AF837128

PERSPECTIVES IN OPERATIONS MANAGEMENT
ESSAYS IN HONOR OF ELWOOD S. BUFFA

PERSPECTIVES IN OPERATIONS MANAGEMENT

ESSAYS IN HONOR OF ELWOOD S. BUFFA

Rakesh K. Sarin, Editor
Paine Professor of Management
Anderson Graduate School of Management, UCLA
Los Angeles, California, U.S.A.

Kluwer Academic Publishers
Boston Dordrecht London

Distributors for North America:
Kluwer Academic Publishers
101 Philip Drive
Assinippi Park
Norwell, Massachusetts 02061 USA

Distributors for all other countries:
Kluwer Academic Publishers Group
Distribution Centre
Post Office Box 322
3300 AH Dordrecht, THE NETHERLANDS

Library of Congress Cataloging-in-Publication Data

Perspectives in operations management : essays in honor of Elwood S. Buffa / Rakesh K. Sarin, editor.
p. cm.
Includes bibliographical references.
ISBN 0-7923-9236-1 (alk. paper)
1. Production management. I. Buffa, Elwood Spencer, 1923- . II. Sarin, Rakesh K.
TS155.P462 1993
658.5--dc20 92-38951
CIP

Copyright © 1993 by Kluwer Academic Publishers

All rights reserved. No part of this publication may be reproduced, stored in a retrieval system or transmitted in any form or by any means, mechanical, photo-copying, recording, or otherwise, without the prior written permission of the publisher, Kluwer Academic Publishers, 101 Philip Drive, Assinippi Park, Norwell, Massachusetts 02061.

Printed on acid-free paper.

Printed in the United States of America

ELWOOD S. BUFFA

TABLE OF CONTENTS

INTRODUCTION

In the fall of 1992 a conference honoring Elwood S. Buffa was held at the Anderson Graduate School of Management of the University of California, Los Angeles. This book is a collection of the work presented at that conference.

The scholars who gathered to honor El are the prominent researchers in the field of Operations Management. Their collective work published in this book represents the richness of the field and provides the reader with valuable insights into its important issues and problems.

While any grouping of the articles by these distinguished scholars will be arbitrary, I have organized the book in four sections.

In the first section the articles dealing with the strategic issues in Operations Management are compiled. The articles deal with continuous improvement, quality, services, supply chain management, and creating value through operations.

The articles that explore the interface of Operations Management with other functional areas, e.g. engineering and marketing, are grouped in the second section.

The third section of the book contains articles that attempt to model some important planning problems that arise in the management of production and operations. Some of the papers in this section provide state of the art reviews of selected topic areas.

Finally, the fourth section contains articles that deal with future directions for Operations Management. The authors offer several insights into the future evolution of the field.

The book begins with the keynote address given by El Buffa at the start of the conference on November 2, 1991.

What is missing from the book is the tribute that every participant in the conference made to El and his illustrious career in the opening remarks of their respective speeches. There are some authors represented in this book who were El's students but virtually everyone was influenced by his work. El's book Modern Production Management, published in 1961, defined the field of Operations Management in new ways. Literally thousands of academics and practitioners learned about the field through this classic. For many years it remained the dominant text used in Production and Operations Management courses throughout the country and around the world.

El has published nine books that collectively have gone through 26 separate editions. OM Problems and Models, published in 1963, introduced analytical techniques in the field. His books span a wide range of topic areas in our field--from production planning and control to manufacturing strategy.

In addition to his professional contributions through his books and numerous research papers, El has provided exemplary service to UCLA for forty years. During his career, he has held a number of important positions within AGSM and at the University level. Some of his past administrative roles include:

--Associate Dean and Chairman of the Department,

--Chairman of the university-level committee that reviews all appointments and promotions on the campus--somehow then called the Budget Committee,

--Chairman of the UCLA Academic Senate.

He was one of the principal architects of the Executive MBA program and its first director. This program has grown to support a large number of doctoral students, and it provides summer support for dozens of faculty.

More recently, El founded the Center for Technology Management that has aided several faculty members and doctoral students in their research endeavors. The Center has also created a two-way communication between faculty and practitioners in the industry.

While many have benefitted professionally through reading his books, some of us have been particularly fortunate. These are the people who have had the opportunity to be his colleagues and friends. We who work him especially cherish his open door policy and his kind and helpful nature. We have all learned a great deal about good human values from El, and he has been a source of inspiration and a role model to us. We hope that this volume will bring his guiding influence to many others as well.

Rakesh K. Sarin
UCLA

ACKNOWLEDGMENTS

This book was made possible by generous financial support from the Center for Technology Management at UCLA's Anderson Graduate School of Management.

My thanks are due to the contributors to the book who diligently met the deadlines and format requirements and produced truly first-rate contributions. I am also appreciative of all of the participants in the conference who came to UCLA to express their gratitude to El.

Finally, I am thankful to the doctoral students in the Operations and Technology Management area at AGSM who helped in arranging logistics during the conference, and especially to Harriet Katz who coordinated the entire product life cycle of this book and kept it on schedule.

R.K.S.

KEYNOTE ADDRESS

Elwood S. Buffa
November 2, 1991

I am deeply honored by the mere presence of ***the*** key scholars in our field, let alone the fact that this conference carries my name. I would have been honored simply to attend a conference with such a stellar cast.

I am retiring after nearly forty years at UCLA. I am retiring early - I am in that cohort that faces no mandatory retirement age. I had always said that I would never retire, that they would have to carry me out. But things change. The University offered an incentive to people in my category - for me it meant a twelve percent increase in my pension check, plus a signing bonus that will buy me a rather expensive car. Both my take home and gross pay will actually increase. By retiring instead of dying in office, I guarantee to my wife the same life annuity granted me. But more than all these fiduciary reasons, I felt that it was time to step aside for new and more vigorous leadership and let the fine new young staff show the way to the next plateau, and I have faith that they will do it.

We have created this "thing," called operations management, after more than one identity crisis.

When I came on the scene there was a field referred to as Industrial Management. It was all inclusive, spanning personnel management, elements of finance and accounting, marketing, distribution, and so on. Much of the present day curriculum was hacked out of that broad definition of industrial management, creating the first identity crisis.

What was left after this process was a highly descriptive field which came to be known as production management, which gained little respect. Following World War II, with the development of operations research, it was obvious to many of us that aspects of production could be quantified and modeled. We could gain respect by being "scientific," outdoing Frederick Taylor, realizing his dream of Scientific Management. Taylor would have been ecstatic for he did not dream of such powerful analytical methodologies.

But the quantification of the field carried with it the seeds of the next identity crisis for what soon broadened in scope to become the field of operations management we now know. Was there a field separate from the analytical methodology? If we taught a course in operations research, did we need anything else, for most of the examples used in OR courses were from production? Was there really a functional field called operations management, such as was recognized in finance and marketing? If we taught a quantitative operations management course, were we leaving out anything of great importance? Was methodology sufficient?

Though this debate continues, I believe that it is now clear that the confusion between the two has been good for neither management science/operations research nor operations management. Management science/operations research should be providing methodology for management in the general sense, not simply concentrating on production problems. The largest curricular failure of MS/OR in my judgment has been the fact that finance, marketing, and other areas have ignored it.

Operations management should continue to develop as a functional field, using MS/OR and ***other*** methodologies as appropriate to the problems at hand. To restrict ourselves to problems where only modeling methodologies are useful is ultimately suicide, ignoring some of the most important problems in the field.

One of the most effective things we could do would be to get out of our offices and into the field. We must take the emphasis off solving problems that we concoct and on to solving the problems faced by managers. I am not suggesting a lack of rigor, rather being challenged by the **real** problems. It would help in both the teaching and research functions. Doctoral students especially need to spend some time with real systems in order to formulate their dissertation research appropriately.

Then, there is another bad omen. At this late date, we still do not have a seat at the table where the big important decisions are made by management. Finance does, marketing does, but OM doesn't. Perhaps we haven't shown the strategic significance of our work. The study of operations strategy is relatively new, and it may be that we simply have not made our dent yet. Or is it that we have been doing work that is important to the strategy of the firm, but we haven't communicated appropriately? I think that both factors are important, but I am inclined toward the communication breakdown as the primary source of our problem, recognizing that strategy analysis hasn't been a centerpiece of our work in the past.

But we need to make strategy studies a prime objective. Technology is at the center of operations strategy, and indeed often of the strategy of the firm, and we are the ones who study it and understand it. Recognizing the strategic importance of technology, and explicitly integrating it into our work and the connections we have to other fields, and to general management, might be the key to greater acceptance by our colleagues and by practicing managers. I note that both Dick Rosenbloom and Arnoldo Hax will discuss aspects of technology strategy.

I see an opportunity arising from a new curricular movement in the MBA. At least here at UCLA there is a growing demand for illuminating inter-functional issues. I observe that OM is one of the few courses that treats inter-functional issues already through the materials and cases used. As you know I'm sure, the marketing course is pure marketing and finance is pure finance, recognizing that nothing else exists in the firm. Pressing our advantage could be worthwhile. There may be opportunities in both teaching and research. Note that both Warren Hausman and Uday Karmarkar focus on inter-functional issues.

It seems unlikely to me that the present model for MBA education will survive. For students to come back to school for two years to gain the knowledge for a career in management seems to take issues out of the context of the real problems faced by managers. Perhaps the other extreme is on-the-job training as practiced by Japanese industry. Rotating beginning manager talent through a series of organization functions provides a breadth and depth not available through formal classroom work. But, on-the-job training lacks the intellectual depth that can be provided by formal education. The emerging model that may combine the advantages of both may be the Executive MBA, where students continue to work at managerial jobs while taking formal classes. Even the Executive MBA format leaves a great deal to be desired, perhaps because we have largely adapted campus courses to be used in these programs. The challenge for the OM field is to define a new course format that takes advantage of the greater experience of the Executive MBA student backgrounds. It seems to me that this new course format should emphasize operations strategy.

As I review the titles of the papers to be presented here, virtually every one deals with issues that have strategic significance, and some are centered in strategy studies. The range of topics is in itself impressive. For those in the former category, I hope that you will highlight the strategic significance of your work.

Another indication that we have a problem is that recruiters are not looking for OM types of MBAs. Students interested in operations management have a difficult time finding good opportunities - recruiters simply do not come here looking for our product, and when they do, the students complain that they offer below par salaries. Companies seem to be filling their OM needs from other sources. I understand that a large fraction of the first graduates from one of the premier programs in the country went into consulting or some other function.

As I look back over the last forty years here at UCLA I am struck by how happy I have been. I have been allowed to do just about whatever I wanted: teach, do research, write, engage in administrative activities, take part in faculty self-governance through the Academic Senate, and do some consulting work. The deans and faculty chairmen have been supportive of just about anything I wanted to do as I floated in and out of these various activities. I feel privileged to have spent my working existence in a golden age of university life that we all enjoy. I hope that the golden age continues, but I am not sanguine.

Let me thank you all for coming to take part in my passage. I cannot express adequately how warm it makes me feel. I also want to thank my close colleagues who have been so supportive and effective in building our group: Reza Ahmadi, Bob Andrews, Sriram Dasu, Don Erlenkotter, Gordon Shirley, Chris Tang, Bill Yost.

And most important, I wish to thank Rakesh Sarin whose idea it was to have this conference. He organized it, invited all the speakers, and arranged all the details including our ship-board dinner.

Thank you!

STRATEGIC ISSUES IN OPERATIONS MANAGEMENT

CREATING VALUE THROUGH OPERATIONS: THE LEGACY OF ELWOOD S. BUFFA

Morris A. Cohen
Matsushita Professor
of Manufacturing and Logistics

Paul R. Kleindorfer
Universal Furniture Professor
of Decision Sciences and Economics

The Wharton School
University of Pennsylvania
Philadelphia, PA USA

INTRODUCTION

Operations management has undergone significant changes since the original publication of Elwood Buffa's Modern Production Management in 1961. We aim here to trace briefly some of the main currents which have characterized these changes. We also wish to illustrate, with three case studies with which we have been intimately involved, the nature of the challenges which operations management now faces in its role as a full partner in creating value for shareholders and customers.

When operations management entered the limelight as a result of the pioneering work of Professor Buffa and others in the '50s and '60s, its focus was primarily in the area of cost management and production control for manufacturing enterprises. The '70s saw the awakening of the world to international competition in the manufacturing sector and the critical importance of operations as a source of distinctive competence. Not only cost management but also quality management and flexibility through new technologies became the visible requirements for success. The obvious competitive advantages of excellence in operations acted as the driving force for a reexamination of strategy to incorporate better the value-creation potential of operations. Moving beyond the earlier focus on productivity, operations was seen successively as the key source of internal quality, as the key inhibiting and contributing factor toward flexibility and customer service, and finally as a full partner, with marketing, accounting, finance and human resources, in promoting value-creation and long-run competitive advantage. Today, whether in service or goods sectors, major strategy initiatives occupying the attention of senior executives worldwide are centrally concerned with operations. These initiatives include:

- Total Quality Management
- Time-Based Competition
- Improved Customer Service
- Improved Capital Utilization
- Improving the Technology Base

But, interestingly, initiatives such as these also imply two other important management trends: *cross-functional integration* and *increased importance of knowledge-based technologies*.

Concerning cross-functional integration, quality was the original driver linking sales and marketing to operations. More recently, time-based competition has further highlighted the importance of integrating marketing, sales, engineering, operations and accounting, in achieving speedier new product development and in reengineering business processes to decrease time and inventory waste in supply and distribution channels. Other initiatives indicated above have similar implications for cross-functional integration. Strategic emphasis has shifted away from pure functional strategies to achieving lean operations, with clear links to internal and external customers in the value chain, and with cross-functional teams working together for value creation at the enterprise level.

Concerning knowledge-based technologies, a cursory glance at the above list, or at Buffa [1983], indicates a shift from simple control strategies to more complex, information-based strategies to support both cross-functional integration as well as detailed, market-focused operations and services.

In the three case studies that follow, we will illustrate the above themes. We apologize in advance for the focus on our own work, with the humble excuse that we can thereby make our point more knowledgeably and more precisely.

We return at the end of the paper to a brief summary of our debt to Elwood S. Buffa whose singular example and scholarship helped set the stage for the increase in visibility and importance of operations management and related fields.

AFTER SALES SERVICE

The support of products after they reach the consumer is an increasingly important component of a product's value. The rapid provision of after sales service has become a strategic necessity for high technology products. In this section, we report on a project which led to the successful development of a highly effective spare parts logistics system for IBM. The project and its results are reported here since they illustrate a number of the key points of this paper.

In particular, the problem of supplying spare parts to support the field service function for mainframe computers is one of immense operational complexity. It is characterized by a high degree of uncertainty, both with respect to the demand and supply processes. In addition, spare parts are distributed through a facility network which contains thousands of locations (since each customer location and field engineer can be used as a stocking point). Finally, there are hundreds of thousands of stock keeping items and thousands of end products which require support. Effective management of this class of logistics system will be achieved by the adoption of control procedures which are based on a thorough knowledge of the operational processes of the system. Such operational knowledge requires that a scientific, analytic perspective be adopted.

The spare parts logistics problem also illustrates the central role of knowledge and information as a critical ingredient for effective management of operations systems. In this case, information on the location and attributes of each stock item, and of each installed machine are required in order to determine stocking policies for the spare parts. The objective is to use these policies to reach a point on the cost/service tradeoff frontier which is consistent with IBM's strategic objectives for customer service, and efficiency. The management of the spares logistics system therefore requires procedures to specify decision rules for every part/stocking location combination (of which there are millions), which can respond to changing environmental conditions (e.g. the number and distribution of installed machines, the performance of parts under use, changes in operating costs and technology developments). The methods and procedures for developing such rules will be described here. We will also discuss the implementation process which was carried out by IBM.

By its nature, after-sales service is a multi-functional management problem. Marketing, manufacturing, distribution, customer engineering and product engineering are all involved in the provision of service to keep customers' main frames up and running on a continuous basis. The trade-off between logistics costs and customer service, which is fundamental to this problem, is, of course, one of the basic paradigms of the operations management field. It is important to note that the concept of service used in this context pertains to function of the end product and not to the availability of the parts. When a machine fails in the field, spare parts will be required to fix it. The availability of spare parts (the traditional service measure) facilitates the completion of the repair job, but the correct service measure is the "job completion rate", i.e. probability that the machine will be repaired. In other words, <u>all</u> parts required for a job must be available.

Project Background

Computers are an example of high technology products which require a high level of after-sales support. This is so since, due to the complexity of their design, it is highly likely that parts will fail, in the field from time to time. In addition, the costs of down-time can be extreme. The trends in this industry have been to the development of more products, with greater performance and complexity. At the same time, consumer expectations for quality have escalated. As a result of the dramatic developments in electronic circuits, parts have become modules of high cost, with multiple functions. It is no longer feasible to train customer engineers to diagnose faults at the component level. Rather, repair involves selective module replacement, and as a result, the spare part is also a diagnostic tool. These trends have led to a requirement for very high levels of parts availability and extremely high costs for many parts (e.g. a mainframe CPU board can cost tens of thousands of dollars). Moreover, these parts are distributed via a very complex, multi-echelon distribution facility network. Finally, we note that the number of distinct parts is proliferating and their demand is becoming more erratic (both due to increased reliability and lower volume use).

The project was initiated, within IBM, by the realization that their existing methods and systems for managing spare parts were becoming uncontrollable. Inventory investment in spare parts was increasing at a rate which was higher than the growth in customer demand. At the same time, achieved levels of customer service were not improving. As a result more and more assets were being invested to maintain the spare parts logistics systems. A number of attempts to control this growth and to re-allocate inventory assets in order to maximize their benefit all met with failure. The reasons given for these failures included the complexity of the problem and the strategic necessity to provide service which led to a reluctance to experiment with alternative management control systems.

There are three levels of decision making in the spare parts logistics system. The basic operational decisions are concerned with determining the stock control policies for each item at each location. The objective of these decisions is to minimize inventory and logistics related costs in a manner which meets the service targets for end product up-time performance. Over a longer planning horizon, management must also consider changes to the design of the distribution network. This involves selection of locations for new stocking facilities, analysis of decisions to close existing locations and specification of the supply/demand linkages within the network. These structural decisions, of course, must be consistent with the stocking policy in force. Finally, management must set service strategy objectives. As noted above, customer service is defined in terms of end product performance and is facilitated by spare part availability. Customer service targets can vary with product, location and market segment.

The stocking policy decision problem can be best understood by considering the case of one product and a given stocking location. This location serves a market region in which there is an installed base of the product at various customer locations. The product is also characterized by the complement of parts which it contains (i.e. the Bill of Material). The random failure of these products in the field, leads to the demand for spare parts.

If the part stocking policy is specified by a stock-up-to level, S_i, for part i at the location, then an inventory cost function for part i can be determined, as $G_i(S_i)$. Management's problem is to minimize the sum (over i) of $G_i(S_i)$, in a manner which meets a product level service target (see Figure 1). The model which specifies $G_i(S_i)$, is based on our research in stochastic inventory theory (see Cohen, Kleindorfer and Lee [1988]).

Figure 1

SERVICE CONSTRAINTS

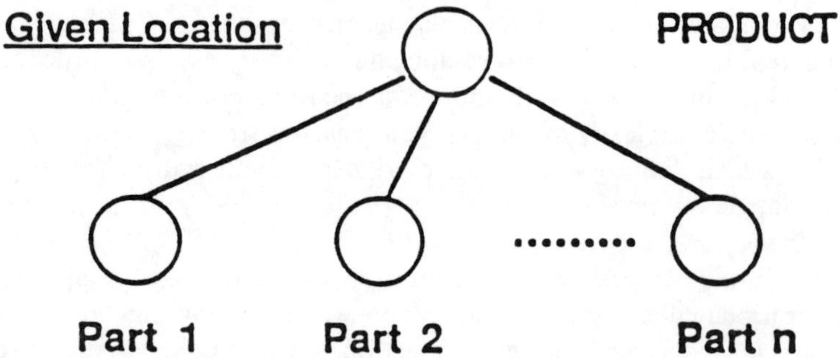

Let $G_i(S_i)$ = Expected Cost, Part i
$P_i(S_i)$ = Expected Fill Rate, Part i

Problem: $\text{Min }[G_1(S_1) + \ldots + G_n(S_n)]$

$$\text{s.t. } a_1 P_1(S_1) + \ldots + a_n P_n(S_n) \geq P$$

$$a_i = \frac{\text{Demand (i)}}{\text{Total Demand }(1, \ldots, n)}$$

Determine $(S_1, \ldots, S_n)$ at this location AND PASS UP EXCESS DEMAND TO NEXT LEVEL. Good Results achieved.

The principal challenge in determining the collection of $\{S_i\}$ for every stocking location is the *linkage* between these decisions over locations and over end products. One particularly interesting aspect of this linkage is the relationship between the demand process faced at an intermediate location, which acts as a supply source to many, lower echelon locations, and the stocking policies at each of these supplied locations. It is clear that the so-called "pass-up" demand is not simply the aggregate of the market demand faced by the lower echelon. Our research to support this project required the development of explicit procedures for analyzing the pass-up demand specification problem.

It was also necessary to develop optimization algorithms which consider the interactions between location stocking policies. A variant of the bottom-up, multi-echelon decomposition method, was used successfully here, (see Cohen, Kamesam, Kleindorfer, Lee, and Tekerian [1990]). Decisions at the lowest (first line of support) level are made. These decisions then affect the demand observed at the next higher level. Stocking decisions for the second level are then made. After all locations have determined their inventory control policies, it is possible to estimate the offered lead time from each supply node to each of its customer nodes in the network. These lead times will, in turn, affect the choice of stocking policy at the lower echelon locations. It is thus necessary to iterate over levels to achieve a stable, near-optimal stocking policy for each item and location.

IBM's National Service Distribution System

We note some of the specifics of IBM's spare parts system. At the time of the study (mid 1980s), there were over 1,000 active end products requiring support. The installed population of these machines in the United States was counted in the tens of millions of units. There were over 200,000 active part numbers and over 15,000 customer engineers were used to repair these machines in the field. The total inventory investment was valued at over a billion dollars (selling price) and there were five or six identifiable echelons in the system.

Figure 2 illustrates the nature of the stocking policies derived by the procedure which we developed. Stocking policies which differentiate by the item's value, demand (and other attributes) are generated. These policies differ significantly from the polices computed by the incumbent system which stocked items at all locations that they passed through. It is important to note that the algorithms used to derive these policies were a component in an

Figure 2

STOCKING POLICIES (UNCONSTRAINED)

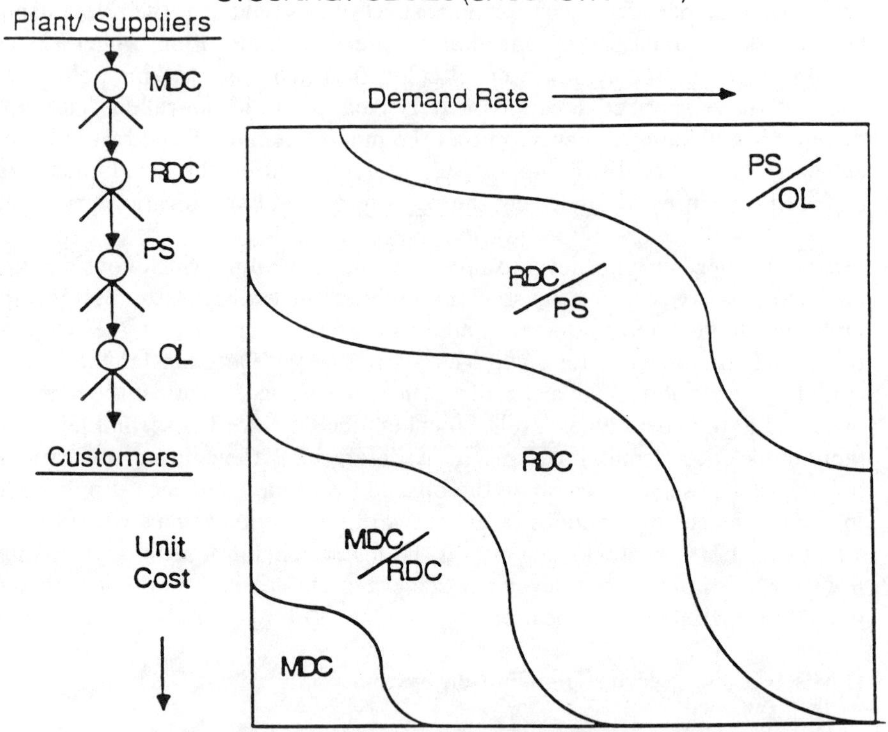

OL = Outside Location
PS = Parts Station
RDC = Regional Distribution Center
MDC = Major Distribution Center

extensive software system which was installed over a five year period. Many technological innovations (like a hand-held communication computer device for each customer engineer) and organizational changes accompanied the installation of this system.

The overall impact of the system was positive. Efficiency improved and in particular IBM was able to reduce its investment in spare parts inventories by over 25% (a $250 million one time reduction). Annual operating expenses were reduced by $20 million and customer service improved by 10%.

At a more qualitative level, the new system improved responsiveness and simplified management, which in turn freed up human resources. Finally IBM used the multi-echelon stocking model to analyze alternative network configurations and based on the results of this analysis, implemented significant reductions in the number of stocking locations.

As noted at the beginning of this section, this project illustrates some of our key points. We note in particular that this project indicates the strong linkage between functions and the strategic impact associated with operations systems. The role of technology was key. This included Information Technology, modelling tools and algorithms. Finally, it is an example of a market driven improvement which is based on a deep understanding of the cost service tradeoffs associated with the logistics function.

GLOBAL OPERATIONS

Operations has truly become an international business function in recent years. Supply, production, distribution, service and product design for many companies involves activities and resources which span national boundaries. At this point in history, when large, multi-country trading blocks are emerging and where global trade issues are central to economic prosperity, it is clear that coordination of operations activities on an international scale is of critical importance to many firms.

The fundamental question is <u>where</u> should value be added in the production of goods and services. Inputs in many industries can be sourced, both domestically and abroad. Similarly, both product and process technology can be acquired from and transferred to foreign locations with increasing ease. The time and distance impediments to global coordination are shrinking. At the same time real and persistent barriers to global operations exist. They include tariffs, import quotas, local content requirements and the like.

There is also intense political pressure on companies to behave as responsible "corporate citizens" in every country in which they operate. The current public debate on automobile imports from Japan is just the latest example of how global manufacturing has become a key issue for the '90s and beyond.

The management of global operations is one of our current research priorities. This problem is particularly relevant to the Buffa conference and to the theme of this paper. Coordination of operations activities on a global scale requires massive information transfer, understanding of operations processes and an understanding of the linkages between decisions at all stages of a firm's supply chain. Of course, in this area, many functions, besides manufacturing and distribution, are involved. In particular marketing (through global advertisement campaigns and product standardization) and finance (through management of foreign exchange risk via a combination of operational and monetary hedges) are heavily involved in issues of global coordination. As a result, a truly multi-function approach must be taken if research on this problem is to make a significant contribution.

We note that there is a distinct development pattern for the globalization of operations. Many companies began with the export of products to international markets. Over time, companies began to produce some portion of their product lines in certain markets. This led to a multi-country approach in which each country acted as an autonomous unit. In some cases, only a limited portion of the products' total value was added in the non-domestic locations. In other cases, replications of the entire system were built. The next stages of this process which leads to true global management of the firm, are full integration of production/sourcing and globalization of Research and Development activities. It is fair to say that most companies today do little to integrate the operation of their research, manufacturing and distribution facilities on an international basis.

Consequently there is a real opportunity for firms to attain a competitive advantage through the achievement of true "global management". Examples where this has occurred are instructive. The Sony company, for example, operates an extensive network out of Singapore which incorporates production, precision engineering, systems design, sales, logistics, and after-sales service. These activities are distributed throughout Southeast Asia and are operated in an integrated fashion. (See Figure 3).

Recently, we have developed a normative model framework for the global operations problem. The international supply chain perspective considers both structural (location and capacity) decisions and tactical decisions affecting product mix, material flow (both inbound and outbound), and cash flow (borrowing, lending, hedging). As noted above, there are a number of factors which are unique to global operations networks. As material crosses borders, cash flows between entities of the company. Duties and tariffs act to increase the unit transportation cost as a result of such international flows. Quota and content requirements, on the other hand, act to constrain the set of feasible material flows. In addition, we must also consider factors which are unique to the global manufacturing problem, such as, country tax rate differences, transfer price rules, and exchange rate fluctuations. The resulting problem is thus much more than a simple variant of the domestic supply chain management formulation.

It is useful to consider the decisions associated with the global operations problem from the perspective of each major management function. For example production and distribution are concerned with plant or distribution center location, as well as choice of technology and specification of the charter of each plant (which determines its product mix on a long term basis). Supply management in the global context must make vendor selections and be especially responsive to currency and tariff issues. Marketing in conjunction with product engineering confront the fundamental issue of product design adaptation to local markets. The pros and cons of standardization vs. customization must be considered in the context of the entire international supply chain. Some companies have noted that significant benefits can be derived by building flexibility into product designs so that local market customization can be carried out at a central manufacturing site or at remote distribution points. Finally, finance and accounting are very much involved with issues which range from cost allocation (the transfer price problem) and currency risk management.

It is our premise that these functional decisions are all inter-connected and significantly influenced by the fact that material and cash flows occur on an international basis. Figure 4 illustrates the structure of our model framework. Note that it includes a variety of operational sub-models, such as a supply chain network model (a stochastic program), a currency exchange rate process model, and a financial flow model. The model framework is driven by a

Figure 3

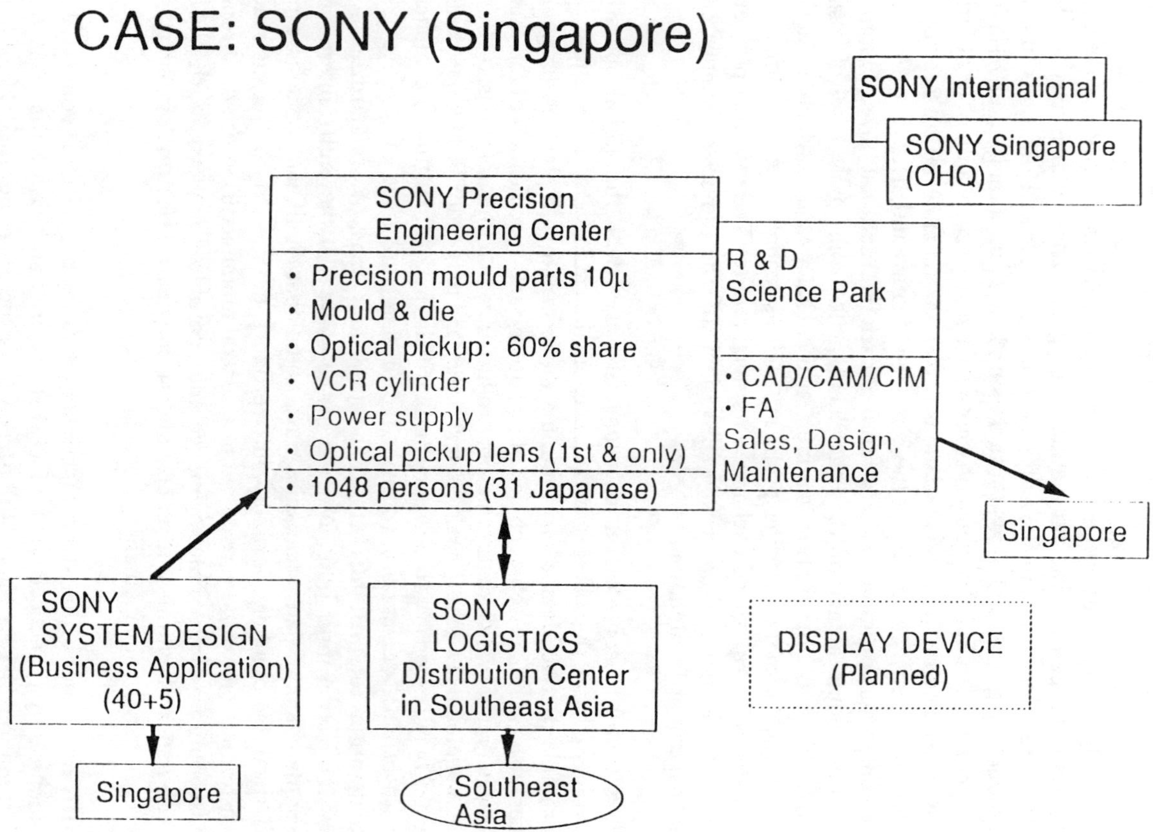
CASE: SONY (Singapore)
SONY International
SONY Singapore (OHQ)
SONY Precision Engineering Center
• Precision mould parts 10µ
• Mould & die
• Optical pickup: 60% share
• VCR cylinder
• Power supply
• Optical pickup lens (1st & only)
• 1048 persons (31 Japanese)
R & D Science Park
• CAD/CAM/CIM
• FA
Sales, Design, Maintenance
Singapore
SONY SYSTEM DESIGN (Business Application) (40+5)
Singapore
SONY LOGISTICS Distribution Center in Southeast Asia
Southeast Asia
DISPLAY DEVICE (Planned)

Figure 4
Global Manufacturing Strategy Model Framework

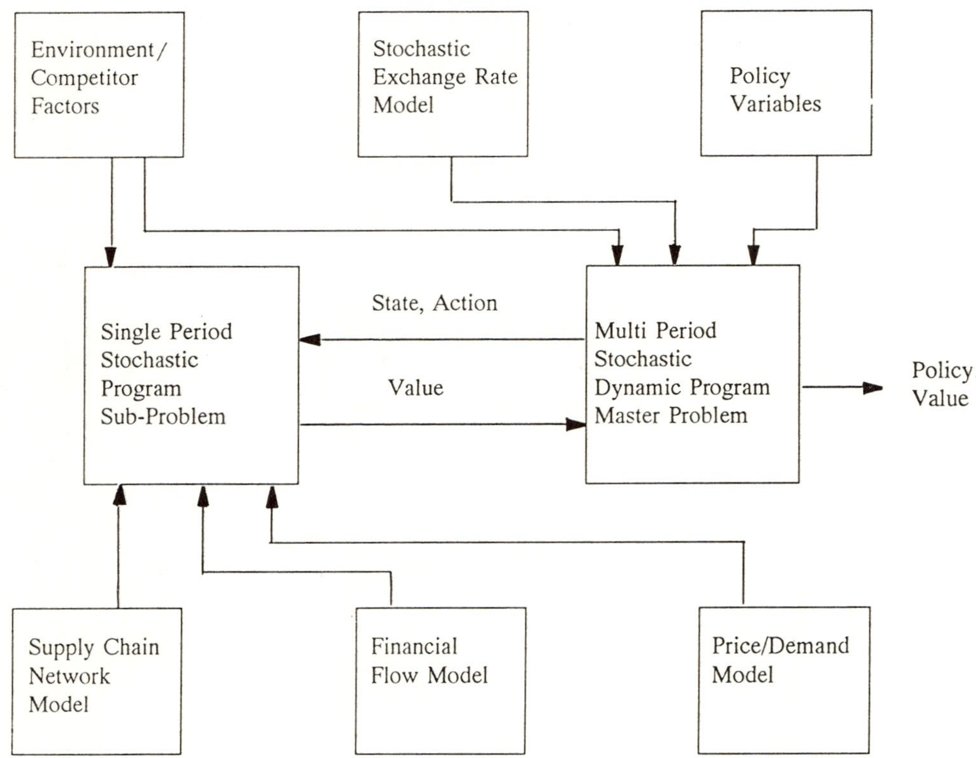

multi-year stochastic dynamic program which is concerned with the major structural decisions noted above. The within period (year) tactical decisions, which determine both material and cash flows are considered in the single period stochastic program sub-problem. This program links the supply chain network formulation and the financial flow model to generate a comprehensive analysis of the firm's global, after-tax profit. Various versions of this framework have been implemented and solved.

This research is ongoing. An early test implementation of the framework was carried out with Apple Computer Company (see Cohen and Lee [1989]). In that implementation a model of Apple's manufacturing operations in the U.S., Ireland and Singapore was linked to a detailed representation of both their inbound sourcing and outbound distribution systems. The model's outputs were used to contribute to the analysis of a decision to locate a new plant. Issues of where and when to build such a plant were considered in the context of the predicted impact on existing operations and on market strategies. The results included projections of plant production plans, global material flows and cash flows. Attention focused on global, after-tax profit, customer service by segment and region (fill rate) and plant utilization (for assembly and sub-assembly processes). The results of this analysis played an important part in the company's policy debate on this key strategic issue. The analysis also quantified the value of global flexibility which is derived from Apple's investment in a coordinated global manufacturing network.

More recently, Cohen and Huchzermeirer [1991] have explicitly incorporated uncertainty into the analysis. A tractable stochastic model has been developed to account for the fluctuation of currency exchange rates and their impact on prices and demand within each country. This research focuses on the role of the global supply chain network as an operational hedge. Long term protection against uncertainty impacts like currency fluctuations, market shocks or political changes cannot be effectively achieved through financial instruments alone. As random factors (e.g. country to country exchange rates, market prices or demand levels, plant productivity gains, etc.) are realized, a firm which has made the prior investment in an integrated global supply chain (i.e. has purchased the operational option), can adjust, on a contingent basis, by shifting production, altering plant mix, changing supply sources or by adjusting the organization of other value-adding activities. Thus the global network must be managed in a manner which explicitly considers the linkages between material and cash flows on an international basis. The research is currently focusing on applications to different company situations.

Some of the lessons learned from this research project are particularly relevant to the themes of this conference. We saw, in the Apple application study, that an international supply chain perspective is a necessity for a firm if it wishes to account for the global, after-tax impact of modifying either the design or use of its production distribution system. As noted above, coordination over a multi-country network is key for absorbing uncertainty and the optimal response to changes can lead to high profits in the long run. Since many companies are not fully exploiting the full potential of coordinated global operations, it is clear that the costs and constraints for global coordination are not well understood. Our research is concerned with analyzing such barriers, which certainly include the firm's organizational structure, its size and the scope of its product line and operations processes.

QUALITY AND PRODUCTIVITY IMPROVEMENT FOR POSTAL SERVICES

Our final example is from the service sector. As noted in Buffa [1983, Chapter 14], the service sector presents equivalent challenges to those in manufacturing in creating value through operations. Indeed, since the service sector accounts for about two-thirds of gross national product in industrialized countries, the value-added opportunities through increasing quality and productivity in this sector are clearly substantial. The particular area with which we will be concerned here is postal service.

Mail is perhaps the one service of central government that touches every citizen on a daily basis. Postal services are large organizations in most major economies with many complicated managerial and economic problems. Despite the importance of postal services to the economy and despite the interesting problems they present, however, very little has been written on the efficient management of postal service until recently.[1]

Postal or mail service in its simplest form is the collection and delivery of written, typed or printed communications. Ostensibly it is a rather low-tech activity. Indeed, for mail service, around 75% of the total cost (now nearing $50 billion per annum) of the United States Postal Service is labor (with now over 700,000 employees). Despite its low-tech character, however, demand

---

[1] For an introduction to recent work on postal services, see Crew and Kleindorfer [1991a].

continues to grow even in a high-tech information age.

Mail service (we concentrate here only on "letter" delivery service) involves a sequence of steps. A letter is placed in a mailbox whose contents are then taken to the originating sorting office or mail processing facility (MPF). Mail processing at the MPF includes a number of operations such as facing (getting all the mail the same way up with the stamp in the top right corner), canceling (postmarking and canceling the stamp), and sorting according to end destination, which may take place either manually or by various automated means. Following a number of sorting and other processes it is then transported to the destinating office, where it is sorted again for delivery by the postman (carrier) to the individual residence or business. Priority of processing is further determined by mail class. Preferential mail (like 1st class letters) receives priority treatment relative to non-preferential mail (such as 3rd class bulk mailings). Given the multi-stage nature of postal service, it is not surprising that the managerial problems of improving quality and productivity for postal services are very much akin to those of flow-shops, with multiple priorty classes of jobs corresponding to the multiple service classes of mail.

Beyond the usual problems of flow-shop scheduling, however, modern mail service faces a number of additional challenges. First, given the public sector character of postal service and the importance and magnitude of its economic effects, it is a heavily regulated activity. Pricing policies and service quality standards must therefore stand up to the scrutiny of the public, and typically (as in the U.S.) also regulatory authorities. The nature of the theory underlying recent trends in postal service is evident in Crew, Kleindorfer and Smith [1990]. This theory builds on the traditional theory of regulated monopolies in prescribing optimal prices (P) and service quality levels (R) as those which maximize an expected welfare function W(P,R) defined as

$$W(P,R) = WTP(P,R) - C(P,R),$$

which is the difference between total consumer willingness-to-pay (WTP) for services minus the total cost (C) of producing these services, with both WTP and C evaluated at the prices P and quality levels R of the services offered. The actual modelling of WTP and C for postal services is complicated, but the prescriptive outcome of maximizing W(P,R) can be summarized rather intuitively:

Pricing Rule: Price for each mail service offered should be based on the marginal cost of producing that service.[2]

Service Quality Rule: Service quality for each mail service offered should be set so as to equate the marginal benefits of increasing service quality (as measured by consumer WTP such increases) to the marginal cost of such increases.

In addition to the prescriptions embodied in these rules, technology and capacity choice are governed by cost minimization subject to meeting the demands and service quality requirements implied by the above rules. As noted by Crew and Kleindorfer [1991b], implementing either of the above rules in practice involves, as with our other examples, a strong interaction between operations and other functions, including marketing, accounting and finance. Consider, for example, pricing. For both preferential and non-preferential mail, determining marginal costs involves very detailed modelling of operations, both to assure cost-minimizing operations, but also to estimate the marginal impact on cost at each stage of mail processing of increased mail volumes. Such marginal cost estimates for one service, in turn, are highly dependent on the volumes of other services, since mail of different service classes uses many of the same facilities. These marginal costs also depend on the required service quality levels for each service, the optimal settings for which require marketing information on the benefits to consumers of increased quality levels.

While the above description would be complicated enough in a labor-intensive monopoly, recent trends have added even further difficulties for theory and practice in the form of rapid technological change toward a high automation environment. Advances in pattern recognition and other AI-based approaches to optical character recognition are being used to sort mail. Nearly universal

---

[2] Most national postal authorities operate as public enterprises with a breakeven constraint, usually exercised over a 3-4 year tariff period. Prices must therefore usually be marked up above marginal costs to recover sufficient revenues to meet breakeven constraints. Such mark-up procedures begin at marginal cost and follow well-understood principles (see, e.g., Crew and Kleindorfer [1991a]).

barcoding of letters is expected over the next decade, making it is possible to have almost all letter mail sorted by machines, vastly speeding up the sorting of mail and substituting capital for labor. The scheduling and capacity planning problems of the resulting more capital intensive environment will have significant payoffs for utilization and efficiency of postal equipment and facilities. As in our other examples, the resulting increased information intensity and automation of the environment of postal services for the '90s will also imply a tighter coordination between operations, marketing and accounting activities and a greater commitment to knowledge-based technologies as the key to efficient delivery of high quality postal services in the future.

CONCLUDING COMMENTS

The above case studies underline the themes we announced in the introduction:

(1) The centrality of operations management in creating value in the '90s, in both services and goods;

(2) The trend toward linking operations to other functional areas to achieve improved quality, productivity and flexibility not just of operations, but of the entire value-added chain;

(3) The trend toward increasing complexity, both model-based and data-based, and the resulting increased reliance on systems and knowledge-based technologies.

It is fair to say that Elwood Buffa anticipated all of these trends in his work and reinforced these trends with an encompassing sense of scholarship in defining the field of operations management. There can no doubt that he has been the voice of the profession for these past decades in declaring the centrality of operations. More than anything, his dedication to detail and to systems approaches to operations problems will be a lasting legacy of Elwood S. Buffa for the operations manager of the future.

REFERENCES

Buffa, Elwood S. (1983). *Modern Production/Operations Management*, 7th Edition. John Wiley & Sons, Inc.

Cohen, Morris A., and Huchzermeier, Arnd (1991). *Valuing Global Manufacturing Flexibility under Exchange Rate Uncertainty.* Decision Sciences Working Paper 91-08-01, The Wharton School, University of Pennsylvania.

Cohen, Morris A., Kamesam, Pasumarti V., Kleindorfer, Paul R., Lee, Hau L., and Tekerian, Armen (1990). *OPTIMIZER: IBM's Multi-Echelon Inventory Systems for Managing Service Logistics.* Interfaces 20, January-February, pp. 65-82.

Cohen, Morris A., Kleindorfer, Paul R., and Lee, Hau L. (1988). *Service Constrained (s,S) Inventory Systems with Priority Demand Classes and Lost Sales.* Management Science, Vol. 34, No. 4, April, pp. 482-499.

Cohen, Morris A., and Lee, Hau L. (1989). *Resource Deployment Analysis of Global Manufactuing and Distribution Networks.* Journal of Manufacturing and Operations Management, Vol. 2, pp. 81-104.

Crew, Michael A., and Kleindorfer, Paul R., (Eds) (1991a). *Competition and Innovation in Postal Services.* Boston: Kluwer.

Crew, Michael A., and Kleindorfer, Paul R. (1991b). *Quality and Price Differentiation in Postal Service*, in Walpurga Speckbacker (Ed.), Die Zukunft der Postdienste in Europa. Berlin: Springer-Verlag.

Crew, Michael A., Kleindorfer, Paul R., and Smith, Marc A. (1990). *Peak Load Pricing in Postal Service.* Economic Journal, 100, September, pp. 793-807.

MODELLING IN SUPPORT OF CONTINUOUS IMPROVEMENTS TOWARDS ACHIEVING WORLD CLASS OPERATIONS

Edward A. Silver
Faculty of Management
University of Calgary
Calgary, Alberta T2N 1N4 CANADA

ABSTRACT

This paper is concerned with the relevance of mathematical modelling for organizations striving to achieve World Class Operations Management (WCOM) where the latter is viewed as a management philosophy that fosters rapid development and delivery, at reasonable cost, of high quality products and services. Continuous improvements are an integral part of WCOM. Existing mathematical models are reviewed and it is seen that almost all are of a conceptual nature as opposed to being in a form that would facilitate easy use by an organization in choosing among proposed improvement alternatives. Some suggestions for research topics are presented.

INTRODUCTION

Much has been stated and written about the strategic importance of a number of related topics that have the common theme of striving for continuous improvement in the delivery of goods and services. These topics include World Class Manufacturing, Just-in-Time Systems, Time-Based Competition, Continuous Improvement and Total Quality Management. However, consider an organization faced with selecting one or more of several proposed specific improvement actions. From that perspective most of the aforementioned material, if it uses mathematical modelling, is of a general conceptual nature, thus of limited value for helping select among specific alternatives. Other portions of the material may be helpful, but these make very limited use of mathematical modelling; rather they are either procedural and qualitative in nature (e.g., a list of steps to be followed in reducing setup times) or empirical (survey) material (e.g., what policies, actions, etc., seem to lead to improved overall performance). Examples of conceptual and procedural writings are Hall (1983) and Shingo (1985), respectively. Empirical studies include Hayes and Clark (1985), Krafcik (1988) and Schmenner (1988).

This paper attempts to address this relative lack of mathematical modelling useful for choosing among specific suggested improvement actions. One of the underlying themes, as has been stated elsewhere (Silver (1990)), is that, if quantitative models are to be more useful as aids for managerial decision making, they must represent more realistic problem formulations, particularly permitting some of the usual "givens" to be treated as decision variables. Moreover, such models must ultimately be in an operational form such that the user can understand the inherent assumptions, the associated required input data can be realistically obtained, and the recommended course of action can be provided within a relatively short period of time. Whybark (1988) has expressed much the same viewpoint, viz:

> In recent discussions with practitioners and in several of the conferences that have taken place over the last few years it is very clear that the drum beat in the class room is quite different from that on the factory floor. In our classes we still approach problems from the standpoint of, "given the following constraints, what is the optimal resource expenditure to achieve maximum profits?" The reverse is occurring in practice, "how can we remove these constraints in order to improve our performance?"

Consistent with the philosophy underlying the aforementioned topics (World Class Manufacturing, etc). and to maintain a reasonable scope of coverage, we are restricting attention in this paper to continuous improvement actions or alternatives that do <u>not</u> involve large up-front investments. In particular, we do not consider alternatives such as robotics, automated guided vehicles, flexible manufacturing systems, and so on. Rather we concentrate on changes that can be accomplished with relatively low cost expenditures. Moreover, the objective is to provide guidelines for the use of mathematical models to aid in choosing among change options as opposed to prescribing specific models to deal with particular situations.

In the next section we introduce the term World Class Operations with an emphasis on its continuous change aspect. A listing is provided of a variety of usual "givens" that can possibly be changed (i.e., possible improvement actions). The subsequent section is concerned with a fundamental question, namely "are mathematical models really needed to make continuous improvement decisions?" Next, we present some <u>simple</u> models that illustrate the impact of improvement actions. Then, a summary is presented of some of the relevant models that have already appeared in the literature. We subsequently point out some of the problems in applying conventional investment evaluation techniques in the context of World Class Operations. Finally, we provide some suggestions for future research that should facilitate greater utilization by practitioners of decision aids in selecting among potential continuous improvement actions.

WORLD CLASS OPERATIONS

Schonberger (1986) coined the term World Class Manufacturing. He and others (e.g., Billesbach and Schniederjans (1989) and Inman and Mehra (1991)) have pointed out that the underlying concepts are equally applicable in non-manufacturing contexts. Thus, we prefer to use the terminology World Class Operations to emphasize the broader scope. Specifically we view World Class Operations Management as a management philosophy that fosters rapid development and delivery, at reasonable cost, of high quality products and services. World Class Operations are achieved through the elimination of all activities that do not add value to the product or service (particularly unnecessary movement, storage or inspection). This elimination is achieved through continuous improvement actions, "kaizen" in Japanese (Imai (1986)). Those advocating Time-Based Competition (see, for example, Blackburn (1990), Schmenner (1988), Stalk and Hout (1990), Suri and de Treville (1991), and Vesey (1990)) argue that improvements in overall performance are best achieved by focussing attention on reducing the time to develop and deliver the product or service. In contrast, the proponents of Total Quality Management (see, for example, Crosby (1979), Deming (1986), Feigenbaum (1983), and Juran and Gryna (1980)) stress a central theme of achieving higher and higher levels of quality. Zipkin (1991) sounds a word of caution against over-zealous, revolutionary means of achieving World Class Operations.

There are a wide variety of possible improvements to undertake (equivalently, usual "givens" to change) in the operations of organizations. These include:

- improve understanding of customer needs,
- lower the time to design the product or service (e.g., by using concurrent engineering),
- fundamentally change the nature of the product/service or the way it is produced/provided (i.e., reengineering concept),
- reduce the number of distinct parts and components of services or products (e.g., through greater use of commonality),

- delay product differentiation during the production/distribution process (see, for example, the paper by H. Lee in this book),
- increase the degree to which the same activities can be repeated (e.g., through the use of Group Technology),
- ensure much higher quality levels (typically through procedures of a more preventive nature),
- use multiple machines instead of one very large (possibly flexible) machine,
- modify packaging, containers, identification mechanisms, etc. (for operational, not just marketing, benefits),
- provide tighter linkages of consecutive stages in multistage operations (by reducing both buffer inventories and spacing),
- reduce the mean and/or variability of the production changeover times/costs,
- reduce the mean and/or variability of the unit processing time of products and services,
- develop higher levels of employee involvement (e.g., through more effective training, more open communication, etc.),
- encourage learning from problem instances instead of just fire fighting (e.g., recording the cause of each work slowdown or stoppage),
- reorganize the workplace (e.g., maintaining specific storage areas for each process input and keeping the inputs close to their point of use),
- improve layout of facilities (e.g., use of cellular rather than functional layout),
- change the nature of demand patterns (e.g., levelling peaks and valleys by providing customer incentives),
- improve the accuracy of forecasting,
- consider the use of other transportation options for the movement of both raw materials and finished goods,
- reduce the mean and/or variability of replenishment lead times at least partly through more rapid information exchange (e.g., through the use of Electronic Data Interchange),
- reduce ordering costs (i.e., the fixed costs associated with purchasing orders),
- lower the minimum permitted size of the replenishment quantity,
- diminish inventory levels (through other changes that reduce or eliminate their causes),
- provide more slack time (i.e., equipment and labour buffers),
- lower the frozen period in the scheduling of operations,
- increase the frequency of planning,
- alter production rates, in particular, moving away from running individual processes at maximum capacity,

- ensure more reliable operation of equipment (so-called Total Preventive Maintenance).

Elsewhere (van Beek and van Puttin (1987) and Ramasesh (1990)) have given the name strategic (or policy) variables to what we have called the usual "givens". They then refer to related shorter-term decision variables as being operational in nature. For example, the fixed cost per setup and the economic order quantity would be linked strategic and operational variables, respectively.

ARE MATHEMATICAL MODELS REALLY NEEDED FOR MAKING CONTINUOUS IMPROVEMENT DECISIONS?

Where the author received a doctorate in Operations Research, the posing of the above question may be viewed by some as an act of heresy. However, the answer is not necessarily obvious and there are a number of individuals who see very little, if any, need for mathematical models to assist in continuous improvement decision making. For example, Schonberger and Schniederjans (1984) state "JIT does not call for a new quantitative formula but, more properly, a direction for action: namely a never ending succession of inventory reductions." Elsewhere, Schonberger (1986) has written

> "A whole generation (or more) of industrial engineers and management science people have been educated to use complex models as tools to unravel the complexities in places like factories. It never worked very well and so was frustrating. Our practising IEs and other plant management people seem happy to move on to visual models, which yield more satisfying results".

By a visual model Schonberger means a simple, manual visual simulation model or even using the actual system itself for experimentation. Moreover, in arguing that the changes in World Class Manufacturing are obtainable, the reasons he gives include "simplicity and overwhelming logic - no need to wait for studies to be done to provide proof that the concepts are right." The lack of applicability of <u>existing</u> mathematical models has also been conveyed to the author by other applied researchers and practitioners.

In contrast, there have been a number of individuals who have advocated the need for <u>new forms</u> of mathematical modelling for use in such decision areas as continuous improvement. Sample quotes include:

> "... the challenge of contribution to industrial productivity will lead OR/MS in new directions. These will include: finding methods for

extracting decision making information out of massive amounts of automatically collected data; helping to shrink the time required to develop and install new products and processes; and empowering frontline workers in organizations to improve their effectiveness .... Another opportunity lies in supporting the goal of simultaneity and time compression. We need to develop tools for assisting cooperative work. If lead times are to be shortened, many persons will have to work on different aspects of the same problem at the same time but with close coordination. What are the models and systems that will help make this possible?" *(Little (1991))*

"This raises the question of whether operations research has a significant role to play in the new manufacturing and if so, what that role might be .... There are many critical issues that we have not addressed in the past where our skills could provide an important contribution. Carving out this broader and more central role for operations research in manufacturing would require a different style of research than we have followed recently — a style that emphasizes close observation of real world phenomenon and careful reflection on how modeling and analysis can provide insights." *(Fisher (1990))*

"In many product/market combinations a wide range of logistics opportunities exists to improve on their performance, sometimes they are labelled with slogans such as 'Zero Inventory' or 'Just in Time' ... Quantification of the combined effects of simultaneous changes in several of these factors is almost never easy but is nevertheless necessary as a basis for assessment of what should be improved and to what extent. Quantification is also needed to be able to compare planned and actual improvements." *(Centre of Quantitative Methods (1986))*

"A key phrase in the description of kaizen is 'constantly looking for opportunities for improvement'. The speed with which models can be set up, and the ease of evaluation of many alternatives, make rapid modelling technology (RMT) an ideal tool to support kaizen." *(Suri and de Treville (1991))*

It is the author's contention that carefully constructed mathematical models do indeed have two important roles to play in supporting continuous improvements in operations. The first is to provide conceptual insights as to the general nature of policies to be followed, whereas the second role is to quickly evaluate whether or not to initiate a specific improvement action under consideration.

SIMPLE MODELLING ILLUSTRATIONS OF THE IMPACT OF IMPROVEMENT ACTIONS

In this section we illustrate, with two relatively simple models, the probable impacts of improvement actions. The first shows a result, likely to frequently hold in practice, namely that appreciably changing one of the usual "givens" of a problem situation, without any rigorous attempt at optimization, will have a much more important effect than optimizing subject to the fixed given. Other such illustrations (in the areas of inventory management and quality assurance) can be found in Silver (June 1990, 1992). The second example shows that improvement actions may have to be jointly considered in that the impact of a particular action may be highly dependent upon whether or not other actions are taken. This interaction effect has also been discussed by van Putten and van Tongeren (1986.)

a). Preventive Maintenance Example

One of the key dimensions of World Class Operations Management is Total Preventive Maintenance, the ensuring of very reliable operation of equipment. Here we show the effects of changing a usual given in a classical preventive maintenance problem.

Consider a piece of equipment having a discrete probability distribution of the operating time (t) until breakdown if there was no preventive maintenance done. Specifically, let

$$p_t(t_o) = \text{prob}(t = t_o) \qquad t_o = 1,2,3,...$$

We're also given that

m_b is the mean time to repair a breakdown,

m_p is the mean time for preventive maintenance,

c_b is the mean cost (of repairs and lost production) associated with a breakdown,

and c_p is the mean cost for a preventive maintenance.

Moreover, we assume that a repair or maintenance puts the equipment back into the same good operating condition. A clock is initialized at that time and a preventive maintenance is subsequently undertaken if an operating time T elapses without a breakdown. The usually-stated problem is to select the (integer) value of T so as to minimize the expected total relevant costs per unit time, denoted by ETRC(T). Use of renewal theory permits us to write

$$ETRC(T) = \frac{\text{expected cost per cycle}}{\text{expected duration of a cycle}}$$

$$= \frac{c_p p_{t>}(T) + c_b p_{t\leq}(T)}{(T + m_p)p_{t>}(T) + \sum_{t_o=1}^{T} (t_o + m_b)p_t(t_o)}$$

where $p_{t\leq}(T) = \text{prob}(t \leq T)$
and $p_{t>}(T) = \text{prob}(t > T)$.

Numerical Example:

Suppose $m_b = 2$, $m_p = 1$, $c_b = 200$ and $c_p = 50$. Furthermore, the lifetime distribution is given by

t_o	1	2	3	4	5	6	7
$p_t(t_o)$	0.1	0.1	0.15	0.2	0.2	0.15	0.1

The expected total relevant costs as a function of T are shown in the upper curve of Figure 1.

Now suppose we institute one or more corrective actions that reduce c_b from 200 to 100 (e.g., streamlining repair procedures, having other flexible equipment/workers that can take over the production during the downtime, etc.). The resulting ETRC(T) function is shown in the lower curve of Figure 1. It is seen that $ETRC(T/c_b=100)$ is lower for a wide range of T values than the best point on the curve for $c_b = 200$.

Of course, there may be significant costs associated with the action(s) that reduce c_b (although there have been many practical illustrations of changes made at negligible cost). Suppose that there was an up-front expense of x and on-going costs of y per unit time. Moreover, assume that after the change the best associated T value would be used. Then, one could use a discounted cash flow (present value) type of analysis to determine if the change was attractive to make. Specifically

$$\text{DCF(of cost change)} = x + \sum_{j=1}^{H} \frac{(y + 15.8 - 25.3)}{(1 + r)^j} \tag{1}$$

where 15.8 and 25.3 are the approximate low points on the two curves of Figure 1, r is the desired rate of return, and average period by period costs are assumed to occur at the end of each period out to an horizon of H periods. A simple what-if analysis could be carried out to ascertain if the change was attractive for various plausible values of x, y, r and H.

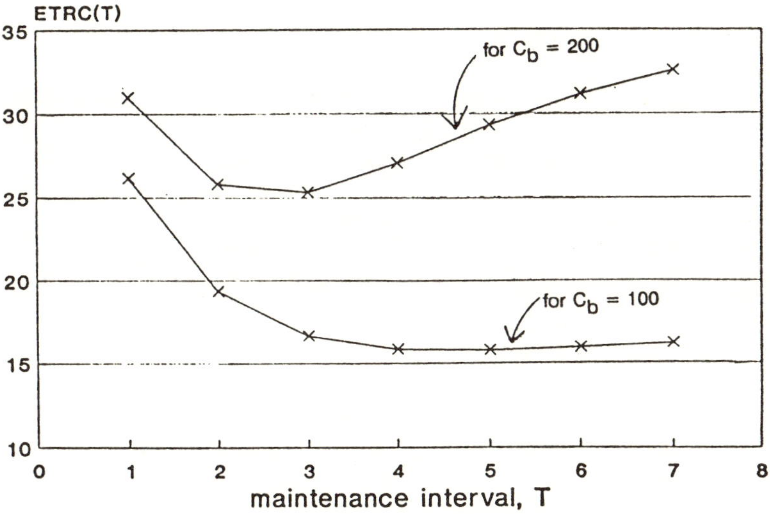

Figure 1

b). <u>Inventory Example</u>

Consider a periodic review (R periods), order-up-to-level (S) control system where the value of S is selected so as to provide a desired customer service level (fraction, P, of demand satisfied directly from stock). Specifically (see, for example, Silver and Peterson (1985)) we set

$$S = \hat{x}_{R+L} + k\sigma_{R+L} \tag{2}$$

where L is the replenishment lead time,

$\hat{x}_{R+L}$ is the forecast (or expected) demand over an interval of duration R+L,

σ_{R+L} is the standard deviation of forecast errors over such an interval,

and k is the safety factor.

For the case of forecasts (over R+L) having normally distributed errors k is chosen so as to satisfy

$$G_u(k) = \frac{RD}{\sigma_{R+L}}(1 - P) \tag{3}$$

where D is the average demand per period

and $G_u(k) = \int_k^{\infty} (u_o - k) \frac{1}{\sqrt{2\pi}} \exp(-u_o^2/2) du_o$ is the partial loss function of the unit normal distribution.

Note that the expected value of the replenishment quantity will be RD so that the average inventory level will be

$$\bar{I} = RD/2 + k\sigma_{R+L} \tag{4}$$

Numerical Example:

Consider an item with the following parameter values: R = 1 period, L = 16 periods, D = 10 units/period, σ_1 = 8 units and P = 0.99. Furthermore, assume that $\sigma_t = \sqrt{t}\sigma_1$. Using (3) and (4) one can show that

$$\bar{I} = 5 + 77.8 = 82.8 \text{ units}$$

Suppose that a proposed change is to lower the review interval to ½ period. This leads to

$$\bar{I} = 2.5 + 83.9 = 86.4 \text{ units,}$$

i.e., the average inventory level rises because the cycle stock decrease is more than offset by an increase in the safety stock necessary to maintain the desired service level (there are now more frequent opportunities for stockouts).

Suppose that a proposed second change action will markedly reduce σ_1 to 2 units (e.g., improved forecasting perhaps by better information sharing with customers). This change alone reduces $\bar{I}$ to 20.4 units but, in addition, its use, together with the previous change, produces an $\bar{I}$ of 19.7. The results are summarized in Table 1 where we see that reducing R is attractive only when σ_1 is also reduced.

Table 1 — Illustration of Joint Effects (on Average Inventory) of Changing Givens

		forecast inaccuracy	
		8 units	2 units
review interval	1 period	$\bar{I}$ = 82.8 (base case)	20.4
	½ period	86.4	19.7

SOME OF THE MODELS ALREADY REPORTED IN THE LITERATURE

There is a rapidly growing literature on modelling the effects of changing the "givens" in Operations Management decision situations. Table 2 provides a brief summary of a portion of this literature with the references simply shown in alphabetical order. Most of the models include the cost of the change, usually amortized as part of total relevant costs per unit time.

PROBLEMS IN APPLYING CONVENTIONAL INVESTMENT EVALUATION TECHNIQUES

Continuous improvement alternatives can be viewed as investment options in that typically there are associated costs and benefits spread over time. Conventional investment evaluation approaches, such as DCF, ROI and so on, compute summary measures of these costs and benefits. If the majority of the relevant costs and benefits of a continuous improvement action can be estimated, then it is plausible to use DCF, etc. (as was illustrated in equation (1)). Unfortunately, the usual conservative approach used has been to include only the costs and benefits that can be reasonably estimated, thus implicitly ignoring the more intangible factors. This approach has been criticized in the literature as indicated by the following sample quotes, which are primarily related to investments in manufacturing technology, but are also relevant to continuous improvement actions involving substantially smaller capital outlays:

Table 2 - Literature Models of Changing "Givens"

Reference	Improvement Actions (Givens Changed)	Context	Additional Comments
Axsäter and Olhager (1985)	Setup times, unit production time	Several identical items with Poisson demand sharing a single machine	
Baker (1985) and Baker et al (1986)	Commonality of parts	Assemble-to-order situation with two items having three parts, one of which is common	
Billington (1987)	Setup cost	Economic production quantity	
Bourland et al (1991)	Knowledge of actual demand	Two echelon serial system with and without use of electronic data interchange	
Fine (1986)	Quality (conformance) level	Dynamic behavior of a single item considering costs of production, prevention and appraisal	Incorporates quality-based learning effects to show that optimal quality level is higher than the usual economic conformance level
Fine (1988)	Quality (control levels)	Process is either in or out-of-control with probability of defective depending upon the state	A 2-state, partially observable Markov process model is used. Transition probability to out-of-control state is reduced by learning
Fine and Freund (1990)	Production flexibility (economies-of-scope)	Considers acquisition costs of product-flexible capacity in context of uncertain product demand	Modelled as a two-stage stochastic program
Fine and Li (1988)	Production flexibility	Multiple products whose life cycle patterns differ	Deals with major capital investments
Fine and Porteus (1989)	Setup cost, quality level	Single item EOQ context but including rework costs	Employs stochastic dynamic programming model of continuous improvement process

Freeland et al (1990)	Setup cost	Multiple, but independent, items in EOQ context	Develops guidelines for sequencing the setup cost reductions of the different items
Gallego and Moon (1992)	Setup times	Multi-item economic lot scheduling on a single machine	Setup times are reduced at increased cost through externalizing setup operations
Gerchak and Henig (1986, 1989)	Commonality of parts	Assemble-to-order situation	Looks at special cases of multiperiod problem
Gerchak and Parlar (1991)	Variability of replenishment lead time	Single item continuous review inventory model	
Graves (1988)	Production rate and mix flexibility	Primarily for a single-stage process, but with some suggestions for multistage extensions	Shows effects on required safety stocks. Costs are not considered.
Hill and Khosla (1990)	Customer response time, replenishment lead time	Single item make to order or make to stock situation	Unit value and demand rate are both analytic functions of the response or lead time
Jönsson and Silver (1989)	Commonality of parts	Assemble-to-order situation with budget constraint on components	Heuristics and MILP formulation
Jönsson et al (1992)	Commonality of parts	Assemble-to-order situation with budget constraint on components	Uses scenario aggregation
Karmarkar and Kekre (1987)	Production flexibility	Two product, economic lot sizing situation	One large flexible machine versus two separate dedicated machines
Mefford (1989)	Quality, lower inventories	Single item EOQ and quality cost context	Incorporates inefficiency effects of inventories and lower quality levels and their interaction. Presentation is conceptual in nature, but does deal with dynamic effects.

Moon and Gallego (1992)	Production rates	Cyclic production of a family of items on a single facility	
Porteus (1985)	Setup costs	Single item EOQ situation	Particular analytic forms assumed for setup cost as a function of investment
Porteus (1986)	Setup costs, quality level	EOQ context but where process has a fixed probability of going out of control as each unit in a lot is produced	
Silver (1990)	Production rates	Cyclic production of a family of items on a single facility	
Spence and Porteus (1987)	Setup costs	Multiproduct capacitated EOQ situation with overtime available	
Tang (1990)	Yield rates, demand variability	Single item multistage serial production with random yield at each stage and random end item demand	
van Beek and van Putten (1987)	i). Setup costs ii). Changeover times iii). Lead times iv). Forecast accuracy v). Production rates	i). Single item finite rate EOQ ii). Two items sharing equipment iii). & iv). Single item reorder point, order quantity system v). Two stage serial system with level end-item demand	In each case value of strategic variable is a convenient analytic function of the investment level
Vander Veen and Jordan (1989)	Production flexibility		How many machines to acquire and possessing what flexibilities (in terms of products that each can manufacture)

".... CIM technology provides many additional benefits — better quality, greater flexibility, reduced inventory and floor space, lower throughput times, experience with new technology — that a typical capital justification process does not quantify. Financial analyses that focus too narrowly on easily quantified savings in labor, materials, or energy will miss important benefits from CIM technology .... These benefits are ... much harder to quantify ... The difficulty arises in large part because these benefits represent revenue enhancements rather than cost savings. It is fairly easy to get a ball park estimate for percentage reductions in costs already being incurred. It is much harder to quantify the magnitude of revenue enhancement expected from features that are not already in place." *(Kaplan (1986))*

"It is simply incongruous to use a justification approach based on cost savings to make a decision about turning manufacturing into a strategic competitive weapon. Strategies are, by definition, oriented toward the long term, and any anticipated cost benefits simply may not materialize in the short-term financial horizon" *(Meredith and Hill (1987))*

"Capital investment has always been justified on the basis of cost reduction or capacity increase. These technologies offer only slight improvements on these grounds; their big advantages of market response, economic variety, customization, and so on, simply don't clear the ROI hurdle. Does a new sales promotion have to satisfy ROI criteria? or a new hire in the VP ranks?" *(Meredith (1987))*

"The obvious costs — the amortized cost of the machine and the labor to operate it — only scratch the surface. WCM requires paying heed to some other machine capabilities — which together overshadow the surface costs:
· How fast the machine can be set up,
· How easy it is to maintain the machine and keep it making good product,
· Whether the machine's speed can be adjusted up and down to match up-and-down use rates at next processes and up-and-down final demand rates,

· Whether the machine's price is low enough for multiple copies of it to be bought over time, matched to the growth of demand" *(Schonberger (1986))*

"Just as it is difficult to quantify the costs and benefits of investments in (hard) factory automation, it is also difficult to quantify costs and benefits of a 'soft' technology such as JIT" *(Fine (1992))*

As the above quotes imply, some of the benefits of the continuous improvement philosophy are probably very difficult to quantify and/or convert into a single measure of effectiveness, yet they may be crucial from a strategic viewpoint. For example,

- reducing throughput time can change the whole nature of a company's market, e.g., from complete make-to-stock to complete make-to-order,
- not doing anything probably will not maintain the status quo if competitors reduce throughput times, improve quality levels, etc.,
- there can be learning effects of some continuous improvement actions whose benefits will only be realized in the implementation of later actions.

Finally, flexibility is of central concern in World Class Operations Management, but it is difficult to measure and evaluate its benefits. Sethi and Sethi (1990) define the flexibility of a system as "its adaptability to a wide range of possible environments that it may encounter". They go on to provide a categorization of 11 different types of flexibility.

Where there are important intangibles present at least two approaches are possible for incorporating them. First, as suggested by Kaplan (1986), one can use an indifference argument, simply estimating how large the intangible benefits have to be in order for use of the associated alternative to be justified. Second, to make a choice among competing alternative courses of action one can use so-called scoring models (see, for example, Buffa and Sarin (1987) or Evans et al (1990)) where a score on each of several factors is assigned for each alternative and the alternatives are then compared by means of weighted sums of the scores.

SUGGESTIONS FOR FUTURE APPLIED RESEARCH

From what has been presented to this juncture it is clear that mathematical modelling is currently playing a very limited role in supporting the continuous improvement philosophy at the level of the individual organization. It is felt that the following suggestions regarding applied research should help ameliorate the situation:

1. Development of relatively simple heuristic decision rules, recognizing that the philosophy of continuous improvement implies many possible changes, each of which has to be evaluated quickly and inexpensively. As Suri and de Treville (1991) have stated "... the need for quick answers supersedes the need for minute precision".
2. More use of what-if, as opposed to optimization, analyses in that fewer, questionable assumptions are usually needed if optimization is avoided. See Ward et al (1991) for a discussion of why many theoretical decision models fail to meet the needs of practical decision–making. One likely fruitful form of what-if analysis would involve a relatively simple, (perhaps visual interactive) simulation model. The rapid modelling technology of Suri and de Treville (1991) is of this type.
3. Where possible, consider using existing models to answer more relevant questions than have been typically addressed by them. For example, the EOQ model, rather than finding the order quantity for a given value of the setup cost, can be used to find by how much the setup cost has to be reduced in order to economically produce in small lots (Freeland et al (1990), Hill (1991)).
4. Greater use of indifference approaches as opposed to attempting to pin down precise quantitative benefits of important intangibles.
5. Development of modelling tools more explicitly oriented to service situations, e.g., Little's point made earlier, viz. "... empowering frontline workers in organizations to improve their effectiveness".
6. Utilization of mathematical tools, such as parametric programming, stochastic ordering, etc. that facilitate variation of the values of the parameters in models (Gallego (1991)).
7. Development of company-specific scoring models that could then be easily used for repetitive decision making.

SUMMARY

In this paper we have reviewed some of the existing mathematical models that are relevant to implementing the continuous improvement philosophy. Most of these models are of a conceptual nature, as opposed to being directly useful to a specific organization in selecting among several possible change actions. Some recommendations for future research directions have been presented.

ACKNOWLEDGEMENTS

The research leading to this paper was supported by the Natural Sciences and Engineering Research Council of Canada under Grant No. A1485. A special acknowledgement is made to Elwood Buffa whose pioneering work in the field of Operations Management has had a profound positive effect on the activities of so many individuals including myself. I would also like to thank a number of individuals who provided writeups of related material as well as comments on an early outline of the paper. These people include Sven Axsäter, Joe Blackburn, Charles Fine, Marshall Fisher, Jim Freeland, Guillermo Gallego, Stephen Graves, Arthur Hill, Robert Mefford, Jack Meredith, Evan Porteus, David Pyke, Tom Rohleder, Gary Scudder, Suresh Sethi, Jim Smith, and Paul van Beek. This paper is dedicated, in loving memory, to my son, Norman Ronald Silver.

REFERENCES

Axsäter, S. and Olhager, J., "The Impact of Capacity Investments on Work-in-Process and Inventories", Engineering Costs and Production Economics, Vol. 9, 1985, pp. 119-124.

Baker, K. R., "Safety Stocks and Component Commonality", Journal of Operations Management, Vol. 6, No. 1, 1985, pp. 13-22.

Baker, K. R., Magazine, M. J. and Nuttle, H., "The Effect of Commonality on Safety Stock in a Simple Inventory Model, Management Science, Vol. 32, No. 8, August 1986, pp. 982-988.

Billesbach, T. J. and Schniederjans, M, J., "Applicability of Just-in-Time Techniques in Administration", Production and Inventory Management Journal, Vol. 30, No. 3, 1989, pp. 40-45.

Billington, P. J., "The Classic Economic Production Quantity Model with Setup Costs as a Function of Capital Expenditure", Decision Sciences, Vol. 18, 1987, pp. 25-42.

Blackburn, J. D., "Time-Based Competition", Chapter 10 in Strategic Manufacturing: Dynamic New Directions for the 1990s (editor: P. Moody), Dow Jones-Irwin, 1990, pp. 189-206.

Bourland, K. E., Powell, S. G. and Pyke, D. F., "Electronic Data Interchange and the Management of Inventories", Working Paper #266, Amos Tuck School of Business Administration, Dartmouth College, August 1991.

Buffa, E. S. and Sarin, R. K., Modern Production/Operations Management, Eighth Edition, John Wiley & Sons, New York, 1987, pp. 583-586.

Centre for Quantitative Methods, "Decision Support for Logistics Improvements", Information Sheet No. 13, Philips, Eindhoven, The Netherlands, March 1986.

Crosby, P. B., Quality is Free, McGraw-Hill, New York, 1979.

Deming, W. E., Out of the Crisis, Massachusetts Institute of Technology Center for Advanced Engineering Studies, Cambridge, Mass., 1986.

Evans, J. R., Anderson, D. R., Sweeney, D. J. and Williams, T. A., Applied Production and Operations Management, Third Edition, West Publishing, St. Paul, MN, 1990, pp. 171-2.

Feigenbaum, A. V., Total Quality Control: Engineering and Management, Third Edition, McGraw-Hill, New York, 1983.

Fine, C. H., "Developments in Manufacturing Technology and Economic Evaluation Models", Logistics of Production and Inventory, (editors: S. C. Graves, A. Rinnooy Kan and P. Zipkin), North Holland, in press.

Fine, C. H., "A Quality Control Model with Learning Effects", Operations Research, Vol. 36, No. 3, May-June 1988, pp. 437-444.

Fine, C. H., "Quality Improvements and Learning in Productive Systems", Management Science, Vol. 32, No. 10, October 1986, pp. 1301-1315.

Fine, C. H. and Freund, R. M., "Optimal Investment in Product-Flexible Manufacturing Capacity", Management Science, Vol. 36, No. 4, April 1990, pp. 449-466.

Fine, C. H. and Li, L., "Technology Choice, Product Life Cycles, and Flexible Automation", Journal of Manufacturing and Operations Management, Vol. 1, 1988, pp. 372-399.

Fine, C. H. and Porteus, E. L., "Dynamic Process Improvement", Operations Research, Vol. 37, No. 4, July-August 1989, pp. 580-591.

Fisher, M. L., "Opportunities for Operations Research in the New Manufacturing", Operational Research 90: Selected Papers from the 12th IFORS International Conference on Operational Research (editor: H. Bradley), Pergamon Press, Oxford, England, 1990, pp. 577-588.

Freeland, J. R., Leschke, J. P. and Weiss, E. N., "Guidelines for Setup-Cost Reduction Programs to Achieve Zero Inventory", Journal of Operations Management, Vol. 9, No. 1, January 1990, pp. 85-100.

Gallego, G., private correspondence, June 1991.

Gallego, G. and Moon, I., "The Effect of Externalizing Setups in the Economic Lot Scheduling Problem", Operations Research, in press.

Gerchak, Y. and Henig, M., "Component Commonality in Assemble-to-Order Systems: Models and Properties", Naval Research Logistics, Vol. 36, 1989, pp. 61-68.

Gerchak, Y. and Henig, M., "An Inventory Model with Component Commonality", Operations Research Letters, Vol. 5, 1986, pp. 157-160.

Gerchak, Y. and Parlar, M., "Investing in Reducing Lead-Time Randomness in Continuous-Review Inventory Models", Engineering Costs and Production Economics, Vol. 21, 1991, pp. 191-197.

Graves, S. C., "Safety Stocks in Manufacturing Systems", Journal of Manufacturing and Operations Management, Vol. 1, No. 1, 1988, pp. 67-101.

Hall, R. H., Zero Inventories, New York, Dow Jones-Irwin, 1983.

Hayes, R. H. and Clark, K. B., "Explaining Observed Productivity Differentials Between Plants: Implications for Operations Research", INTERFACES, Vol. 15, No. 6, 1985, pp. 3-14.

Hill, A. V., private correspondence, August 1991.

Hill, A. V. and Khosla, I. S., "Models for Optimal Lead Time Reduction", Working Paper 90-6, Department of Operations and Management Science, Curtis L. Carlson School of Management, University of Minnesota, August 1990.

Imai, M., Kaizen: The Key to Japan's Competitive Success, Random House, New York, 1986.

Inman, R. A. and Mehra, S., "JIT Applications for Service Environments", Production and Inventory Management Journal, Vol. 32, No. 3, 1991, pp. 16-20.

Jönsson, H., Jornsten, K. and Silver, E. A., "Application of the Scenario Aggregation Approach to a Two-Stage, Stochastic, Common Component, Inventory Problem with a Budget Constraint", European Journal of Operational Research, in press.

Jönsson, H. and Silver, E.A., "Common Component Inventory Problems with a Budget Constraint: Heuristics and Upper Bounds", Engineering Costs and Production Economics, Vol. 18, 1989, pp. 71-81.

Juran, J. M. and Gryna, F., Quality Planning and Analysis, Second Edition, McGraw-Hill, New York, 1980.

Kaplan, R. S., "Must CIM Be Justified By Faith Alone?", Harvard Business Review, Vol. 64, No. 2, March-April 1986, pp. 87-95.

Karmarkar, U. S. and Kekre, S., "Manufacturing Configuration, Capacity and Mix Decisions Considering Operational Costs", Journal of Manufacturing Systems, Vol. 6, 1987, pp. 315-324.

Krafcik, J. F., "Triumph of the Lean Production System", Sloan Management Review, Vol. 31, 1988, pp. 41-52.

Little, J. D. C., "Operations Research in Industry: New Opportunities in a Changing World", Operations Research, Vol. 39, No. 4, July-August 1991, pp. 531-542.

Mefford, R. N., "The Productivity Nexus of New Inventory and Quality Control Techniques", Engineering Costs and Production Economics, Vol. 17, 1989, pp. 21-28.

Meredith, J. R., "The Strategic Advantages of the Factory of the Future", California Management Review, Vol. 29, No. 3, 1987, pp. 27-41.

Meredith, J. R. and Hill, M. M., "Justifying New Manufacturing Systems: A Managerial Approach", Sloan Management Review, Vol. 28, No. 4, 1987, pp. 49-61.

Moon, I. and Gallego, G., "Controllable Production Rates in the Family Production Context", International Journal of Production Research, in press.

Porteus E. L., "Optimal Lot Sizing, Process Quality Improvement and Setup Cost Reduction", Operations Research, Vol. 34, No. 1, January-February 1986, pp. 137-144.

Porteus, E. L., "Investing in Reduced Setups in the EOQ Model", Management Science, Vol. 31, No. 8, August 1985, pp. 998-1010.

Ramasesh, R. V., "Recasting the Traditional Inventory Model to Implement Just-in-Time Purchasing", Production and Inventory Management Journal, Vol. 31, No. 1, 1990, pp. 71-75.

Schmenner, R. W., "The Merit of Making Things Fast", Sloan Management Review, Vol. 30, No. 1, 1988, pp. 11-17.

Schonberger, R. J., World Class Manufacturing, Free Press, New York, 1986.

Schonberger, R. J. and Schniederjans, M. J., "Reinventing Inventory Control", INTERFACES, Vol. 14, No. 3, 1984, pp. 76-83.

Sethi, A. K. and Sethi, S. P., "Flexibility in Manufacturing: A Survey", The International Journal of Flexible Manufacturing Systems, Vol. 2, 1990, pp. 289-328.

Shingo, S., A Revolution in Manufacturing: The SMED System, Productivity Press, Stamford, Conn., 1985.

Silver, E. A., "Changing the Givens in Modelling Inventory Problems: The Example of Just-in-Time Systems", Engineering Costs and Production Economics, in press.

Silver, E. A., "Changing the Givens in Operations Research Modelling: The Operations Management Example of Just-in-Time Systems", Inaugural Visiting Scholar Lecture, The Institute of Advanced Studies, Vienna, Austria, June 1990.

Silver, E. A., "Deliberately Slowing Down Output Rates in the Family Production Context", International Journal of Production Research, Vol. 28, No. 1, January 1990, pp. 17-27.

Silver, E. A. and Peterson R., Decision Systems for Inventory Management and Production Planning, Second Edition, John Wiley & Sons, New York, 1985, Chapter 7.

Spence, A. M. and Porteus, E. L., "Setup Reduction and Increased Effective Capacity", Management Science, Vol. 33, No. 10, October 1987, pp. 1291-1301.

Stalk, G. and Hout, T. M., Competing Against Time, Free Press, New York, 1990.

Suri, R. and de Treville, S., "Full Speed Ahead", OR/MS Today, June 1991, pp. 34-42.

Tang, C. S., "The Impact of Uncertainty on a Production Line", Management Science, Vol. 36, No. 12, December 1990, pp. 1518-1531.

van Beek, P. and van Putten, C., "OR Contributions to Flexibility Improvements in Production/Inventory Systems", European Journal for Operational Research, Vol. 31, 1987, pp. 52-60.

van Putten, C. and van Tongeren, A. R., "Quantification Makes Integral Logistics Plans More Effective", Logistics News, Vol. 54, O&E Information Centre, Philips, Eindhoven, October 1986, pp. 1-4.

Vander Veen, D. J. and Jordan, W. C., "Analyzing Trade-Offs Between Machine Investment and Utilization", Management Science, Vol. 35, No. 10, October 1989, pp. 1215-1226.

Vesey, J. T., "Meet the New Competitors: They Think in Terms of Speed-to-Market", Industrial Engineering, Vol.22, No.12, December 1990, pp.20-26.

Ward, S. C., Chapman, C. B. and Klein, J. H., "Theoretical Versus Applied Models: The Newsboy Problem", OMEGA, Vol. 19, No. 4, 1991, pp. 197–206.

Whybark, D. C., "The Curriculum Crisis", Operations Management Review, Vol. 6, Spring and Summer 1988, pp. 53-55.

Zipkin, P. H., "Does Manufacturing Need a JIT Revolution?", Harvard Business Review, Vol. 69, No. 1, January-February 1991, pp. 40-50.

DESIGN FOR SUPPLY CHAIN MANAGEMENT: CONCEPTS AND EXAMPLES

Hau L. Lee
Department of Industrial Engineering and Engineering Management
Stanford University
Stanford, USA

ABSTRACT

In recent years, there has been increased emphasis on linking the design of products and processes to cost, quality and efficiency measures in manufacturing. Concepts such as "design for manufacturability," "design for assembly," "design for producibility," and "design for testability," have emerged. The underlying motivation of these concepts is that product costs are largely determined by the design of the product. Hence, functionality and performance are no longer sufficient criteria for excellent design. There are additional important cost and service determinants of product success and profit that go well beyond a narrow definition of manufacturing. Factors such as logistics and distribution costs, inventory investment, product availability, customs duties, serviceability, and flexibility in dealing with unexpected changes, are important considerations which can be affected by the design of the product or its manufacturing process. We define a supply chain as a network including the procurement of raw materials, processing of intermediate and finished products, customization of the product for local market needs, and distribution of the product to customers. This paper describes emerging concepts for product and process design that include a supply chain perspective. Four case studies are presented to illustrate how these concepts can be implemented in real situations.

INTRODUCTION

In recent years, there has been increased emphasis on the incorporation of manufacturing considerations in the design phase of a product. It has been found that the manufacturing cost of a product is largely determined by the design of the product. Consequently, it is important to include consideration of the costs, efficiency and quality aspects of manufacturing in the design phase. Concepts such as "design for manufacturability," "design for assembly," "design for producibiliity," "design for testability," and "design for quality" have emerged both in research and in practice (see Whitney, 1988, Dean and Susman, 1989, and Taguchi and Clausing, 1990).

However, manufacturing is only one of the many steps in the order fulfillment cycle of the product. This order fulfillment cycle involves the procurement of raw materials, the transformation of materials into intermediate and finished goods (manufacturing), the transfer of goods from one site to another, the customization of the product to suit the local market needs, and the distribution of the product to customers. We use the term "supply chain" to denote the network of facilities that perform the functions in the order fulfillment cycle (see Cohen and Lee, 1988). Managing a supply chain is a complex task. Lee and Billington, 1992, describe some of the pitfalls and opportunities existing in the management of supply chains. Product design can have a significant impact on factors like logistics and transportation costs, customization strategies, inventory investment, tax and customs duties and flexibility costs to changes. They can affect the operational performance (cost, service, flexibility) of the supply chain. We use the term "design for supply chain management" to describe the concept of design products and processes that would support the management of a supply chain in addressing the above factors. These factors and their cost implications should not be overlooked. For example, logistics costs alone generally constitute 10% to 26% of the total sales of a company (see Ljungdahl, 1991).

The complexity of supply chains and the associated potential opportunities for gaining a competitive edge by designing a product and process to support supply chain management can be significant. For companies that have manufacturing and distribution facilities worldwide, serving a global market, the complexity and potential are even greater (see Cohen and Lee, 1989). We contend that design for supply chain management is a important strategic and operational consideration for companies competing in a global market.

In this paper, some basic concepts related to design for supply chain management are described. This is by no means an exhaustive list. Some examples drawn from industrial cases are used to illustrated. In particular, we describe the issues that need to be resolved in order to have successful implementation of these concepts in practice. The paper is organized as follows. The basic concepts are first introduced. This is followed by a description of the common obstacles that explain why design for supply chain concepts have not been more widely used or implemented, despite their potential great payoffs. Four case examples are then presented to illustrate how such concepts can result in competitive advantages and cost savings. The specific challenges and issued faced by the designers and managers in these cases are also described.

SOME BASIC CONCEPTS

A key concept in design for supply chain management is delayed product differentiation. This is also known as the deferred differentiation strategy, end of line configuration, late point differentiation strategy, or simply a postponement strategy.

In a global market, due to the different local requirements in taste, language, environment and government regulations, multiple versions of a single product are required, each meeting the specific requirements of a local geographical region. For example, computer products for various countries may differ in the power supply module to accommodate local voltage, frequency, and plug conventions. Keyboards and manuals must match local language. Telecommunication products may also be differentiated by the communication protocols supported. In some cases, the need for localized versions of a product results from government-imposed local content requirements.

Even within the same geographical region, there may also be multiple models of the product within a product family that relate to different functionalities or capabilities of the product. These different models may also reflect the differential needs of different market segments, e.g., business, education, personal, or government sector. Hence, it is not uncommon to have significant proliferation of product numbers within a single product family. Delayed product differentiation refers to delaying the point in time when a product assumes its identity, i.e., a particular model in a particular region for a particular market segment.

It is usually much more difficult to forecast the demands of an individual model in a region/market segment. Greater accuracies are usually observed in forecasting total product family demand than forecasting at the individual model/region/market segment level. Inventories held in generic form can be much lower than in individual model form, due to the benefit of risk pooling, to meet the same level of customer service. Hence, delaying product differentiation increases the supply chain's flexibility in meeting customer demands, reduces inventory, and improves customer service performance.

Product designs to allow for delayed product differentiation usually involve modular structure of the product, with a few key modules serving as the differentiator of the product for the different models/regions/market segments. Hence, "design for modularity" is an important strategy (Barkan, 1991). It also calls for the layout or configuration of the product so that the differentiating module(s) can be easily assembled at the end of the supply chain.

The point at which product differentiation occurs could be the end of the traditional manufacturing process, the distribution centers, the distribution channels (such as dealers for computer products), en route or in transit to the customers, or even at the customer site.

The strategy to delay product differentiation to meet different local region requirements is also known as "design for localization," whereas the strategy to meet the needs of different models/market segments is sometimes termed "design for customization" (see Lee et al., 1991).

The most extreme point for delayed product differentiation is to have differentiation occurring at the time when the product reaches the customer. In that case, the product needs to have built-in capabilities to handle the multiple requirements. There are many such examples in the consumer electronics market. For example, products can have a switchable power supply, allowing the customer to choose the particular power voltage at the time of usage. Telephones having both rotary dial and direct tone dial capabilities are another example. Printers that can be used for both Macintosh or IBM-compatible personal computers, requiring a change of cable or a simple switch, are another example. At first sight, it seems that products that provide the flexibility for product differentiation at the consumption level are more costly. This is not always the case. Besides the marketing value of the flexibility, the postponement of product differentiation means that there is no need to have manufacturing or distribution processes inside the supply chain of a company to perform the physical tasks of product differentiation. There can be cost savings and inventory reduction resulting from such standardization.

A concept that is highly related to delayed product differentiation is increased part commonality and interchangeable subassemblies ("design for flexible manufacture.") Part commonality can result in cost savings in part number administration, inventory reduction, and supplier management. A more powerful benefit of part commonality, often ignored in the literature, is that it can be used as a means to achieve delayed product differentiation. The manufacturing process of a product and its associated multiple versions may involve multiple stages, each requiring different input parts and subassemblies. Increasing the level of part commonality at the early stages of the manufacturing process is similar to postponing the differentiation of the products until after these early stages. When used appropriately, part commonality can result in benefits that go beyond the commonly cited ones.

As mentioned earlier, the logistics costs of moving materials and products from one manufacturing site to another, from factory to distribution sites, or from distribution sites to customer sites can be a significant part of the product cost. "Design for logistics" is an emerging concept that is being used at companies such as Digital Equipment and Hewlett-Packard. The idea is to design products that would be easy and cost-efficient to be transported. For example, the packaging of the product and the shape of the product itself should be designed so as to maximize the pallet density, to lower the freight cost of the product. This concept can also complement design for delayed product differentiation. For example, if final packaging can be done at distribution centers instead of at the factory, then the product can be transported from the plant to the distribution center in bulk-packaged form, consequently lowering freight costs.

A related design concept is to design products so that anticipated engineering changes can be easily made. Again, a modular product design structure would make this possible. Engineering design changes are common in new products; sometimes these changes may have to be made to products already sold, or to products in inventory at different parts of the supply chain. A design that allows such changes to be made without complete disassembly of the product, for example, could save costs and resources.

"Product line structuring" is another powerful concept that supports design for supply chain management (see Scott and Tuoti, 1982). It relates to the structuring of the product so that the many models and versions of end products are assembled from relatively independent subassemblies and auxiliary systems. The benefit is that these subassemblies can then be manufactured concurrently instead of serially. This reduces manufacturing lead time and increases flexibility to meet customer demands. Barkan, 1991, discusses the benefits of such approaches. He also cites companies such as General Electric, Chicago Pneumatic and Allen-Bradley as ones who have employed product line structuring with great success (Barkan, 1992).

Product line structuring and product variety reduction strategies are design methodologies which offer potentials in improving supply chain efficiency (Dumaine, 1989, and Suzue and Kohdate, 1990). They also offer rich opportunities and materials for interdisciplinary research and for the development of new design methodology.

Product design can also have significant impact on customs and duties for global companies. Such companies typically have manufacturing and distribution sites that are located globally. Subassemblies, semi-finished products, and finished products may cross continents before they reach customers. To what extent semi-finished products are configured can result in either no, partial, or full customs and duties levied on the semi-finished products when they cross borders. It may not be the primary focus for designers to design products to minimize customs and duties paid, but this is certainly a consideration in a global market.

COMMON OBSTACLES FOR IMPLEMENTATION

There are, of course, technical and creative challenges that designers must meet so that the new design of the product or process can result in superior operational performance (cost, service, flexibility) of the supply chain for the product. An encouraging note should be mentioned here. In the many examples cited in references describing Taguchi's method of design for quality, it is often stated that design changes do not always have to be an expensive and difficult undertaking. There often exist simple design changes that can result in significant reduction in product quality variations. Similarly, there are also many instances in which design changes for supply chain management are not overwhelming in terms of their technical difficulties or costs. Oftentimes, it is the mind-set of the design team that matters, and the associated performance evaluation systems put in place.

Similar to the "design for manufacturability" concept, design for supply chain management requires design engineers to take a broader perspective than product functionality and performance. It also requires enlarging the definition of costs in the evaluation of alternative designs, which often are narrowly defined as simply involving the material costs of the product and direct labor for assembly. Such a broad perspective is neither one that design engineers were trained to use, nor one advocated by traditional performance measurement or incentive systems in their work environment. Linking design with manufacturing is already nontrivial. Linking design with logistics, distribution, customization, customs and duties, and customer service requirements is even more challenging.

Designs that take advantage of the supply chain structure, or that result in bigger savings in the non-manufacturing elements of a supply chain, often require close collaboration of multiple functions within a corporation. These functions may include distribution, sales and marketing, finance, and customer support, in addition to engineering, R&D, and manufacturing. The organizational barriers between these functions can be very high. Collaboration is also difficult when the cultures, objectives, and measurement systems in place at the different functions are different.

Design changes may not be completely free. There will have to be some investment to make it happen. Standardizing components and parts may increase material costs. Delaying product differentiation may require adding special processes at the end of the supply chain. Providing flexibility to customers may also require adding functionality to the product. However, the key is that there will also be significant benefits that may outweigh these cost investments. In that sense, design for supply chain management may be "free," just like Crosby's "quality is free" concept (Crosby, 1979).

One difficulty with the initial cost investment that is required for refining a design to suit supply chain management is that management and the design team could be reluctant to proceed with the new design unless there is a concrete estimate of the corresponding benefits. Logistic costs such as freight, customs and duties, etc., can be easily quantified. The benefits in terms of lower inventory, faster response times to customers, increased availability levels and increased flexibility, are much more difficult to quantify. It is here that inventory theories and models can make a contribution. There will also be benefits, such as increased worker morale, improved quality, and marketing values, etc., which one may never be able to appropriately quantify. The inability to quantify costs and benefits can be an impediment to the development of designs for supply chain management.

DESIGN FOR DELAYED PRODUCT DIFFERENTIATION:
A Computer Disc Drive Example

To illustrate some of the concepts described earlier, and to highlight the significance of the obstacles and issues involved in the implementation of design for supply chain management, four example cases are presented.

The first one deals with a US disc drive manufacturer in a build to order environment. Orders for the products of this manufacturer come from different computer manufacturers (OEMs), the disc drive of each of which is unique. These orders, once placed, are to be met within some given time window. Disc drive manufacturing has a long lead time, since many time-consuming tests are involved. Hence, it is necessary to have in-process inventory so as to shorten the response time to customer orders.

Orders for different disc drives (destined for different OEMs) are highly variable, due in part to the difficulties for the OEMs to accurately forecast their own demands. Consequently, high levels of in process inventory are needed to enable the disc drive manufacturer to provide a high level of reliability to meet the required order lead times.

The overall manufacturing process of disc drive can be divided into two parts. The first part is generic to all disc drives, i.e., disc drives completed in this first part can be used to satisfy demands for any OEMs. In the second part, the disc drive is customized to the specifications of the different OEMs. Hence, the beginning of the second part is where product differentiation begins. The ideal point for in process inventory is thus at the end of the first part of the process, since inventory held at that point can be used to be customized for any of the different disc drives. The manufacturing process, however, was designed in such a way that the first part is relatively short. There are a number of key time-consuming tests that need to be carried out during the process. Printed circuit boards are needed to perform these tests. Different OEM products require different printed circuit boards. Thus, the insertion of these boards also "customizes" the disc drive for the different end products. Consequently, the point at which printed circuit boards are inserted into the disc drives is the point where the second part of the overall manufacturing process begins.

Since the second part of the manufacturing process is relatively long, a high level of in-process inventory (at the point where the first part ends) is required in order to support high reliability of some order lead time targets. It would be greatly desirable to postpone the point of product differentiation, since in-process inventory can then be held much closer to the end of the process. When inventory is held at a later point of the process, less of that is required to meet the same reliability of the order lead time targets. However, inventory at a later point of the process would usually have higher value-added. Hence, although a lower level of safety stock is required, the value of inventory investment may actually be higher.

It would be ideal to design the process so that the point of product differentiation is postponed without increasing the value of inventory at that later point. This would call for reconfiguring the process so that process steps with little value-added that used to be performed after the differentiation point can be made generic, i.e., performed before the differentiation point. For our disc drive manufacturing example, an improvement to the current process can be achieved by the insertion of a "coupon" board (can be viewed as a generic board) so that the testing steps subsequent to product differentiation can now be completed prior to the insertion of the printed circuit board. After the series of tests were completed, the coupon board is then removed. The actual printed circuit board is then inserted at this point. Hence, the point of product differentiation is postponed from the beginning of the tests to after the tests were completed. Since tests do not add significant value to the product, the delayed product differentiation strategy does not incur an increase in the value of inventory investment. This is illustrated in Figure 1 below.

Figure 1: Disc Drive Manufacturing Example Illustrated

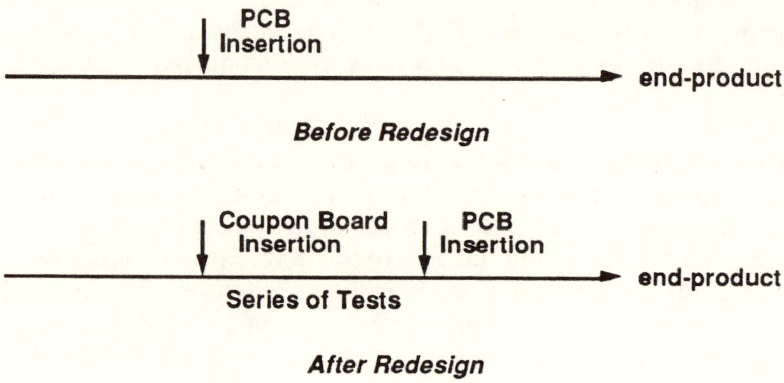

Despite the apparent attractiveness of delayed product differentiation, the implementation of such a concept for the disc drive manufacturer was not without major challenges. The biggest resistance came from manufacturing management personnel who were "Just-In-Time purists." They believed that any non-value added manufacturing steps should be eliminated. The insertion and subsequent removal of a coupon board was viewed as such a step. It is hard to appreciate the value of flexibility resulted from delayed product differentiation. Nevertheless, quantification of the benefits of inventory reduction was a key step to solidify management support for eventual implementation of the process design change.

DESIGN FOR LOCALIZATION:
A Deskjet Printer Example

The Deskjet is one of several printers manufactured by a key computer manufacturer. There are two key stages in the manufacturing process: (1) printed circuit board assembly and test (PCAT); and (2) final assembly and test (FAT). PCAT involves the assembly and testing of electronic components such as ASICs (application-specific integrated circuits), ROM (read-only memory), and raw printed circuit boards to make logic boards and print head driver boards for the printers. FAT involves the assembly of other subassemblies like motors, cables, key pads, plastic chassis and "skins," gears, and the printed circuit boards from PCAT to produce a working printer, as well as the final testing of the printer.

The localization of the Deskjet for different countries involves packaging the appropriate power supply module, with the correct voltage and power cord terminators (plugs), and manual with the printer. In the past, such a step was performed at the factory itself. Hence, the finished products of the factory consisted of printers destined for all other countries. These products were then sorted into three groups for the three distribution centers (DCs): North America, Europe and Asia-Pacific, respectively. Such a manufacturing process is referred to as "factory-localization."

Both the PCAT and FAT stages are located at a plant in the U.S. Outgoing products are shipped to the DCs. The printer industry is highly competitive. Consequently, there has been increasing pressure for the manufacturer to provide high levels of availability at the DCs for the dealers. In response, management has decided to operate the DCs in a make-to-stock mode in order to provide very high levels of availability to the dealers. Hence, the DCs operate as an inventory stocking point with large safety stocks to meet a target off-the-shelf fill rate, where the replenishment of products come from manufacturing.

Manufacturing, on the other hand, operates in a pull mode. Production plans are set to replenish the DCs "just-in-time" to maintain the target safety stocks.

Under "factory-localization," the different versions of the Deskjet (printers with different power supplies and manuals) are shipped to the two non-US DCs by sea, with a transit time of approximately one month. The consequence of such a long lead time is that the DC's ability to respond to fluctuations in demands for the different versions of the product is limited. In order to assure high fill rate service for the customers, the European and Asian-Pacific DCs thus have to maintain high levels of safety stocks.

Localization at the DCs provides an attractive alternative to gain control of inventory and service. Hence, the factory would be responsible for manufacturing a generic Deskjet printer without the power supply module and manual. Generic products are shipped to the DCs. At the DCs, the generic product is then localized to the different specific options as they are needed. From standard inventory theory, such a change would have the benefit of the risk pooling effect, and consequently, less safety stock in finished goods would be needed. Such a strategy is termed "DC-localization." Figure 2 shows the two localization strategies for the Deskjet printer.

To be able to implement DC-localization, some design changes had to be made to the product. For example, it is important to redesign the product so that the power supply module is the last component to be added on and can be added easily at the DC. To minimize the time that it takes to localize the product, the assembly of the power supply module to the product should be a simple operation, e.g., a simple plug-in. Moreover, even if the assembly of the power supply module and manuals is a simple kitting process, some investment is needed to

Figure 2: Printer Localization Example Illustrated

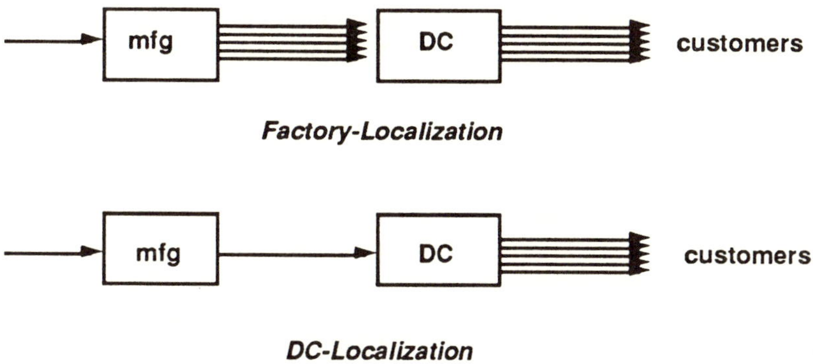

equip the international DCs with such a capability.

In terms of inventory control, in the case of factory-localization, safety stocks of the power supply modules and manuals are kept at the factory only, while the DCs keep stocks of the localized finished product. For DC-localization, the international power supply modules and manuals are kept at the remote DCs. Semi-finished goods, i.e., unlocalized version of the product, are stocked at the European and the Far East DCs. When actual orders are received, a quick operation localizes the generic version into the specific product required.

Analysis of the safety stock requirements indicates that the inventory savings from factory-localization to DC-localization is substantial. In addition to inventory, there are many other factors that should be considered in a comprehensive evaluation of the localization alternatives. First, a localized printer contains localization materials, and so it has a higher value than an unlocalized printer. Hence, the capital tied up in pipeline inventory (inventory in transit) is also lower when localization occurs at the end of the chain, i.e., DC-localization. Second, an unlocalized printer is much less bulky than a localized one, as the localization materials and many of the final packaging materials that are needed for the customers do not have to be bundled with the printer. One can thus ship the unlocalized printers in bulk pallets, and cut the cost of transportation significantly. Third, design changes to the product are needed to implement DC-localization. This requires valuable engineering resources. Moreover, final configuration and packaging capability would now have to be added to the Far East and European DCs. There is, of course, a cost associated with such an implementation.

Finally, there are some other non-quantifiable factors to be considered as well. For example, increasing "local content" and local "manufacturing" presence can make a product more marketable, which supports doing DC-localization at the non-US DCs (see Cohen and Lee, 1989). There is also a need to develop a local supply base of the localization materials for the DCs. The pros and cons of decentralizing or regionalizing supply of materials should be considered (again refer to Cohen and Lee, 1989). Finally, since DC-localization requires the DCs to perform some operations that are traditionally viewed as manufacturing activities, there may be cultural and organizational barriers to overcome.

After considering both the quantifiable and the non-quantifiable factors mentioned above, the manufacturer is now beginning to design or redesign all current and future products to support the DC-localization strategy.

DESIGN FOR COMMONALITY
A Color Printer Example

Our next example concerns a printer manufacturer which is about to introduce a new product, a color printer. Demands for the new color product and the existing mono printer are highly uncertain. It is quite likely the demands for the two products are negatively correlated. Again, product availability is a key factor for competition in this market. To meet high levels of product availability, the manufacturer has to stock high levels of finished products. Yet, with high degree of uncertainties in the market place, it is likely to over- or under-forecast the demands of the two products, resulting in inventory building up for one product while shortages exist for the other one.

The manufacturing processes for the two products are basically very similar, except for the materials used. As before, there are two key stages: printed circuit assembly and test, and final assembly and test. There are two key subassemblies that differentiate the color and mono printers. At the printed circuit assembly stage, two distinct head driver boards are used, one for each product. At the final assembly stage, two distinct print mechanism interfaces are also used. There are also other differences in the bill of materials for the two products, but these are minor and can easily be standardized.

The differences in the head driver boards and the print mechanism interfaces result in two distinct manufacturing processes to meet the demands of the products. There is little flexibility at the factory to respond to changing product mixes. Our earlier concept of delayed product differentiation would call for redesigning the processes so that they are identical except for the final step. That would provide the necessary flexibility in the manufacturing process to respond to demand uncertainties. Recall that color printer demand would likely be negatively correlated with mono printer demand. It is a much easier task to forecast orders for the two products combined, while there could be significant forecast errors for the individual products.

Delayed product differentiation in this case requires the use of a common head driver board and a common print mechanism interface. Once this design change is made, the bill of materials for the two products are essentially the same throughout the manufacturing process, except at the very end. In this case, commonality is a key step towards delayed product differentiation.

Note that if a common head driver board is used in conjunction with two distinct print mechanism interfaces, then delayed product differentiation is still achieved. Here, the point of product differentiation occurs right after printed circuit board assembly, instead of after final assembly achieved by making both the head driver board and the print mechanism interface common. If making the print mechanism interface common for both the color and mono printer is technologically infeasible, then a common head driver board is a good intermediate step towards delayed product differentiation. Figure 3 shows the two alternatives of delayed product differentiation.

Figure 3: Printer Commonality Example Illustrated

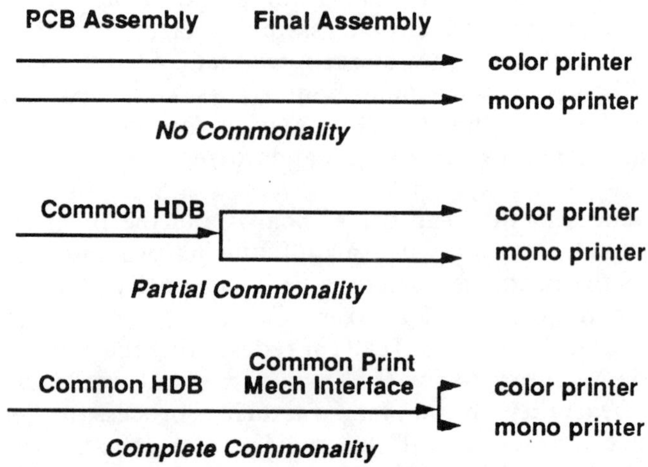

There are several challenges to design for commonality in this case. First, common parts result in higher part costs. A color printer is a product with more functionality, whereas a mono printer is one with less. Having a standardized part for both means that the material cost for the mono printer would probably go up, since a part with greater capability has been used. Second, design for commonality does pose a significant technological challenge. It turns out that it is more difficult to standardize the head driver board, whereas it is relatively easy to standardize the print mechanism interface. Standardizing the print mechanism interface helps to reduce the management cost and the inventory of print mechanism interface as a part. However, it does not help to delay product differentiation without a common head driver board. Once a head driver board for a color printer is used at the beginning of the final assembly stage, the printer is destined to be a color printer. Hence, to implement delayed product differentiation, the technological challenge of designing a common head driver board must be overcome. This problem is being explored by the printer manufacturer.

After the new product introduction phase, it is possible that the demands for the two products begin to stabilize, so that their variabilities decrease. The value of part commonality as a way to gain flexibility (via delayed product differentiation) may then decrease. At that point, one may want to pursue material cost reduction through more unique parts. Hence, the value of design for supply chain management may vary depending on the stage of the product in terms of its product life cycle.

PRODUCT LINE RESTRUCTURING
A Network Printer Example

A European manufacturer has formed a strategic alliance with a Far East manufacturer to produce network printers for the European market. The manufacturing process calls for the European company designing and assembling the main board for the printer, shipped it over to the Far East manufacturer. The Far East partner then integrates the board with the other main body of the printer, which includes building the housing, motors, and other components. The finished product is then shipped back to Europe for distribution.

The long manufacturing and transportation time for the product is a major cause of concern for the European manufacturer. Long lead times mean that very high level of safety stocks are needed at the European DC to meet high level of customer service. Adopting a DC-localization strategy such as the manufacturer described in the previous section would help, but the problem remains. Since the product is quite bulky, shipping the product from the Far East by air to reduce replenishment lead time would not be economical. The other option is to utilize subsidiaries of this Far East manufacturer in Europe. There is a high set up cost associated with such a move. Moreover, the much higher labor cost at these subsidiaries makes this unattractive.

The long manufacturing and transportation lead time for the product is duc to thc sequential nature of the manufacturing process. Board manufacturing has to be done first, followed by integration with the main body of the printer, and finally, final assembly to complete the process. This is a case where product line restructuring would help. Redesigning the process (and part of the product) to allow for simultaneous processing by both the European and the Far East manufacturer is the direction towards design for supply chain management.

The key to product line structuring here is to design the process so that the integration of the board and the rest of the printer can be performed at the end of the process. In that way, the Far East manufacturer can work on the printer itself while the European manufacturer is working on the board. The next step is to have the European manufacturer to perform the integration step, since it certainly is a lot closer to the European DC. Such restructuring of a serial system to an assembly system helps to reduce the manufacturing lead time significantly. Moving integration to Europe reduces the transportation part of the replenishment lead time. This can lead to significant inventory reduction at the DC, as well as increased flexibility for the manufacturer to respond to the uncertainties of demand. Figure 4 describes such a change.

Figure 4: Product Line "Restructuring" Illustrated

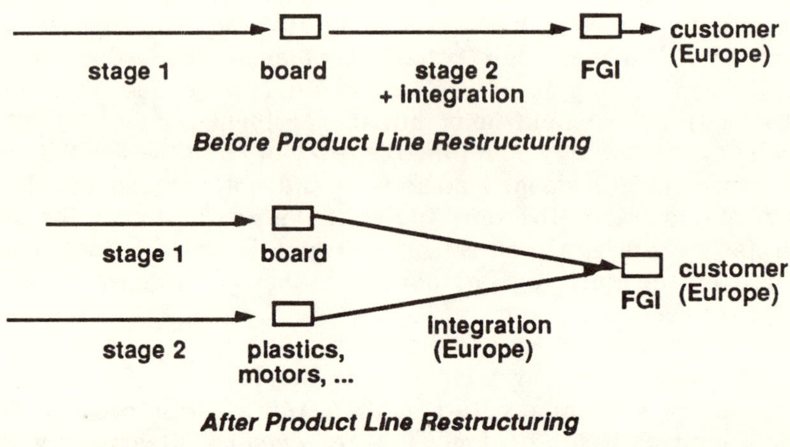

There are several critical issues that need to be resolved to implement the above product line structuring strategy. First, the technological barrier for "delayed" integration needs to be overcome. Second, the Far East partner needs to be convinced of the benefits of such a change. Here, logistics models that help to quantify the benefits to both the European and the Far East manufacturers would help. Third, the impact (benefit) to custom duties needs to be analyzed.

It is gratifying to see that the above product line structuring strategy has indeed been adopted and carried out. Having integration moved back to the hands of the European manufacturer gives a sense of control to the manufacturer. Currently, the manufacturer is working on development projects so as to push the integration step to the DC. This way, integration and localization can be done together at the DC, in concert with the concept of delayed localization and delayed integration.

CONCLUSION

Design for supply chain management can have great potentials for companies to develop a competitive edge in a global market. As companies increase the geographical spread of their customer base, and as the pressure for high customer service continues to be building up, it is critical for companies to look for product and process designs as ways to gain control of their costs of logistics, distribution and inventory, while maintaining high levels of customer service.

Acknowledgment

The author gratefully acknowledge comments and inputs from Professors Warren Hausman and Philip Barkan (Stanford University), Corey Billington, Brent Carter and Debbie Flandro (Hewlett-Packard Company), Herb Shumway (Digital Equipment Corporation), Laird McCulloch, (SUN Microsystems) and Marc Lorrain (Ford Motor Company).

REFERENCES

Barkan, P., "Strategic and Tactical Benefits of Simultaneous Engineering," Design Management Journal, Spring, 1991, 39-42.

Barkan, P., "Product Line Structuring and Design for Modularity," Lecture Notes, 1992.

Cohen, M.A. and H.L. Lee, "Strategic Analysis of Integrated Production-Distribution Systems: Models and Methods," Operations Research, 36, 1988, 216-228.

Cohen, M.A. and H.L. Lee, "Resource Deployment Analysis of Global Manufacturing and Distribution Network," J. of Manufacturing and Operations Management, 2, 1989, 81-104.

Crosby P., Quality is Free, New American Library, New York, 1979.

Dean, J.W. Jr. and G.I. Susman, "Organizing for Manufacturable Design," Harvard Business Review, January-February, 1989, 28-36.

Dumaine, B., "How Managers Can Succeed Through Speeds," Fortune, Feb 13, 1989, 53-73.

Lee, H.L. and C. Billington, "Pitfalls and Opportunities in Supply Chain Inventory Management," Sloan Management Review, Spring, 1992, forthcoming.

Lee, H.L., C. Billington, and B. Carter, "Gaining Control of Inventory and Service Through Design for Localization," 1992, Interfaces, forthcoming.

Ljungdahl, L., "International Manufacturing Logistics as a Value Add," paper presented at the Conference on Manufacturing in Europe '92: The Outlook for US Companies, University of Rochester, September, 1991.

Scott, G.A. and F.X. Tuoti, "Product Line Structuring: A Management Tool for Improving Business Productivity," paper presented at the Third International Conference on Improving the Productivity of Technical Resources, IIT, November, 1982.

Suzue, T. and A. Kohdate, "Variety Reduction Programs," Japan Management Association, Productivity Press, Cambridge, MA, 1990.

Taguchi, G. and D. Clausing, "Robust Quality," Harvard Business

Review, January-February, 1990, 65-75.

Whitney, D.E., "Manufacturing by Design," Harvard Business Review, July-August, 1988, 83-91.

COMPETING THROUGH QUALITY: PERSPECTIVES AND RESEARCH DIRECTIONS

Herbert Moskowitz
Krannert Graduate School of Management
Purdue University

ABSTRACT

Quality has become the mantra for enterprise viability and success. What is quality? How has our view and practice of quality changed? We describe the change in quality perspective and thinking that has taken place in our society and from this suggest some methodological and managerial (applied) research directions.

Introduction

There has been a tremendous change in the thinking, philosophy, and practice of quality management and control in recent years. This has not been due to any breakthrough in the technology or science of statistical quality control and improvement methodologies. Rather, it has been due to the necessity for organizations to compete globally in the manufacture of technology, products, services, as well as education. Quality has become the mantra for survival. The implications of this change in quality perspective has both organizational and methodological implications.

This research was supported, in part, by a grant from the Center for the Management of Manufacturing Enterprises and the Engineering Research Center at Purdue University under NSF Grant No. 85002

We first describe briefly the change in quality perspective that has and is taking place, and from this, suggest some methodological and managerially focused research directions. We shall first trace the history of quality practices to better understand how the change in quality thinking has evolved.

Historical Evolution of Quality Perspectives

From a historical perspective, the United States has been a leader in developing quality concepts and techniques but has been a recent follower in actually implementing those principles. If you examine the history of quality approaches and practices over the past 130 years, you will note that they were invented in the United States, but were exported (e.g., DeVor, Chang, and Sutherland, 1992).

Before the industrial revolution, groups of individual craftsmen controlled and assured quality by being involved in all aspects of the product life cycle and by their pride of ownership. With the industrial revolution and subsequent birth of "Taylorism," pride of ownership was lost. Manufacturing became production oriented with the introduction of the work standard and wage incentive plans. F.W. Taylor's scientific management focused on producing large quantities -- fast. A separate quality control department evolved, which meant that someone other than the production worker was responsible for quality.

Then, in 1925, Dr. Walter Shewhart of Bell Labs, inspired by the statistical research of R.A. Fischer in agriculture (*Business Week*, Jan. 15, 1992, p. 15) introduced statistical process control to continually monitor on-line process variations. It did not catch on in the 1920s because the emphasis was on producing and it was counter to the work (production) standard. Shewhart's work was misunderstood as controlling the product, not the process.

In the 1930s, Dodge and Romig at Bell Labs introduced acceptance sampling methods. In this system, lot-by-lot sampling inspection was used to determine whether products were suitable for shipping to customers. This method, which used a probabilistic approach based on sampling for predicting lot quality, focused on controlling, detecting, and containing defective/nonconforming products. Dodge and Romig's method became immensely popular during World War II, because the need for 100% inspection was eliminated with minimal risk. The focus, however, was on product control (defect detection) not process control (defect prevention).

In the 1950s, Dr. W. Edwards Deming used the concepts of Shewhart to develop a statistical and managerial approach to quality and productivity improvement that focused on the enterprise and emphasized the responsibilities and obligations of top management. During this time, J.M. Juran developed the concept of Total Quality Control (TQC), which advocated applying quality to everybody in an organization, from managers to clerical staff. We in the United States listened to these quality pioneers but not very enthusiastically. As a result, one of our best technologies was exported to Japan, where it played a major role in its transformation into a world-class industrial power after World War II.

The United States finally took notice in the 1970s and '80s, as it was losing market share in the global marketplace by producing poorer quality products than its global competitors, particularly the Japanese. The concepts and approaches of Deming and Juran were reclaimed, along with those of others, such as Crosby and Fiegenbaum, and U.S. industrial leaders began to change their management philosophy. Managers in the United States were also introduced to the concepts and methods of Ishikawa and ofTaguchi -- statistical design of experiments for robust product and process design and the concept of a quality loss function.

Although the concepts and approaches of the quality pioneers differ somewhat, they all have many commonalities. Hence, in the interest of brevity, we shall next focus on Deming.

Deming's Principles and Their Implications

Deming's principles have both managerial and statistical implications. The Deming Prize in Japan, the Malcolm Baldrige Award in the United States, and similar awards in Europe and in individual enterprises embody the managerial aspects of Deming's philosophy and approach (as well as that of the other pioneers). The Baldrige Award, initiated in 1988 to promote quality in U.S. firms, focuses on customer-driven quality, leadership, continuous improvement, actions based on facts and data, and the participation and empowerment of all employees. Unlike traditional management philosophy, the Baldrige Award criteria for quality focuses from the top down, on individuals as assets, and on the customer as the bottom line in terms of profits.

From a statistical perspective, Deming's principles focus on defect prevention rather than defect detection, the goal being to eliminate inspection by reducing variation or noise in the system. Variations are produced by either common causes (system capability, a responsibility of management) or special causes (operator error, for example, due to lack of proper motivation, training). Sources of variation occur in the product design process, manufacturing process, in the operating environment, and its usage over time, i.e., the age and wear of the product.

Inspection and acceptance sampling filter the noise, but they do not reduce the source of the defects. Hence, you end up simply satisfying product specifications, but do not learn anything about the source of the problem. Process control (e.g., employing statistical process control (SPC) techniques) reduces the variation in manufacturing, while robust product and process design (e.g., using classical statistical or Taguchi design of experiments techniques) reduce the amount of noise produced in the design phase, and hence, throughout the life cycle of the product. The notion is to push the quality process upstream into the product design phase. Hence, companies should spend more time on design, resulting in fewer quality problems subsequently. The Japanese have placed much effort into designing and producing a product that will function in (i.e. be insensitive to) a variety of environments as well as its usage over time. U.S. managers have tended to place less effort than the Japanese on quality in the product development and design phase and more effort downstream in manufacture and assembly. This has resulted in more costly redesigns and longer development times.

Where U.S. Firms Stand Today

Garvin (1988) identified four phases of the quality movement:

(1) inspection, (2) statistical quality control, (3) quality assurance, and (4) strategic quality management (Figure 1). Of these four phases, most U.S. manufacturing firms appear to be typically in the second or third phase of their quality evolution. In manufacturing, we have gone from inspecting quality to statistical process control -- a major step. Once this step is accomplished, going to quality assurance - spreading quality across the entire production chain - and the strategic quality management stages are less difficult. Most service industries are lagging behind manufacturing firms in implementing quality concepts. For example, in the areas of financial services, a 1987 survey by the Conference of the Financial Institutions Marketing Association found that:

- 25 percent of customers had contacted their financial institution with a complaint in the past year.
- High-income customers accounted for the highest percentage of complaints (35 percent).
- 15 percent of the customers never obtained a satisfactory resolution.
- 42 percent of customers had changed financial institutions at least once because of service problems.

The importance of quality in the service sector has never been greater. For example, deregulation and international competition in the financial industry have created huge competitive challenges (witness the many institutional failures of banks and savings and loans). At the same time, automation and computerization of operations and a more sophisticated public are increasing demands for quality. There are growing inroads in the service sector by Japan, leading to speculation that the service sector is following manufacturing into an era of strong global competition.

Strategic Quality Management

The concept and practice of quality has undergone a major evolution/transformation. Organizations now speak of Strategic Quality Management -- Stage 4 of the Quality Movement in Figure 1. Increasing foreign competition, product liability lawsuits, government regulations, and the dramatic inroads made by Japanese manufacturers (e.g., in semiconductors, autos, machine

	Stage of the Quality Movement			
Identifying Characteristics	Inspection	Statistical Quality Control	Quality Assurance	Strategic Quality Management
Primary concern	detection	control	coordination	strategic impact
View of quality	a problem to be solved	a problem to be solved	a problem to be solved, but one that is attacked proactively	a competitive opportunity
Emphasis	product uniformity	product uniformity with reduced inspection	the entire production chain, from design to market, and the contribution of all functional groups, especially designers, to preventing quality failures	the market and and consumer needs
Methods	gauging and measurement	statistical tools and techniques	programs and systems	strategic planning, goalsetting, and mobilizing the organization
Role of quality professionals	inspection, sorting, counting, and grading	troubleshooting and the application of statistical methods	quality measurement, quality planning, and program design	goal-setting, education and training, consultative work with other departments, and program design
Who has responsibility for quality	the inspection department	the manufacturing and engineering departments	all departments, although top management is only peripherally involved in designing, planning, and executing quality policies	everyone in the organization, with top management exercising strong leadership
Orientation and	"inspects in" quality	"controls in" quality	"builds in" quality	"manages in" quality

Figure 1. The Four Major Quality Eras

Source: Garvin, D.A., *Managing Quality*, Free Press, 1988.

tools, color TVs) have sensitized top management to the link between quality and profitability -- an increase in quality increases product demand, increases productivity, and reduces costs. Hence, quality is increasingly being viewed by top management as an aggressive competitive weapon from both the demand (e.g. profit, market share) and supply (cost) sides, and requiring its inclusion in the strategic planning process. The key essentials of Strategic Qualty Management or, alternatively, Total Quality Management (credited to Fiegenbaum) are: (1) A strong customer focus -- customer satisfaction as related to competing products, satisfaction formed over the lifetime of a product (not just at time of purchase), and that a composite of attributes is necessary to provide customer satisfaction; (2.) Continuous improvement based on continuous learning; (3) Involvement of everyone in the enterprise;

(4) Disciplined linkages among organizational entities (cross-functional integration).

Operationalization of Strategic/Total Quality Management

To implement and manage quality requires that it first be defined and then measured. Unfortunately, there are as many definitions of quality as there are approaches to achieving it. Some definitions of quality include (Garvin, 1988): (a) Transcendent - cannot be precisely defined; (b) Product-Based - characterized by the attributes of a product and its performance; (c) User-Based - based on the product or service that best satisfies a customer's preferences; (d) Manufacturing-Based - whether the product conforms to design specifications; (e) Value-Based - considers performance and conformance at a reasonable price.

From these various perspectives, it becomes obvious that quality is a complex concept, and is viewed differently by different functional areas in the enterprise. A "quality problem" as viewed by one function may not necessarily be a "quality problem" to another function, suggesting the need for cross-functional integration.

Quality is thus multidimensional, involving a vector of attributes, e.g., performance, features, reliability, conformance, durability, serviceability, aesthetics, and perceived quality (Garvin, 1988). These attributes and their relative importances are heterogeneous, varying across market segments. Hence, it is important to identify various market segments in order to compete effectively. Moreover, quality perceptions/preferences are dynamic, thus requiring their continuous tracking. For example, preferences for automobile characteristics shifted from styling in 1970, to fuel economy in energy-conscious 1975, and then to design quality in 1980 (Takeuchi and Quelch, 1983).

Implementing the concept of strategic or total quality management requires quality integration throughout the enterprise. For example, the concept of quality function deployment (QFD), which relates customer attributes to engineering design characteristics, and engineering design characteristics to manufacturing specifications, is being employed as a concept/tool to integrate the business, engineering, and manufacturing functions in the design process (Hauser and Clausing, 1988). Information systems for tracking, communicating, and decision support, and human resource management for cross functional team building and employee improvement, provide the essential glue to achieve integration.

Implications for Management Education and Research

Although there has been increased emphasis on quality in university management and engineering programs, it is still mostly embryonic. Quality control, operations management, statistics, and engineering design courses cover the classical aspects of statistical quality control and some contemporary aspects of quality concepts and processes; but they generally do not address marketing, human resource, and the broader strategic issues of the enterprise. Many management schools are now incorporating total quality management issues into their degree and executive programs.

To be effective, schools of management need to integrate quality into and across functional management areas. Management schools should seek input from companies to obtain both information about the expectations of the marketplace, as well as which concepts, tools and techniques are important to stress in the classroom. There are both opportunities and pressures in achieving these goals. Corporate recruiters want students to be exposed to quality management concepts in the curriculum. Funding opportunities are available from corporations and foundations to facilitate such implementation. However, education(as well as research) in quality demands a cross-disciplinary team effort, a traditionally weak characteristic of academic institutions. Universities typically have not offered inducements for cross-disciplinary work. Moreover, academics often lag behind industry in adopting new concepts and approaches. Changes are, however, beginning to occur. Management schools are increasingly integrating total quality management into their professional curriculum, and some universities are actually applying such concepts to assess and improve their administrative processes.

There are many research opportunities in quality -- both methodological and managerial. We shall address some of these in the next two sections.

Methodological Research Directions

Some of the methodological research directions to be discussed are as follows: (1) Quality Function Deployment; (2) Continuous Improvement: Learning; (3) Integrating Operations Research (OR) Techniques and Artificial Neural Networks (ANN) for Real Time Scheduling; (4) Statistical Process Control (a. Correlated Data, b. Application of Bootstrapping, c. Learning); (5) Design of Experiments Novice Decision Support System; (6) Supplier Evaluation Models; (7) Warranty Models

Quality Function Deployment

Quality Function Deployment (QFD) is a concept for translating the voice of a customer through the various stages of product planning, engineering, manufacturing, and service (Sullivan, 1986; Hauser and Clausing, 1988). There is a lack of formal modeling of this process. Aa recent formal approach to QFD employs fuzzy concepts and methodologies (Kim, Moskowitz, Dhingra, and Evans, 1991). Multiattribute value theory combined with fuzzy regression and optimization theory are employed to allow the designer to consider tradeoffs among the various customer attributes as well as the inherent fuzziness in the system (Keeney and Raiffa, 1976; Tanaka, Uejima, and Asai, 1982; Tanaka and Watada, 1988). A systematic procedure to evaluate the robustness of the design parameters with respect to varying customer preferences (market segments) is also being investigated, with the aim of developing a mechanism to determine which design parameters are robust to customer tastes, thereby simplifying the product design/manufacturing process through a generic design, whose sensitive parameters can be customized to various market segments (quality niches). A QFD decision support system is also being developed to facilitate the implementation of these concepts.

Continuous Improvement: Learning

Continuous process improvement is now recognized to be an essential activity for a manufacturing enterprise to remain competitive in the long run. While process improvement efforts cost the firm in the short run due to loss of production and other related expenses, these efforts prepare the firm to make productivity gains in the long run. Because of this tradeoff between short term and long term gains, the overall profitability of an enterprise over its life would depend on how it allocates its resources among production activities to meet customer demand and process improvement activities. The optimal allocation of resources to production activities and process improvement activities will depend on such factors as the cost of capital and the time horizon over which the enterprise is going to remain in business. It would be useful to develop a model to determine the optimal process improvement effort, and level of production activities over time (viz, "optimal load path"), given a plant's maximum production capacity. At any time, part of this capacity is used for process improvement activities and the remaining capacity is available for production activities to satisfy customer demand. Process improvement efforts lead to higher process quality, which in turn lead to lower quality related costs (internal and external) and possibly lower per-unit production costs.

The optimal load path has implications for the marketing strategy of an enterprise. An important element of an enterprise's marketing strategy for the products manufactured in a plant is to plan for the number of customer orders to be filled from the production in the plant. It can be argued that it is important for management to recognize that these decisions should be guided by the optimal load path and not by the maximum plant capacity (*Copeland Corporation: Evolution of a Manufacturing Strategy,* 1975-1982, HBS Case #9-686-088). The Japanese practice of planning the shop load such that there is always some excess capacity is consistent with this concept (see, e.g., Schonberger, 1981). The failure to meet customer expectations in the case, *"The Great Nuclear Fizzle at Old B & W"* (Hill, 1989, p. 136-248) can be explained by management's failure to recognize the need for allocating resources to process improvement activities.

Several researchers have addressed the issue of continuous improvement. Fine and Porteus (1988) focused on the problem of investment to improve process quality. They formulated a markov decision model allowing many small investments over time with potential process improvements of random magnitude. Fine (1988) developed a model to characterize optimal inspection policies, where investment in inspection activities leads to process improvement over time. Tapiero (1987) developed a stochastic dynamic programming model to determine an optimal quality control policy when investment in quality control efforts leads to lower manufacturing costs.

Another area for research in continuous improvement involves the application of statistical process control. Consider, for example, economic-based control charts, developed by Duncan (1956) and extended by others (Montgomery, 1980). The design of such models is often affected by such factors as the costs of sampling, the costs associated with investigating out-of-control signals and possibly correcting assignable causes, and the costs of allowing defective products to reach the consumer. These costs have been assumed to be constant, i.e., no learning about the process is considered, which results in cost reductions of these economic factors. It would be useful to relax the assumption of constant cost, and develop economic control charts, where such costs are reduced over time through learning and continuous improvement.

Integrating OR Techniques and ANN for Real-Time Scheduling

Scheduling is a major problem in manufacturing. Managers allocate limited resources to achieve a desired production level in a given period. In some cases, the time allowed for making such decisions is so short that the next action must be taken based on the current state of the system only. This is called real-time scheduling.

The concept of real-time scheduling was first introduced in developing computer operating systems in the late 1960's. Because a computer connects to a number of terminals that may send jobs at any time they are in use, it is extremely difficult for the operating system to plan and schedule the jobs in advance; thus, scheduling must be done on a real-time basis. In the manufacturing domain, many problems have similar characteristics. For example, when a material handling robot is used to move jobs between stations, real-time scheduling is useful when uncertainties in processing time and machine failure exist.

Consider a circuit-board manufacturing process, which must be processed in a series of chemical tanks. The processing time in each tank is determined by the product type. Under- or over-processing spoils the product. In the manufacturing process, robots are used to move the boards from tank to tank. The real-time scheduler controls the robot to transport the jobs, say, with known processing times through the tanks. The objective is to maximize the throughput of good (conforming) products in the process (i.e., the number of items produced x the % nondefective).

Traditionally, human experts are required to interact with a computer system to control a robot. This poses several potential problems. Human experts often take considerable time to train, and their performance may be inconsistent and/or biased. Recently, research has focused on developing an automated computer-based scheduler. The automated scheduler analyzes the current state of the production line and decides the next action, which is then sent to proper devices for execution.

Two major approaches have been employed in designing automated schedulers: (1) optimization, such as mathematical programming (Shapiro and Nuttle, 1988) or a semi-Markov model (Yih, 1990b); and (2) an expert system that incorporates heuristic rules into computer systems (e.g., Fox and Smith, 1984; Thesen and Lei, 1986, 1990; Yih, 1990a). These approaches have their obvious limitations -- restrictive assumptions (optimization), optimal rules from a semi-Markov approach cover only a subset of entire state space, the quality of expert system rules are contingent on knowledge of the expert, etc. These limitations suggest a hybrid approach that integrates computer simulation, semi-Markov optimization, and ANN for automated knowledge acquisition and real-time scheduling (Liang, Moskowitz and Yih, 1991). Computer simulation is used to collect on-line scheduling decisions for use as training data for the ANN. This allows scheduler's decisions to be obtained in a non-intrusive way. Once sufficient data are collected, a semi-Markov decision model can be applied to filter inconsistent and biased scheduling decision data on a subset of the entire state space encountered. The purified training data can then be used to build an ANN model to automate the real-time scheduling process and apply it to the entire state space. Such an integrated OR-ANN approach has potential applicability to a broad array of problem classes and structures, which need to be identified and investigated.

Statistical Process Control

Notwithstanding the considerable amount of research on statistical process control (SPC), there are several developments worth pursuing, such as, (1) SPC models for correlated product output data from a process, and (2) nonparametric models using, for example, bootstrapping.

SPC models with correlated data. Traditional SPC charts assume that observations are independent and normally distributed about some process parameter, e.g., the mean. However, often in industrial practice, for discrete as well as continuous production processes, observations are not independent, but are serially correlated, such as in some machining and forging operations. Here, traditional SPC charts may be inappropriate for monitoring process quality, and thus more sophisticated correlated models probably should be considered. Such correlated models might also be useful for predicting future quality. With powerful personal computers, use of such more sophisticated SPC models appears practical, making it possible to automate SPC modeling and charting (Wardell, Moskowitz, and Plante, 1990, in press).

Nonparametric (bootstrapping) models. The most commonly used methods in SPC are parametric, in the sense that they require assumptions about the statistical properties of the underlying process. Shewhart process control charts, for example, assume that observations are independent, and that the statistic of interest is normally distributed. Often, in practice, these assumptions are violated -- the population distribution of a performance variable may be strongly skewed, in which case the central limit theorem does not hold for small sample sizes, which are typically employed in SPC. Moreover, observations can be correlated. Nonparametric statistical methods, such as bootstra (Efron, 1979), may be used to obviate these problems. Bootstrap is a computer intensive resampling method that does not require any distribution assumptions, a priori. Computer simulations are used to find estimates of the accuracy of the statistics of interest.

Design of Experiments Novice Decision (DOE) Support System

Design of Experiments (DOE) has become one of the most discussed tools of quality management, yet one of the least applied due to its apparent complexity. Because of this, often high-priced consultants rather than in-house process experts are used in developing experimental designs. Thus, it would be useful and economic to develop a novice friendly DOE Decision Support System (DSS) that could be used by in-house experts to design experiments to improve process and product designs. We are attempting to develop such a novice friendly intelligent DOE DSS with an appropriate user interface. Existing DOE software systems require considerable DOE expertise to handle even relatively simple experimental designs (Nachtsheim, 1987). This has limited its application in practice. There is also a need to develop highly efficient fractional factorial designs (FFDs) to facilitate the understanding and modeling of complex system phenomena (having many independent variables and vectors of possibly correlated dependent variables) associated with manufacturing process research and application.

Warranty Models

Most research on warranty policies, and those employed in practice, have been concerned with a single attribute, e.g., a warranty period (time). More theoretical and applied research is needed on multiattribute warranty policies. One such line of investigation has focused on the two attribute warranty policy, e.g., product age and mileage in a y-year, m-mile protection plan in the automobile industry (Moskowitz and Chun, 1991). For example, two types of such policies have been developed: (1) a fixed warranty policy in which all customers are offered the same plan, and (2) a flexible policy in which customers are permitted to choose any warranty plan among the numerous plans furnishable by the producer. Two attribute warranty plans (as opposed to single attribute plans) have the desirable characteristic of protecting both producer and consumer, with the flexible plan being more mutually equitable than the fixed plan for the same price, thus providing a competitive potential advantage.

Managerial Research Directions

There are many research issues that need to be addressed in the area of total quality management, regarding (1) implementation, (2) human resource management, (3) customer satisfaction, and (4) competitive impact. Some of these are outlined briefly below.

Implementation

Empirical studies of the strategies to implement quality programs and their impact on the enterprise would provide generic insights and guidelines, as well as what strategies work best in particular environments. Some specific issues that might be addressed are as follows: (1) Clarification of the role of senior management in leading total quality management (TQM); (2) Problems and strategies in getting senior management to use the processes and data driven approaches of TQM in their own work on core enterprise problems; (3) How do organizations develop effective programs to promote and implement quality? (4) What are the key capabilities (managerial and quantitative skills and tools) that allow an enterprise to attain different levels of success in managing process performance, from achieving *process stability* to achieving *continuous process improvement* ; (5) How can firms develop an effective enterprise quality information system -- one that measures quality, its rate of change, in quasi real- time so that it can monitor the effect of quality improvement programs and take responsive action if quality deteriorates; (6)How are new concepts about quality accepted and integrated into the existing corporate culture? (7) Which approaches to quality improvement are appropriate for which types of organizations - separated, for example, by size or stage of development? (8) How can the enterprise get its senior management to understand and internalize customer attitudes and quality improvement concepts? (9) What impact does technology have on creating quality? (10) How are quality improvement issues and programs addressed in, for example, European, Japanese, vs U.S. enterprises? (11) How can quality initiatives be applied to research and development activities?

Human Resources Management

To incorporate quality throughout an enterprise, all employees must embrace the concept. The following issues might shed light on how this can be achieved: (1) What role and what type of executive and employee compensation/reward schemes promote and create quality in an enterprise? (2) What are the most effective methods of training employees on quality? (3) How can enterprises keep up the motivation and enthusiasm of employees while maintaining high levels of quality? (4) How do you make quality relevant to employees? (5) What is the role and effectiveness of champions in making quality improvement efforts a success?

Customer Satisfaction

If quality is viewed as meeting customer's multidimensional needs and requirements, it is vital to understand what customers want and expect. (1) How can we improve our understanding of the effect very satisfied customers have on brand loyalty, customer retention and repurchase decisions? What customer attributes discriminate very satisfied customers from satisfied customers, and how do they affect purchase decisions? (a) What does it require to move customers from satisfied to very satisfied? (b) What product development functions most significantly contribute to very satisfied customers? (2) What is the relationship between customer perceptions and intentions at the time of a survey versus at the time of the actual purchase decision? (3) What role can artificial intelligence technology play in the management of customer satisfaction, both in terms of problem detection and prevention? (4) What is the linkage between improvements in customer satisfaction and enterprise external (e.g., market share, return on assets) and internal (e.g., number of defectives, machine breakdowns) performance measures? (a) What is the marginal value gained by an enterprise for a unit improvement in customer satisfaction? (b) What level of investment is then required to achieve a unit of improvement in customer satisfaction, particularly when going from, say, 90% to 100% satisfied customers? (5) How can customer feedback be effectively incorporated into the management decision making process: (a) in integrating marketing plans and business strategies, (b) in planning goals for new products, (c) in the product development process, (d) To effect current product customers satisfaction improvements and strategies? (6) What is the relationship between external customer satisfaction and employee satisfaction in the enterprise? (7) Should (or perhaps how should) customer satisfaction performance be incorporated as a motivator in managerial annual performance measurement and compensation schemes?

Competitive Impact

Studies that directly tie quality improvement strategies with enterprise performance measures, such as profit and market share, would be valuable for senior management in determining what kind of quality strategy to choose. (1) What is the link between various measures of quality to enterprise performance measures? (2) What are the key characteristics of successful quality programs? What are the key leverage points of success? (3) What impact has the Malcolm Baldrige Award (and/or similar award programs) had on quality and enterprise performance; for winners as well as losers?

Many (if not all) of the menu of managerial issues raised are currently being addressed by both practitioners in organizations as well as academicians in many schools of management, often working as research partners.

Summary

We have described the change in quality perspective that has been evolving in the practice of total quality management. This change in perspective as well as practice has been motivated by survival; namely, for enterprises to remain viable and compete in the global marketplace. We then described and enumerated some methodological and managerially oriented research directions in quality. Opportunities for interesting and significant research are manifest, and accommodate a portfolio of researchers' tastes and capabilities. Hopefully, this will help to motivate some research in an area of vital impact to our national economic competitiveness.

References

1. Copeland Corporation, "Evolution of a Manufacturing Strategy," Harvard Business School, Case #9, 686-088.

2. Effron, B., "Bootstrap Method: Another Look at the Jackknife," The Annals of Statistics, 7, 1979, 1-26.

3. Fine, C., "A Quality Control Model with Learning Effects," Operations Research, Vol. 36, Iss. 3, May-June, 1988.

4. Fine, C.H., and Porteus, E. L., "Dynamic Process Improvement, Operations Research, Vol. 37, No.4, July-August, 1989.

5. Fox, M.S. and Smith, S.F. (1984), " ISIS -- A Knowledge Based System for Factory Scheduling," Expert Systems Journal, 1, 1, pp. 25-49.

6. Hill, Terry, "Manufacturing Strategy: Texts and Cases," Richard D. Irwin, 1989, pp. 136-248.

7. Hauser, J.R. and Clausing, D., "The House of Quality," Harvard Business Review, 66 (1988), 63-73.

8. Keeney, R, and Raiffa, H., Decisions with Multiple Objectives, John Wiley and Sons, New York, 1976.

9. Kim, K., Moskowitz, H., Dhingra, A., and Evans, G. "Fuzzy Multicriteria Methodologies for Quality Function Deployment," Center for the Management of Manufacturing Enterprises, Working Paper Series, No. 91-10-1, Krannert School of Management, Purdue University, West Lafayette, IN 47907, October 1991.

10. Liang, T.P., Moskowitz, H., and Yih, Y., "Integrating Neural Networks and Semi-Markov Processes for Automated Knowledge Acquisition: An Application to Real-Time Scheduling," Center for the Management of Manufacturing Enterprises, Working Paper Series, No. 91-8-1, Purdue University, West Lafayette, IN 47907, August 1991.

11. Moskowitz, H., and Tsai, H.-T., "Single-Sided Economic Screening Models Incorporating Individual Unit Misclassification Error and Risk Preference," European Journal of Operational Research, (in press).

12. Moskowitz, H., Plante, R. D., and Wardell, D.G., "Use of Run Length Distributions of Control Charts to Detect False Alarms," Center for the Management of Manufacturing Enterprises, Working Paper Series, No. 91-5-1, May 1991.

13. Moskowitz, H., and Chan, Y.-H., "A Poisson Regression Model for the Two-Attribute Warranty Policy," Center for the Management of Manufacturing Enterprises, Working Paper Series, No. 91-6-2, June 1991.

14. Nachtshim, C.J., "Tools for Computer-Aided Design of Experiments," Journal of Quality Technology, Vol. 19 (1987), pp. 132-160.

15. Pettit, R.E., "Vendor Evaluation Made Simple," Quality Progress, March 1984, 19-22.

16. Schonberger, R. J., "Operations Management," Business Publications Inc., 1981.

17. Shapiro, G.W., and H. L. W., Nuttle (1988), "Hoist Scheduling for a PCB Electroplating Facility," IIE Transactions, 20, 2, pp. 157-167.

18. Sullivan, L.P., "Quality Function Deployment," Quality Progress (June 1986), p.39.

19. Tanaka, H., Uejima, S., and Asai, K., "Linear Regression

Analysis with Fuzzy Model," IEEE Transactions on SMC, SMC-12 (1982), 903-907.

20. Tanaka, H., and Watada, J., "Possibilistic Linear Systems and Their Applications to the Linear Regression Model," Fuzzy Sets and Systems, Vol. 27 (1988), 275-289.

21. Tapiero, C. S., "Production Learning and Quality Control," IEE Transactions, Vol. 19, Iss. 4, December 1987.

22. Thesen, A., and Lei, L. (1990), "An Expert Scheduling System for Material Handling Hoists," Journal of Manufacturing Systems, 9:3, pp. 247-252.

23. Tsai, H.-T., and Moskowitz, H., "A One-Sided Single Screening Procedure Using Individual Unit Misclassification Error," Technometrics, (submitted and revised).

24. Wardell, D. G., Moskowitz, H., and Plante, R.D., "Control Charts in the Presence of Data Correlation," Management Science, (in press).

25. Wardell, D. G., Moskowitz, H., and Plante, R. D., "Run Length Distributions for Special-Course Control Charts for Correlated Processes, Center for the Management of Manufacturing Enterprises, Working Paper Series, No. 90-3-1, March 1990.

26. Yih, Y. (1990a), "Trace-Driven Knowledge Acquisition (TDKA) for Rule-Based Real-Time Scheduling Systems," Journal of Intelligent Manufacturing, 1, 4, pp 217-230.

27. Yih, Y. (1990b), "Learning Scheduling Rules for FMS from the Optimal Policy of User-Based Semi-Markov Decision Processes," Proceedings of the Fourth International Conference on Expert Systems in Production and Operations Management, Hilton Head Island, South Carolina, May 14-16, 1990.

SERVICE QUALITY DEPLOYMENT: QUALITY SERVICE BY DESIGN

Ravi S. Behara
School of Management
University of Manchester
Institute of Science and Technology (UMIST)
Manchester, United Kingdom

and

Richard B. Chase
School of Business Administration
University of Southern California
Los Angeles, USA

If we designed cars the way we seem to design services, they'd probably come with one axle and five wheels.

ABSTRACT

The service-dominated and product-dominated sectors are just awakening from a service slumber to find a new generation of quality activities that call for systematic approaches to design and improvement. This paper proposes a methodological framework called *Service Quality Deployment*, to enable the creation of quality services by cross-functional efforts geared to incorporating the customer's views. The approach is a synthesis of the Japanese Quality Function Deployment design system and the SERVQUAL instrument for measuring service quality. It is customer-driven in that perceived quality is directly coupled to the factors that affect the corresponding service quality attributes. The approach lends itself to the redesign of existing services for continuous improvement in quality, and to designing-in quality in new services.

INTRODUCTION

The widespread interest in manufactured product quality has resulted in a variety of frameworks and tools to build in quality at the design stage. Of particular interest to virtually all manufacturers is creating methodologies which incorporate the voice of the customer in product specifications and identify organizational actions to execute these specifications. Service organizations, as well, have been actively engaged in developing quality programs, and have often sought to build in customer preferences at the design stage. However, services, unlike manufacturers, need tools that enable them to simultaneously consider data on customer treatment, server requirements, and back office support which underlie effective service encounters. In this paper, we present an approach which addresses these requirements by combining a standard tool for measuring customer satisfaction with service provision (SERVQUAL) with a standard tool of product design (Quality Function Deployment) to develop a "houseof service quality."

QUALITY FUNCTION DEPLOYMENT

Quality Function Deployment (QFD) is defined as "a system for translating customer requirements into appropriate company requirements at every stage, from research through production design and development, to manufacture, distribution, installation and marketing, sales and services" (ASI, 1987). QFD arises from the need to specify the product/service and the process at the design stage. It is implemented by the use of a sequence of matrices and charts resulting in a House of Quality, that links customer requirements and characteristics of the product or service, along with competitive benchmarking, to assign quality related tasks to appropriate departments (Kogure and Yoji, 1983). QFD is most widely used in manufacturing, where its full scale application drives the voice of the customer into each phase of the production-distribution cycle. In particular, it is seen in product planning, process planning, production planning, and parts deployment.

Observed benefits of QFD include improvements in both the product and the process. Product improvements include increased reliability, improved designed-in quality, reduced warranty claims, and lower design and production costs. Process improvements include greater functional integration, improved marketing opportunities and decision making, and a more customer- oriented work force (Burns, 1990). In addition, a study of some U.S. QFD projects in manufacturing showed qualitative improvements in collaboration and communication at the strategic level as measured by improved marketing strategies, improved analysis, and building a common vision (Griffin, 1989).

SERVICE QUALITY DEPLOYMENT

Our approach to Service Quality Deployment is based upon a general framework we call "qualityimprovement by design"which explicitly recognizes the relationships suggested by the "service quality lever." We will now briefly discuss these concepts, as well as SERVQUAL, and then develop the specific features of the House of Service.

Quality Improvement by Design. Figure 1 illustrates how design and redesign links the inputs of customer perceptions, service failures, service firm processes and personnel, and competitive benchmarking to drive quality improvement actions. SQD, in essence is merely a systematic way of characterizing this process for subsequent implementation. The key point of this model is that quality improvement actions must affect both service firm processes and personnel and customer perceptions of the process. This is in contrast to manufacturing quality improvement where the customer is concerned mainly with the end product, not in the details of how it was made.

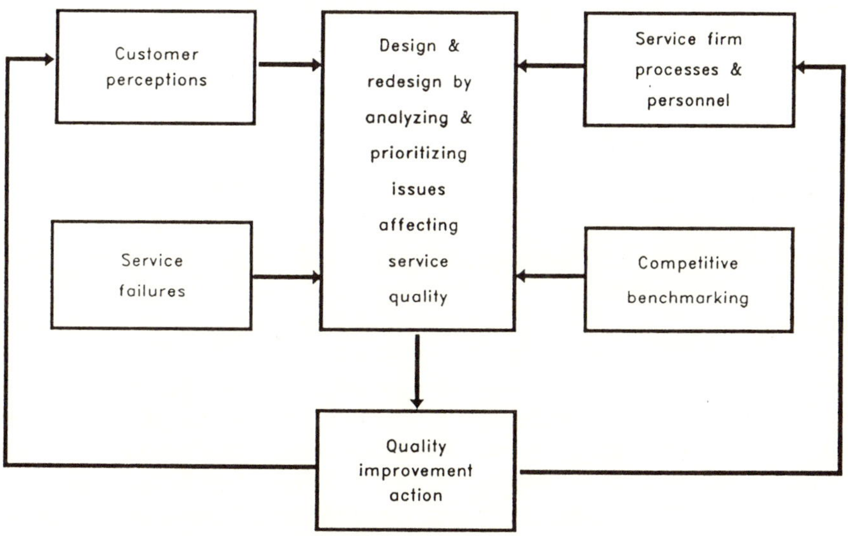

Figure 1: Quality Improvement by Design

The Service Quality Lever. SQD like QFD in manufacturing arises from the need to make the voice of the customer heard, with minimal attenuation, throughout the firm. Also like QFD it is of critical importance to get customer feedback at the design stage. The service quality lever shown in Figure 2 (modified from ASI, 1987) highlights this point indicating the relative impact of designed in service quality versus remedial actions. Here for example, customer complaints during delivery are highly visible but at best could initiate an effective recovery. Moving customer inputs to the design stage, before errors are made, however, substantially reduces the risk of visible (to the customer) service failures.

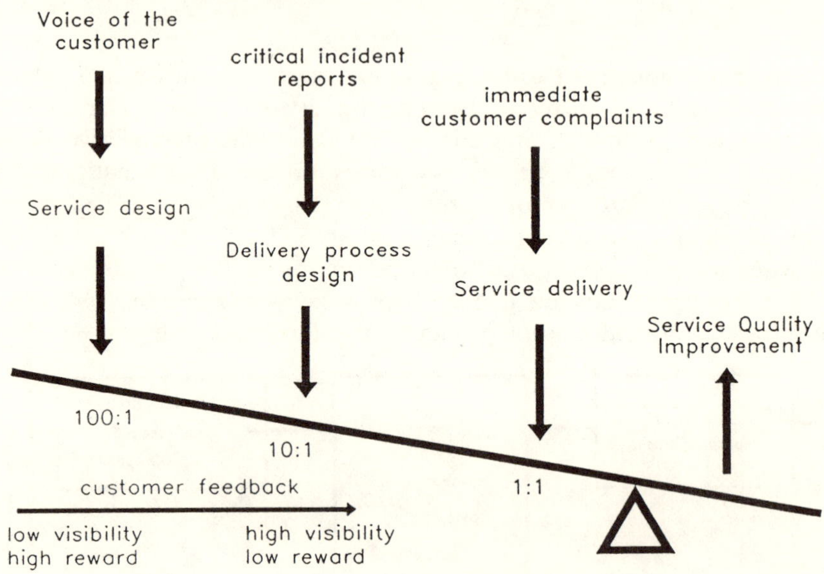

Figure 2: The Service Quality Lever

SERVQUAL

SERVQUAL, developed by Parasuraman, et al (1988), is a widely used survey instrument to measure customer perceptions of service quality. It consists of twenty-two statements grouped into five dimensions: Tangibles, Reliability, Responsiveness, Assurance, and Empathy. Assessing the quality of service using SERVQUAL involves computing the difference between the ratings customers assign to paired expectation perception statements. This difference is referred to as "Gap 5." (Four other gap scores can also be calculated to determine the causes of this gap.) The scores for the five dimensions are averaged to yield an unweighted SERVQUAL score. This score is then adjusted by customer supplied importance weights to give a weighted SERVQUAL score. By virtue of its broad application (and scientific validation) in services, SERVQUAL is an ideal tool to incorporate into an SQD house of service.

BUILDING A HOUSE OF SERVICE

Quality Relationship Matrix

The formulation of the *quality relationship matrix* is fundamental to the subsequent development of the House of Service. The relationship matrix (Figure 3) comprises three distinct but related sections: *customer quality criteria* (what customers perceive), *service company facets* (how these criteria are created by the firm), and the *relationship grid* (how the two are related).

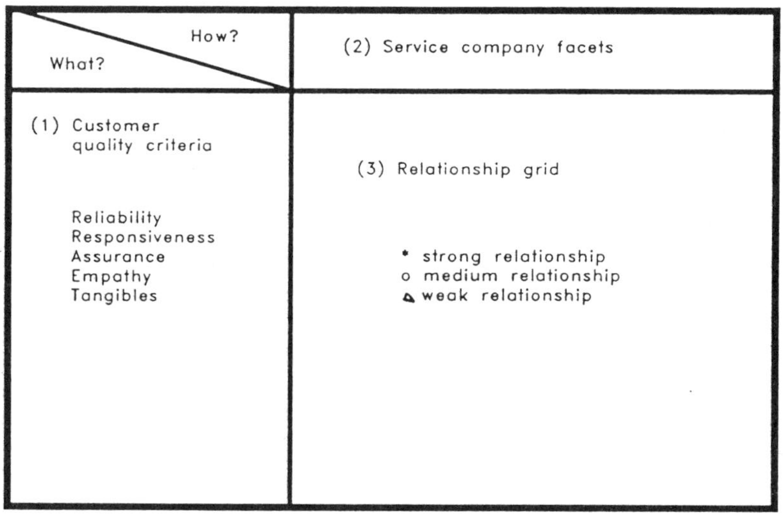

Figure 3: Service Quality Relationship Matrix

(1) **Customer Quality Criteria.** The nature of services dictates that the customer can usually provide only tertiary (or detailed) expectations and perceptions. In addition, these would tend to be partial and ill-

defined. However, research using focused groups has consolidated this information. It has shown that customers generally have similar primary (and secondary) criteria for evaluating service quality, independent of a specific service. These are identified in the SERVQUAL measuring instrument, and forms the basis for this section.

(2) **Service Company Facets.** Service quality is determined by the interaction of the customer with a range of products, processes and people of the service firm. These multiple facets of the company would also have generic primary categories such as planning, procedures and personnel. Service-specific secondary and tertiary sub-categories should, however, be developed by a cross-functional employee group.

(3) **Relationship Grid.** The relationships between every customer criterion and the factors that contribute to it are established in this matrix. The relationships are categorized (subjectively and symbolically) as being strong (*), medium (o) or weak (Δ). The grid is best developed by a focus group consisting of customers and company personnel.

A section of a completed *quality relationship matrix* based on a study of an auto-service network (Behara and Lemmink, 1991) is shown in Figure 4. This illustrates how the primary customer quality criterion of reliability can be disaggregated to secondary and tertiary quality determinants using SERVQUAL scores derived from a sample of customers across six dealerships. The primary company facets of planning, procedures, and personnel are further disaggregated into secondary auto-service related factors. These are drawn from preliminary analyses of the activities by Behara and Lemmink. The relationship grid couples the quality determining criteria and the factors that affect them.

House of Service

The House of Service is developed by extending the symbolic quality relationship matrix to include quantitative information. The use of SERVQUAL scores instead of symbols to couple customer perceived quality to company facets is fundamental to this approach. The sections involved in the construction of the House of Service (Figure 5) are described below. The (1) *customer quality criteria* and (2) *service company facets* are sections identical to those developed for the *quality relationship matrix*.

				Planning				Procedures							Personnel		
What? / Customer quality criteria — How? / Service company facets			Relative importance	Layout	Resources (eqpt.)	Resources (per)	Systems capacity	Housekeeping	Customer handling	Car handling	Information handling	Routine/nonroutine Situations	Inventory	Job/personnel scheduling	Selection	Skills training tech/interpersonal	Attitude/morale
Prin	Secondary	Tertiary															
Reliability	Accuracy	Correct problem diagnosis			*	*						○			○	*	△
		Work right first time			*	○					*	○		*	○	*	△
		Correct billing			○											*	
	Maintaining Records	Clear statement of work done			○						*	○				*	
		Update maintenance book									*					*	○
	Honoring promises	Do work mentioned by customer			*	*			*		*		△		△	○	△
		Car ready at promised time			*	*	*				△		△	*		△	○
		Keep promises/ appointments							*		*					*	○

Figure 4: Section of Relationship Matrix for Auto-Servers

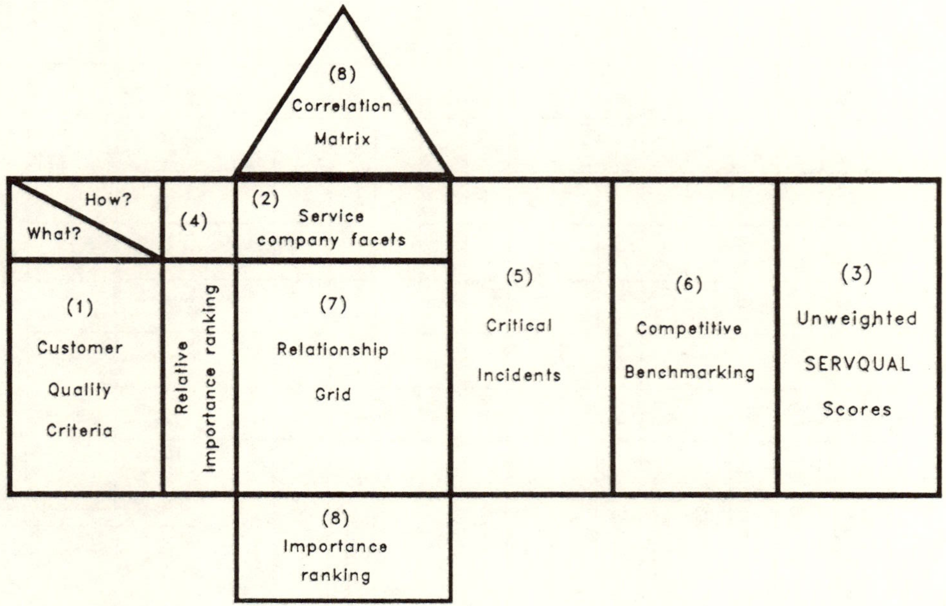

Figure 5: House of Service

(3) **SERVQUAL Scores.** The survey instrument is applied to all tertiary quality criteria and the unweighted average "Gap 5" score for each is recorded. These scores form the basis for the grid scores and are given in the last column in Figure 6. (A negative sign is provided only to indicate that perceived service was, on average, below the expected service. It is not used in the calculations.)

(4) **Relative Importance Ranking.** The customers' quality criteria are ranked and weighted in order of relative importance. Research by Zeithaml et al. (1990, p. 28) of the relative importance (to the customer) of the five SERVQUAL dimensions showed reliability [5] being the most important. This was followed by responsiveness [4], assurance [3] and empathy [2], with tangibles [1] being of least relative importance. This ranking is independent of a specific service. The ranks are used in calculating weighted grid scores.

(5) **Critical Incidents.** This section tabulates the number of customer complaints of service failure (for a given period) corresponding to the tertiary customer perceived quality dimensions. If no complaints for a given criterion are reported, the cell is left blank and omitted from the calculation of the grid scores. The significance of complaints is reflected by using these values as weights in determining the associated grid scores.

(6) **Competitive Benchmarking.** In our approach, we follow the QFD convention of having a benchmarking team score best in class competitors relative to our company on a one to ten scale. We then take the difference in these scores and enter them in the matrix (if that difference favors the competition.)

(7) **Relationship Grid.** The symbolic relationship grid is quantified here. Strong relationships are assigned a value of 9, medium relationships a 3, and weak relationships a one. Each relationship is then multiplied by the negative of the SERVQUAL score, the relative importance ranking, the critical incident score, and the competitive benchmarking score of the associated quality criterion. The negative of the SERVQUAL score is used so that the resultant number is a positive measure of importance. The reason for the unusual scoring of critical incidents was because the multiplicative nature of the calculation makes it susceptible to zero values. A blank, however, will prompt discussion of the reason behind the absence of reported incidents. For instance, the company may have made it difficult for customers to complain about this facet. Also, only a relatively small portion of customer complaints are actually reported. Therefore, the

Figure 6: House of Services:
Auto Services

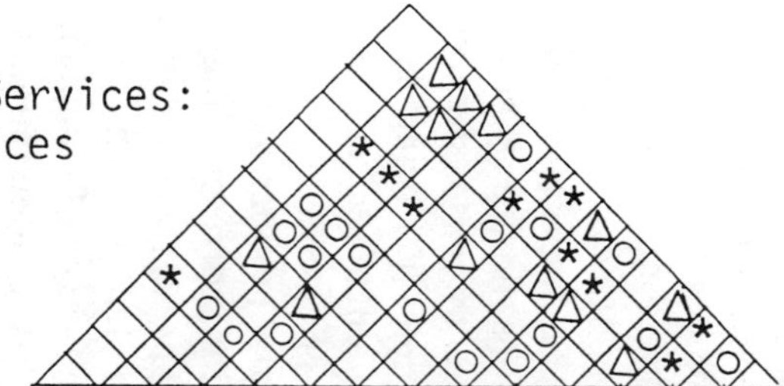

Pr.	Secondary	Tertiary	Relative importance	Layout	Resources (eqpt.)	Resources (per)	Systems capacity	Housekeeping	Customer handling	Car handling	Information handling	Routine/nonroutine situations	Inventory	Job/personnel scheduling	Selection	Skills training tech/interpersonal	Attitudes/morale	Critical incidents	Competitive benchmarking	SERVQUAL Scores
Reliability	Accuracy	Correct problem diagnosis	5		76	76						25			25	76	8	2	3	-.28
		Work right first time	5		917	306					917	306		917	306	917	102	7	3	-.37
		Correct billing	5		130						389					389		6	3	-.48
	Maintaining records	clear statement of work done	5		138						414	138				414		5	4	-.46
		Update maintenance book	5								77					77	26	3	-	-.57
	Honoring promises	Do work mentioned by customer	5		558	558			558		558		62		62	186	62	7	3	-.59
		car ready at promised time	5		149	149	149				17		17	149		17	50	5	3	-.22
		keep promises/ appointments	5						478		478					478	159	6	3	-.59
		absolute importance		-	1968	1089	149	-	1036	-	2850	469	79	1066	393	2554	407			
		relative importance			3	4	10		6		1	7	11	5	9	2	8			

Sample cell value calculation (row 1, col. 2):
5 x (* = 9) x 2 x 3 x .28 ≅ 76.

absence of critical incidents does not necessarily indicate the absence of problems.

(8) **Importance Ranking.** The total weighted scores for each company facet is evaluated from the relationship matrix. The vertical sum of the columns will indicate the *absolute importance* of each of the factors affecting overall quality. The *relative importance* ranking of these absolute scores will prioritize the various factors that need improvement.

(9) **Correlation Matrix.** The service company facets are compared against each other. The relationship between factors can be used to highlight the importance of cooperation between different functions of the firm. Once again, relationship strengths are symbolically represented.

The section of the auto-service *quality relationship matrix* is extended using the above concepts in the construction of the (partial) *House of Service* shown in Figure 6.

A brief interpretation of the *House of Service* is as follows: Only one customer criterion, reliability, is being considered in the example. Hence all the tertiary criteria have a relative importance rank of 5, since reliability is the most significant criterion. The number of critical incidents in a given period of time, and the normalized benchmarking values are also indicated. The total value for each *company facet* column, its *absolute importance*, is an indication of the significance of that facet in creating a quality service as perceived by the customer. It also indicates the extent to which that facet needs to be improved to achieve customer satisfaction. The *relative importance* ranking of these total absolute scores is a means to prioritize quality improvement actions. Hence an absolute importance score of 2850 for the information handling facet of the company makes it the top priority issue to be addressed. The *roof* of the house, showing functional interdependency, is indicative of the need for cooperation and communication between the different units of the service company.

CONCLUSION

The foundation of the Service Quality Deployment approach presented in this paper is that services should be designed to reflect the customer's expectations and desires employing validated customer satisfaction metrics. The conceptual map of the House of Service provides the means to implement these metrics through structuring cross-functional cooperation and coordination requirements. Advantages of the systematic and structured framework presented include:

(a) Enabling a direct customer-driven approach to service design.
(b) Incorporating intangible customer treatment aspects in analyzing services.
(c) Identifying and prioritizing areas for improvement.
(d) Analyzing processes and activities based on contribution to overall quality.
(e) Minimizing waste due to ignorance of customer needs.
(f) Incorporating quantitative assessments to support monitoring of continuous improvement.
(g) Minimizing quality gaps at the design stage by improved and effective interfunctional communication.

Further work on Service Quality Deployment might include:

(a) Developing subsequent service quality deployment charts to explore service design in greater detail and examine issues involving service delivery process planning and control. These can be addressed by the use of appropriately structured house of quality charts based upon the one developed above.
(b) Refining the relationship grid score calculation through normalizing the weighing schema.
(c) Extending the service company factors that affect customer perceived service quality to cover a wider range of corporate activity.
(d) Testing the practicality of the approach not only for improving existing services but for new services as well.

REFERENCES

ASI (1987). *Quality Function Deployment*, Executive Briefing, American Supplier Institute.

Behara, Ravi S., and Lemmink, Jos (1991). *Q-Matrix: A Multidimensional Approach to Using Service Quality Measurements*. Proceedings of the Workshop on Quality Management in Services. European Institute for Advanced Studies in Management, EIASM, Brussels, May, pp. 55-65.

Burn, G. R. (1990). *Quality Function Deployment*. In B. G. Dale and J. J. Plunkett (eds.), Managing Quality. Hemel Hempstead, UK: Philip Allan.

Griffin, Abbie (1989). *Improving the New Product Development Process: Contributions of Quality Function Deployment*. MSI Conference: Enhancing the New Product Development Process (Report No. 89-122). October, pp. 10-12.

Karabatros, N. (1989). *Quality in Transition: Account of the '80s*. Quality Progress, December, pp. 63-73.

Kogure, Masao, and Akao, Yoji (1983). *Quality Function Deployment and Company Wide Control in Japan*. Quality Progress, Vol. 16, October, pp. 25-29.

Parasuraman, A., Zeithaml, V. A., and Berry, L. L. (1988). *SERVQUAL: A Multiple Item Scale for Measuring Consumer Perceptions of Service Quality*. Journal of Retailing, Vol. 64, No. 1, Spring, pp. 12-40.

Zeithaml, Valarie A., Parasuraman, A., and Berry, Leonard L. (1990). *Delivering Quality Services: Balancing Customer Perceptions and Expectations*. New York: Free Press.

STRATEGIC ROLE OF INFORMATION IN SERVICES

James A. Fitzsimmons
Department of Management
University of Texas at Austin
Austin, Texas 78712

ESTABLISHING SERVICE MANAGEMENT AS A FIELD

Within the past two decades a number of seminal articles and textbooks have been published that established service operations management as a field of study embracing all service industries. A marketing professor from Harvard (Levitt, 1972) is credited with the observation that service operations could be viewed from a manufacturing perspective. Using McDonald's as an example of "manufacturing in the field" he proposed that services could adopt manufacturing concepts to insure the delivery of consistent service. The "customer contact model" (Chase, 1978) was the first and still most useful theory of service operations. More than a classification scheme, this theory with its "front and back office dichotomy" unlike manufacturing recognizes the customer as a participant in the delivery system. The implications for marketing and human resource management now became apparent.

Several service marketing textbooks (e.g. Lovelock, 1984) were written addressing issues such as intangibility, positioning, managing demand, and the customer perception versus expectation of service delivered.

The Harvard service management course and the publication of a casebook (Sasser, Olsen, Wycoff, 1977) followed by a textbook (Fitzsimmons and Sullivan, 1982) began the initiation of service operations management courses in the business curriculum.

Service operations management as an academic field was recognized by the Decision Sciences Institute with a separate track at its 1987 Boston meeting and all subsequent meetings.

The International Journal of Service Industry Management was inaugurated in 1989. The First International Service Research Seminar in Service Management was held in France in 1990 with participants from the fields of operations management, marketing, and organizational behavior. Finally, with this conference and an earlier textbook (Lovelock, 1988) the multi-disciplinary nature of service management was recognized by dropping the adjective "operations" and the field is now referred to as service management.

FUTURE OF SERVICE MANAGEMENT

However, the fundamental driving force behind the explosive growth of services in the world economy has not been fully realized except when it is referred to as a transition to an information economy. As the industrial revolution changed an agricultural economy into an industrial economy, today's industrial economy is being changed into a service economy by an information revolution.

The availability of computers and global communications has created industries for the collecting, processing and communicating of information. It is possible for everyone on the globe to be in instant communication with everyone else, and this revolution is changing world society in many ways. Consider the impact of the emerging private satellite network industry that provides uplinks and downlinks for personnel training, product introductions, credit checks, billing, financial exchanges, and overall telecommunications.

K mart was among the first retail giants to establish a private satellite network using the new small-dish antenna VSAT (Very Small Aperture Terminal) placed on store roofs for receiving and transmitting masses of date. The VSAT at each K mart store is linked to the company's Troy, Michigan data center via a satellite transponder leased from GTE Spacenet. The communication network has allowed K mart to better coordinate its multisite operations and realize substantial benefits such as improved data transmission about rate of sales, inventory status, product updates, and most importantly credit authorizations for customers. The instant accessibility to credit histories can significantly lower the risk of nonpayment the credit-card companies face, thus lowering the discount rate that reverts back to the retailer. This savings alone will eventually pay for the cost of the satellite network[1]

THE COMPETITIVE ROLE OF INFORMATION

For service management, information technology is helping define the competitive strategy of successful firms. Figure 1 captures the different roles in which information technology can support a service firm's competitive strategy. We shall explore each of these roles in turn with illustrations from successful applications.

Competitive Use of Information

Strategic Focus	On-line (Real Time)	Off-line (Analysis)
External (Customer)	<u>Create Barriers to Entry</u> Reservation System Frequent User Club Switching Costs	<u>Data Base Asset</u> Sell Information Develop Services Micromarketing
Internal (Operations)	<u>Revenue Generation</u> Yield Management Point of Sale Expert Systems	<u>Productivity Enhancement</u> Inventory Status Data Envelopment Analysis (DEA)

Figure 1 Strategic Roles of Information in Services

[1]From Bernie Ward, "Microspace, Maxiprofits," *Sky*, December 1990, pp. 22-31.

Create Barriers to Entry

James L. Heskett has argued that barriers to entry can be created using economies of scale, market share, creating switching costs, investing in networks, and using data bases and information technology to strategic advantage.[2] We will discuss three uses of information for creating barriers to entry; reservation systems, frequent "flyer" or similar programs to gain customer loyalty, and developing customer relationships to increase switching costs

Reservation systems A barrier to entry can be created by developing online reservation systems that are provided to sales intermediaries such as travel agents. The American Airlines Saber System is an example of the economy of information that is created by a comprehensive information system. United and Delta have duplicated this reservation system at great cost but most smaller carriers use these systems for a fee. The competitive importance of online reservation systems became evident in late 1982. At this time the Civil Aeronautics Board (CBA) and the Department of Justice began a joint investigation of possible antitrust violations by airline reservation systems. In this investigation, Frontier Airlines filed charges accusing United Airlines of unfairly restricting competition in the use of its Apollo computerized reservation system.[3]

Frequent user club It was a small step for American Airlines, given its massive reservation system, to add passenger accounts to accumulate travel credit for frequent flyers awards. These programs, that award free trips, create strong brand loyalty among travelers, in particular, business travelers not paying their

---

[2]James L. Heskett, "Operating Strategy - Barriers to Entry," Chapter 6 in *Managing in the Service Economy*, Harvard Business School Press, Boston, MA, 1986.

[3]For specific allegations, see "Frontier Airlines, Inc.(A)," HBS case no. 9-184-041, Boston: HBS Case Services, Harvard Business School, 1983.

own way. Thus, discount fares on a new competitor have no appeal. Mr. Layer, a travel consultant has been quoted as saying, "It's one of the most anticompetitive programs ever erected."[4]

Alfred Kahn, the father of deregulation, while heading the CAB in the late 1970's did not foresee how airlines would create reservation systems and frequent flyer plans to stifle competition. He is quoted as saying, "Nobody recognized all the ways in which a carrier could insulate itself from competition".[5]

Switching costs Information technology has been used in the medical supplies industry to directly link hospitals to the suppliers distribution network by on-line computer terminals. Both American Hospital Supply and McKesson, the drug distributor, have installed their on-line terminals in hospitals permitting the purchase of supplies and drugs when needed. Significant switching costs are built into this arrangement because the hospital is able to reduce inventory carrying costs and has the convenience of on-line ordering for replenishments. The supplier benefits by a reduction in selling costs knowing how difficult it is for a competitor to entice away a customer co-opted into their system[6]

Revenue Generation

Real time information technologies can play a competitive role with a focus on internal operations to increase revenue opportunities. The concept of yield management is best understood as a revenue maximizing strategy to make full use of service capacity. Advances in micro-computers has created

---

[4]Robert L. Rose and Jonathan Dahl, "Skies Are Deregulated, But Just Try Starting A Sizable New Airline," *Wall Street Journal*, July 19, 1989, pg. A1.

[5]ibid., pg. A8.

[6]From Harold S. Bott, "Information for Competitive Advantage," *Operations Management Review*, Fall 1985, pg. 35.

opportunities for innovative point of sale devices. The use of expert systems tied to (800) numbers allow for increased customer service.

Yield management Using information available in their Saber reservation system, American Airlines was first to realize the potential of what is now called "yield management." Constantly checking the status of their upcoming flights and competitors on the same route, American makes pricing and allocation decisions on unsold seats. Thus, the number of Supersaver fares allocated to a flight can be adjusted to insure remaining empty seats have a change of being sold but not at the expense of a full fare seat. This real time pricing strategy maximizes the revenue for each flight by insuring that no seat goes empty for want of a bargain seeking passenger.

Thus, yield management is the application of information to improve the revenue generated by a time perishable resource (airline seats, hotel rooms, etc.). The success of yield management for American has not gone unnoticed by other services industries, for example Marriott Hotels has installed a nationwide system to increase occupancy rates. American Airlines is capitalizing on its innovation by selling the yield management software to noncompetitive industries such as the French national railroad.

In addition, yield management can be used as an effective barrier to entry against start-up airlines attempting to offer discount fares. By selectively matching a competitors low fares only on certain routes, the national carrier sells a few seats at bargain rates and can subsidize the loss with earnings elsewhere in its network. This strategy was not lost on Mr. Donald Burr, CEO of People Express whose failing airline was bought by Texas Air in 1986, when he said "I'am the world's leading example of a guy killed by a computer chip"[7]

Point of sale Wal-Mart has discovered a new toy for the discount shopper, called a VideOcart. As the shopper pushes the

---

[7]op. cit. Robert L. Rose and Jonathan Dahl, pg. A8.

VideOcart through the store, information about the department at hand flashes on the video screen attached to the cart. The cart also helps customers find items in the store by listing hundreds of products by department and then displays a map of the store for ease in location. The company supplying the cart claims that it has increased sales by $1 per visit in trials at supermarkets.[8] For another example, consider the use of palm held microcomputer transmitters. With this device a waitperson in a restaurant can transmit an order directly to the kitchen monitor and simultaneously the bill to the cashier. Thus, saving unnecessary steps and allowing more time for suggestive selling.

Expert systems Otis Elevator Company is using an expert system together with laptop computers in the hands of maintenance staff to speed repair in the field. Collecting information on the behavior of its elevators over the years has lead to a knowledge base incorporated in the expert system. A repairmen in the field using a laptop computer can call up the system and receive diagnostic help in identifying the source of the problem. The result is elevators placed back in service quickly and the need for fewer repairmen. Some of the earlier applications of expert systems have been in the medical field and conceivably could be accessed for a fee by physicians. An oil exploration expert system was able to identify promising drilling sites for a major oil company.

Data Base Asset

James L. Heskett observed that the data bases a service firm possess can be a hidden asset of strategic importance. The expense of assembling and maintaining a large data base is itself a barrier to entry for competitors. But more importantly, the data base can be mined for profiles of customer buying habits and

---

[8]From Kevin Helliker, "Wal-Mart's Store of the Future Blends Discount Prices, Department-Store Feel," *Wall Street Journal*, May 17, 1919, pg. B1.

opportunities for developing new services.[9]

Sell information Dun & Bradstreet created a business of selling access to its data base of business credit information. American Home Shield, a provider of service contracts for individual home heating, plumbing, and electrical systems, discovered it had a valuable asset in its data base accumulated over many years of repair experience. Manufactures are now invited to access this data base to evaluate the performance patterns of their products. American Express, with detailed access to the spending habits of its card holders, now offers breakdowns of customer spending patterns to its retail customers.

Develop services Sears, Roebuck and Co. has become a financial services giant by exploiting its customer data base. Unlike American Express, Sears was in a position to use its credit card and customer buying pattern data base for the delivery of new services. With its long standing insurance presence in Allstate, Sears has acquired firms such as Dean Witter & Company and Coldwell Bankers to establish itself as the one stop financial services center with retail hours.

Micromarketing Achieving a truly focused service strategy targeted at customers at the micro level is now possible. Bar coding and checkout scanner technology has created a wealth of consumer buying information that can be used to target customers with precision. Analysis of this data base can allow marketers to pin point their advertising and product distribution. To increase sales, Borden Inc. has used this information to select stores in which to feature its premium pasta sauce. Kraft USA saw its sales of cream cheese increase when it targeted its flavors to the tastes of a particular store's shoppers.[10]

---

[9] op. cit. James L. Heskett, pg. 114.

[10] Michael J. McCarthy, "Marketers Zero in on Their Customers," *Wall Street Journal*, March, 18, 1991, pg. B1.

Productivity Enhancement

New developments in the collection and analysis of information has increased our ability to manage multi-site service operations. Using laptop computers retail inventory can be managed on a daily basis to make better use of shelf space by matching products with sales. Information collected on the performance of a multi-site units can be used to identify the most efficient producers. Productivity is enhanced system wide when the sources of these successes are shared with other sites. The foundation for a learning organization is then established.

Inventory Status Using a handheld computer, Frito-Lay sales representatives have eliminated paper forms and instead daily download via telephone to the Plano, Texas headquarters the data collected on their route. The company use this data to keep track of inventory levels, pricing, promotions, and state or returned merchandise. With daily updates on sales, manufacturing and distribution are able to keep fresh products moving through the system matching consumer demands. For a perishable product like potato chips, having the right product, at the right place, and in the proper amount is critical to the success of Frito-Lay. A Frito-Lay spokesman said the company saved more than $40 million in its first year from reduced paperwork, reduced losses from stale products and route consolidation.[11]

Data Envelopment Analysis DEA is a linear programming technique developed by A. Charnes, W. W. Cooper, and E. Rhodes to evaluate nonprofit and public sector organizations. It has subsequently found applications in for profit service organizations. DEA compares each service delivery unit with all other service units for a multi-site organization and computes an efficiency rating based on the ratio of resource inputs to outputs. Multiple inputs (i.e. labor hours, materials, etc.) and multiple

---

[11]Peter H. Lewis, "Looking Beyond Innovation, an Award for Results," *The New York Times*, June 23, 1991, pg. 8.

outputs (sales, referrals, etc.) are possible and desirable in measuring a units efficiency. Taking this information, the linear programming model determines the efficiency frontier based on those few units producing at 100 percent efficiency. By comparing the operating practices of efficient units with less efficient units, areas of improvement can be identified. Sharing management practices of efficient units with less efficient units provides an opportunity for the later's improvement and total system productivity enhancement. Repeated use of DEA can establish a climate of organizational learning that fuels a competitive strategy of cost leadership.

Banker and Morey applied DEA to a 60 unit fast food restaurant chain and found 33 units to be technically efficient. In the analysis three outputs (food sales for breakfast, lunch, and dinner) and six inputs (supplies and materials, labor, age of store, advertising expenditures, urban/rural, and existance of drive-in window) were used. It is interesting to note that the inputs included both discretionary and uncontrollable variables (e.g. demographic variable urban/rural and whether the unit had a drive-in window).[12]

LIMITS IN THE USE OF INFORMATION

Only the positive benefits of using information as a competitive strategy have been addressed. Some of the strategies raise questions of fairness, invasion of privacy, and anti-competitiveness, which if abused could result in harm to consumers.

In an attempt to created barriers to entry, the use of reservation systems and frequent user programs have been identified as potentially anti-competitive. For example, how should a frequent flyer free trip award be considered, in particular, when the passenger has been travelling on business at

---

[12]Banker, Rajiv D. and Richard C. Morey: "Efficiency Analysis for Exogenously Fixed Inputs and Outputs," *Operations Research*, vol. 34, no. 4, July-August 1986, pp. 518-519.

income in kind and corporations believe the free ticket is theirs. However, the long run implication is the removal of price competition in air travel.

It has been said that the easiest way to start a riot is begin asking passengers on a flight how much their ticket cost. Under yield management, ticket prices can change every hour and thus the price paid is a moving target and the ticketing process becomes a lottery. At the extreme, is yield management fair and equitable to the public? Or has every service price always been negotiable and now customers are becoming aware of their buying power?

The concept of micromarketing has the potential to create the most violent backlash from consumers because of the perceived invasion of privacy. Having a record of your every purchase at the local supermarket shared with eager manufacturers could lead to very manipulative sales practices. Lotus Development Corporation recently felt the sting of consumer displeasure after announcing the availability of its MarketPlace household data base to anyone with PC and modem. Lotus received more than 30,000 requests from irate persons wanting to be removed from the data base. Lotus subsequently withdrew its offer of general availability but still sells access to the data base to large corporations.[13]

CONCLUSIONS

Research in Service Management is recognized as an eclectic undertaking. At a minimum the fields of Marketing, Operations Management, and Organizational Behavior must be included in any study of service management. To this set of disciplines I would add Information Technology.

The strategic role of information in service strategies is organized into four categories; create barriers to entry, data base

---

[13]. "How Did They Get My Name," *Newsweek*, June 3, 1991, p.41

asset, revenue generation, and productivity enhancement. Information based competitive strategies were illustrated for each category.

Yield management, pioneered by American Airlines, is the most comprehensive use of information for strategic purposes and best illustrates the integrated nature of services. Using forecasting methods from operations management, pricing strategies from marketing, and consumer psychology from organizational behavior, American Airlines has developed a computer based method to individually price airline tickets to maximize the revenue for any given flight. This innovation directly attacks the classic service dilemma of matching supply and demand.

In conclusion, the discussion of the limits to information use suggests that service managers again must always be sensitive to the perceptions of their actions on the consuming public.

REFERENCES

Banker, Rajiv D. and Richard C. Morey: "Efficiency Analysis for Exogenously Fixed Inputs and Outputs," Operations Research, vol. 34, no. 4, July-August 1986, pp. 513-521.

Bott, Harold S.: "Information for Competitive Advantage," Operations Management Review, Fall 1985, pp. 30-42.

Buffa, Elwood S.: Meeting the Competitive Challenge, Irwin,Homewood, IL, 1984.

Chase, Richard B.: "Where Does the Customer Fit in a Service, Operation," Harvard Business Review, November-December 1978, pp. 137-142.

Hayes, Robert H. and Steven C. Wheelwright: Restoring Our Competitive Edge, Wiley, New York, 1984.

Heskett, James L.: Managing in the Service Economy, Harvard Business School Press, Boston, MA, 1986. Heskett, James L., W. Earl Sasser, and Christopher W.L. Hart: Service Breakthroughs, The Free Press, New York, 1990.

Levitt, Theodore: "Production-Line Approach to Service," Harvard Business Review September-October 1972, pp. 41-52.

Lovelock, Christopher H.: Services Marketing, Prentice-Hall, Englewood Cliffs, NJ, 1984.

Lovelock, Christopher H.: Managing Services: Marketing Operations, and Human Resources, Prentice-Hall, Englewood Cliffs, NJ, 1988.

McFarlan, Warren F.: "Information Technology Changes the Way You Compete," Harvard Business Review May-June 1984, pp. 98-103.

Porter, E. Michael: Competitive Strategy, The Free Press, New York, 1980.

Porter, E. Michael and Victor E. Millar: "How Information Gives You Competitive Advantage," Harvard Business Review July-August 1985, pp. 149-160.

Sasser, Earl W., Paul R. Olsen, and Daryl D. Wycoff: Management of Service Operations, Allyn and Bacon, Boston, 1978.

Sherman, David H.: "Improving the Productivity of Service Businesses," Sloan Management Review, Spring 1984, pp. 41-23.

OPERATIONS MANAGEMENT INTERFACES

THE MANUFACTURING/MARKETING INTERFACE: CRITICAL STRATEGIC AND TACTICAL LINKAGES

Warren H. Hausman
Department of Industrial Engineering
and Engineering Management
Stanford University

and

David B. Montgomery
Robert A. Magowan Professor of Marketing
and Business School Trust Fellow
Graduate School of Business
Stanford University

Presented at the Conference Perspectives in Operations Management in honor of Prof. Elwood S. Buffa, Nov. 2-3, 1991, at UCLA. Research support from the LEK Partnership, the Stanford Graduate School of Business, and the Stanford School of Engineering is gratefully acknowledged. Dr. Sara Beckman of Hewlett-Packard provided valuable conceptual suggestions during the formative stages of this project. Timothy Hough and Timothy Scholes of LEK furnished valuable suggestions at a later stage of this project.

INTRODUCTION

There has been much conceptual discussion of the interfaces between manufacturing and marketing (Hausman and Montgomery, 1985) and manufacturing strategy and tactics (Hayes and Wheelwright, 1984); however there has been a dearth of empirical evidence. This paper presents empirical evidence on these two issues: the linkages between marketing strategy and manufacturing strategy, and the linkages between manufacturing strategy and manufacturing tactics. Data obtained from several groups of executives are presented regarding both of these linkages, as well as data on each executive's view of his/her business unit's current and desired manufacturing strategy priorities. Analysis of the data shows that there is a substantially consistent view of manufacturing/marketing strategic linkages and of tactical/strategic manufacturing linkages.

MANUFACTURING STRATEGY AND MARKETING STRATEGY

The now-classic manufacturing strategy literature (e.g., Hayes and Wheelwright, 1984) typically deals with the dimensions of manufacturing priorities listed in Table 1. While none of these priorities can be ignored, usually some are more critical than others for a given competitive environment; indeed, many case studies in manufacturing strategy courses require the reader to discern some rank ordering of these priorities.

TABLE 1

MANUFACTURING PRIORITIES

Cost (Lowest total cost)
Quality (Conformance to Specifications)
Dependability (Meeting delivery/availability commitments)
Short-Term Flexibility (adapting to changing product volume and mix requirements)
Innovation (rapid introduction of new products and features)

On the marketing side, Table 2 lists six dimensions on which customers evaluate products and services relative to the competition. Specific product-market segments may have additional competitive dimensions; e.g., computer software or hardware may have an ease-of-use dimension which could be classified as a "feature" but which should perhaps be evaluated separately.

TABLE 2

MARKETPLACE COMPETITIVE DIMENSIONS

Price
Quality (reliability, durability; not features)
Availability (response time mean and variance)
Variety (width of product line; options, customization)
Features (Product attributes; vs. competition; first to market)
Post-Sales Service (including spare parts availability)

The dimensions in Table 2 should answer the question, Why would a customer buy from us? Similar to the manufacturing priorities, the marketplace competitive dimensions allow a business unit to determine relative priorities among the various dimensions. An interesting exercise is to name some product or service offerings which focus on each of the six marketplace competitive dimensions (e.g., Volkswagen for low-price transportation, Maytag for high reliability, Federal Express for rapid and assured delivery, Burger King for customized burgers, Sony for product features, and Caterpillar for post-sales service). Market driven priorities may be established for the marketplace competitive dimensions by conjoint analysis as suggested and illustrated in Hausman and Montgomery (1990).

What are the critical linkages between the manufacturing strategy elements (Table 1) and the marketplace competitive dimensions (Table 2)? We asked participants in four 1991 executive programs to consider each manufacturing strategy priority and assess the impact that priority was likely to have on each of the six strategic marketing priorities, on a 1 to 5 scale (5 = highest impact, 1 = lowest impact). The four programs were: Manufacturing Strategy for Competitive Advantage (MSCA), Effective Management of Production Operations (EMPO), the Stanford-American Electronics Association Executive Institute (AEA), and a Management of Technology program sponsored by Kyoto Research Park (KRP). Participants in the latter program were Japanese managers and executives; participants in the first three programs were predominantly (i.e., over 90%) managers and executives from the U.S.

Table 3 presents the mean results by program. Linkages significantly greater than 3.0 at the 0.05 level and beyond are in bold face and italic[1]. Since 3.0 is the scale midpoint value indicating "moderately favorable impact", the significance tests in Table 3 reflect executive assessments of linkages which are significantly greater than "moderate." Thus the important and statistically significant linkages are highlighted in Table 3.

The manufacturing priority of cost was seen as being the most important driver of the marketing priority of price in all four executive samples, with quality being the second most important in all four programs. All four cost results were significant and three of the four quality results were also. The marketing priority of quality was uniformly seen as being most strongly driven by manufacturing quality, while the manufacturing priority of innovation was uniformly seen as the major manufacturing influencer of the marketing priority of product features. Manufacturing flexibility was seen as the second most important driver of features for three of the four executive groups. Note that the Japanese see four of the five manufacturing priorities as having substantial impact on features, in contrast to the US groups. The Japanese appear to have a broader view of the manufacturing priorities affecting features.

The marketing priority of availability was seen as being most influenced by manufacturing flexibility, while dependability was seen as being the second most important (the order of these was reversed for the Japanese sample). The three US executive samples agreed that variety in the marketplace was most influenced by manufacturing innovation, while manufacturing flexibility was seen as being second in importance in delivering variety to the marketplace. Once again, the Japanese executives reversed the marketing impact of these two manufacturing priorities. Finally, the manufacturing priorities of quality and dependability are seen as having the two most important manufacturing linkages to the marketing priority of post-sales service in all programs, although only four of the eight results were statistically significant.

The Japanese executives placed greater importance on the Cost-Features and Quality-Features links than did the U.S. executives; and they placed less importance on the Innovation-Variety and Quality-Post Sales Service links. When asked about the former emphases, a few Japanese participants responded that lower product costs enable companies to provide more features; the Quality-Features link remained unexplained, and may have resulted from an association between marketplace Quality and manufacturing Quality priorities.

---

[1] The linkages with a t ratio greater than 2.1 are highlighted in bold face and italics. This indicates significance at the 0.05 level and beyond as $t = 2.1$ reflects the critical 0.05 level for the program having the smallest sample size (i.e. MSCA with $n=18$). In fact 31 of the 39 significant results in Table 3 have a t ratio greater than 3.0 and reflect significance at beyond the 0.01 level.

TABLE 3

Manufacturing Priorities:	Price	Quality	Features	Availa-bility	Variety	Post-Sales Service
MSCA (n=18)						
Cost	***4.50***	2.94	3.22	2.61	2.89	2.67
Quality	***3.61***	***4.78***	3.11	2.94	2.67	3.50
Depend.	3.28	3.17	2.78	***4.00***	2.83	3.33
Flexibility	2.67	2.28	2.94	***4.28***	3.44	2.78
Innovation	3.50	2.94	***4.22***	3.11	***3.72***	2.78
EMPO (n=53)						
Cost	***4.71***	2.89	2.61	2.61	2.24	1.97
Quality	***3.82***	***4.92***	2.74	2.92	2.18	***3.61***
Depend.	3.26	***4.03***	2.39	***3.42***	2.26	***3.68***
Flexibility	3.05	2.82	***3.50***	***3.97***	***3.82***	2.42
Innovation	3.21	2.97	***3.89***	2.97	***3.89***	2.29
AEA (n=82)						
Cost	***4.41***	2.52	2.15	2.14	2.26	2.41
Quality	***3.49***	***4.81***	2.51	2.58	1.94	***4.10***
Depend.	2.89	3.23	2.40	***4.06***	2.09	***3.54***
Flexibility	2.62	2.37	***3.41***	***4.22***	***3.91***	2.73
Innovation	3.16	2.62	***4.44***	2.93	***4.11***	2.07
KRP (n=25)						
Cost	***4.92***	2.60	***3.52***	2.32	2.36	2.04
Quality	3.36	***4.76***	***3.52***	2.04	2.24	3.12
Depend.	2.76	2.52	3.24	***4.40***	2.40	3.20
Flexibility	2.40	2.08	***3.56***	***4.24***	***3.68***	2.28
Innovation	3.20	2.44	***4.44***	2.28	2.84	1.80

Marketing Priorities: (column headings Price through Post-Sales Service)

Although there are some differences between the executive groups as noted above, there was a substantial amount of agreement among groups in terms of their perceptions of the impact of manufacturing priorities on marketing priorities. In particular, for each of the six marketing priorities, the various programs were quite consistent in their relative ranking of the impact of the five manufacturing priorities. With six marketing priorities and four programs there are twenty-four possible first ranks. In twenty (83%) of the cases, the four programs ranked the same manufacturing priority first. Similarly, in eighteen instances (75%) of the cases the second choice was also identical. If a match is taken to be an overlap of manufacturing priorities ranked first or second, fully 94% overlap was found among the executive groups in their assessment of manufacturing priorities.

Further evidence of the inter-program agreement on the manufacturing/ marketing linkages may be found by correlating, for each pair of programs, their respective vectors of mean manufacturing linkages for each specific marketing priority. The results are shown in Table 4. All but three of the correlations (92%) are significant at the .05 level or beyond, the exceptions being MSCA with EMPO and AEA for Features and with KRP for Variety[2]. Three quarters of the program correlations are greater than 0.9. Hence there was substantial, although not total consistency between the programs in their views concerning the important linkages between manufacturing and marketing priorities.

TABLE 4

Program Correlations: Manufacturing Priorities - Marketing Priorities

	Price	Quality	Features	Availa-bility	Variety	Post-Sales Service
MSCA/EMPO	0.93	0.92	0.72	0.97	0.97	0.98
MSCA/AEA	0.97	0.97	0.77	0.99	0.98	0.95
MSCA/KRP	0.97	0.98	0.98	0.94	0.72	0.94
EMPO/AEA	0.98	0.97	0.96	0.95	0.99	0.92
EMPO/KRP	0.96	0.87	0.85	0.86	0.85	0.96
AEA/KRP	0.99	0.96	0.88	0.94	0.83	0.96

---

2 The critical correlation at the 0.05 level for a one tailed test with five observations is 0.8054 (Fisher and Yates,1957,pg.53).

The aggregate results from combining all four programs are presented in Table 5, with linkages significantly greater than 3.0 at the 0.01 level in boldface and italic type. The aggregate results show the highest linkage between Quality (mfg.) and Quality (market). There are also very strong (mean greater than 4.0) linkages between Cost-Price, Innovation-Features, and Flexibility-Availability. There are also quite strong linkages (mean greater than 3.5) between Quality-Price, Dependability-Availability, Innovation-Variety, Flexibility-Variety links, Quality (mfg.)-Post Sale Service, and Dependability-Post Sales Service.

TABLE 5

Total Sample (n=178)	Marketing Priorities:					
Manufacturing Priorities:	Price	Quality	Features	Availa-bility	Variety	Post-Sales Service
Cost	***4.58***	2.68	2.57	2.37	2.33	2.29
Quality	***3.54***	***4.80***	2.78	2.64	2.12	***3.76***
Depend.	2.98	***3.32***	2.53	***3.92***	2.23	***3.47***
Flexibility	2.68	2.42	***3.37***	***4.14***	***3.74***	2.57
Innovation	3.21	2.72	***4.28***	2.88	***3.80***	2.18

DO FIRMS HAVE DIFFERENT MANUFACTURING PRIORITIES?

This question was explored by asking the same executives to rank their current and desired manufacturing priorities (1 = lowest priority,..., 5 = highest priority). Figure 1 presents the mean results for current priorities for each of the four executive groups. They are generally similar across groups, with Quality, Dependability, and Cost being ranked first, second and third respectively and Flexibility and Innovation being ranked in last place, except for the MSCA group placing Cost above Quality. Similarity was again measured by correlating the mean ranks for all pairs of programs; all but one correlation were significant at the 0.05 level. Only the MSCA/AEA correlation between the mean ranks is not significant at the 0.05 level (it is significant at the 0.06 level). The mean correlation between programs was 0.89.

FIGURE 1. CURRENT MANUFACTURING PRIORITIES

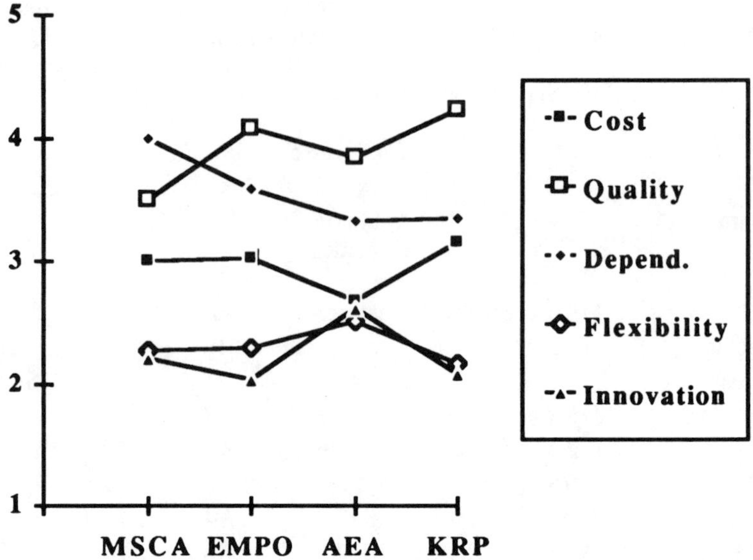

Figure 2 presents similar results for the desired manufacturing priorities. Again, the results are reasonably similar across groups; indeed, the results are not too different from Figure 1's current manufacturing priorities, except that Cost is ranked somewhat lower (except by the Japanese KRP program), Flexibility continues to rank low (except by the EMPO program), and Innovation is ranked somewhat higher (by all but EMPO participants). However, there was not as much inter-program agreement on desired manufacturing priorities as there was for current manufacturing priorities. The average correlation slipped to 0.76 and only three of the six inter program correlations were significant at the 0.05 level. The Japanese in comparison with EMPO and MSCA largely account for the differences.

The degree to which each program's participants are satisfied with their companies' current manufacturing priorities may be seen by correlating the current and desired manufacturing priority rankings across Figures 1 and 2 for each program. All three US executive groups were substantially satisfied with their current manufacturing priorities, with all three correlations being significant at the 0.05 level and the average correlation across the three programs

FIGURE 2. DESIRED MANUFACTURING PRIORITIES

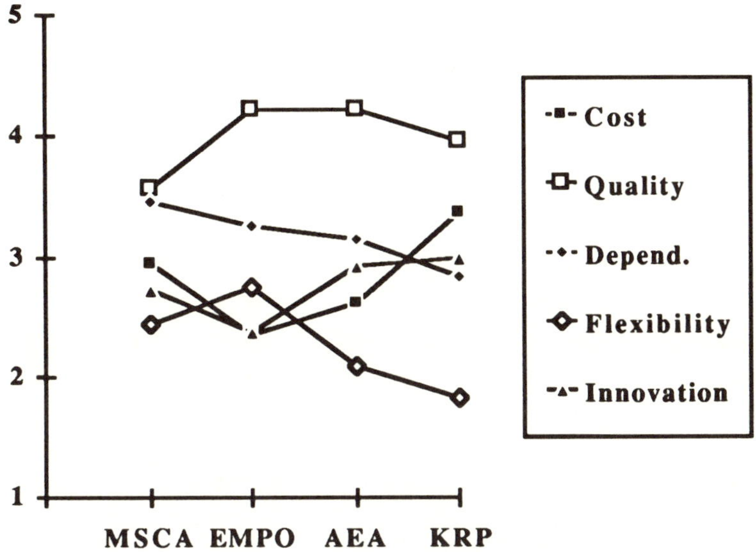

being 0.90. In contrast,the Japanese correlation between the current and desired manufacturing priority ranking was not significant at the 0.05 level and was only 0.76. Interestingly, the Japanese, who have the reputation of having superb manufacturing capability and relentless pursuit of excellence, are the least complacent.

The above results are averages over all participants in each program. In order to see whether or not individual firms face similar manufacturing priorities, we tabulated the number of times each desired strategic priority was listed as "first priority" and the number of times it was listed as "lowest priority"; these results, converted to percentages, are presented in Table 6. The dominant role of quality as a target of manufacturing is further reinforced by the results in this table.

Table 6 also demonstrates that among these 178 executives there are at least seven percent who would rate any one manufacturing strategic priority as "highest", and at least another seven percent who would rate the same manufacturing strategy priority as "lowest"! Thus, in this population, we conclude that individual firms have quite different manufacturing priorities. This observation leads to our final section of the paper; choosing from the many tactical manufacturing programs which are available to manufacturing executives.

TABLE 6

Desired Manufacturing Priorities:	Fraction Ranking it First Priority:				
	MSCA	EMPO	AEA	KRP	TOTAL
Cost	0.17	0.00	0.07	0.12	0.07
Quality	0.33	0.60	0.57	0.44	0.54
Depend.	0.22	0.08	0.11	0.08	0.11
Flexibility	0.06	0.13	0.09	0.12	0.10
Innovation	0.28	0.11	0.15	0.24	0.16

Desired Manufacturing Priorities:	Fraction Ranking it Last Priority:				
	MSCA	EMPO	AEA	KRP	TOTAL
Cost	0.17	0.25	0.24	0.00	0.20
Quality	0.06	0.08	0.07	0.04	0.07
Depend.	0.11	0.08	0.07	0.16	0.09
Flexibility	0.33	0.25	0.41	0.56	0.38
Innovation	0.33	0.28	0.18	0.24	0.24

LINKING MANUFACTURING TACTICS AND STRATEGY

To translate manufacturing strategic priorities into appropriate tactical programs requires an agreed-upon relationship among manufacturing tactics and the manufacturing strategies; additional data were collected to provide evidence on the amount of executive agreement on these linkages and their impact. The executives were presented with ten tactical manufacturing programs and asked to assess the impact of each one on each of the five manufacturing strategy priorities (scale: 5 = high impact, 1 = low impact). The results, by program, are presented in Tables 7-10, with results statistically greater than the scale midpoint (3.0) at the 0.05 level and beyond displayed in boldface and italic.

Recall that results which are statistically greater than 3.0 reflect linkages which are both managerially and statistically significant, since the scale value of 3.0 indicates a moderately favorable impact.

The tactical programs presented were seen as having broad and significant linkages to both manufacturing cost and quality. Fully sixty-eight percent of the possible tactical/strategic linkages with cost and quality across the four programs were seen as being significantly greater than 3.0. This was less true for Dependability, Flexibility, and Innovation which had corresponding significance proportions of 55%, 43%, and 25%, respectively. The results are even more dramatic when one examines the number of linkages above 4.0, reflecting a scale value indicating a very significant impact. Cost and Quality priorities each have 11 out of a possible forty linkages to tactical programs above 4.0, whereas Dependability, Flexibility, and Innovation realized 3, 0, and 2 such values, respectively. In sum, the executive respondents viewed these ten tactical programs as having substantially broader and more significant linkages to manufacturing cost and quality than to the other three manufacturing strategic priorities.

The issue of executive agreement across programs with respect to the tactical/strategic links may be addressed in two ways. An examination of Tables 7 - 10 reveals that in seventy percent of the cases the programs are in complete agreement as to which of the ten tactical programs is the most impactful on a strategic priority. If one examines the agreement in rating the top two tactics for each manufacturing strategic priority, there is seventy-five percent agreement across the programs.

Further evidence relating to consensus among the programs may be obtained by correlating the vector of mean linkage ratings over the manufacturing tactics for each of the manufacturing strategic priorities. The correlation results are presented in Table 11. All programs have correlations significant at the 0.05 level and beyond for Cost, Quality, Flexibility, and Innovation[3]. For Dependability, however, four of the six correlations are statistically insignificant and less than 0.50, which suggests substantial disagreement among the four programs on this dimension. It may be that the definition of Dependability (meeting delivery/availability commitments) dichotomizes respondents into two groups: those in a "to order" environment versus a "to stock" environment, and certain tactics may work quite differently in those two environments. In any case, the lack of consensus on the Dependability dimension suggests that this dimension needs to be refined in future research in order to identify tactics which may be used to impact it.

---

3 The critical correlation at the 0.05 level for a one tailed test with ten observations is 0.55 (Fisher and Yates, 1957, pg.53).

TABLE 7

MSCA
n=18

Tactical Program:	Cost	Quality	Depend-ability	Flexi-bility	Innova-tion
Quality Awareness	3.22	***4.72***	***3.61***	2.33	2.44
Statistical Process Control	***3.67***	***4.50***	***3.83***	2.50	2.39
Just-in-Time	***4.06***	2.72	3.61	3.06	2.22
Design for Mfrability	***4.39***	***4.06***	***3.56***	***3.89***	***3.72***
Concurrent Engineering	***3.89***	***3.67***	2.94	***3.44***	***4.06***
Workcells	***3.44***	***3.67***	3.17	***3.72***	3.11
Self-directed Teams	3.33	***3.56***	***3.67***	***3.61***	3.50
Reducing Mfg Cycle Time	***4.39***	3.33	***4.17***	***3.67***	3.11
Time-based Competition	***3.72***	3.28	***3.78***	3.33	2.89
CAD/CAM	3.33	3.11	2.83	3.28	***3.78***

Strategic Manufacturing Goals: Cost, Quality, Dependability, Flexibility, Innovation

TABLE 8

EMPO
n=53

Tactical Program:	Cost	Quality	Depend-ability	Flexi-bility	Innova-tion
Quality Awareness	***3.71***	***4.89***	***3.82***	2.08	2.42
Statistical Process Control	***3.53***	***4.61***	***3.76***	2.24	2.13
Just-in-Time	***4.37***	***3.63***	3.37	***3.58***	2.55
Design for Mfrability	***4.53***	***4.34***	***3.74***	3.32	3.29
Concurrent Engineering	***3.95***	***3.82***	***3.42***	***3.39***	***3.68***
Workcells	3.26	***3.71***	2.92	***3.63***	2.92
Self-directed Teams	***3.34***	***3.76***	***3.39***	***3.79***	***3.55***
Reducing Mfg Cycle Time	***4.16***	2.89	3.21	***3.53***	3.11
Time-based Competition	3.22	2.84	3.30	***3.57***	3.03
CAD/CAM	3.25	***3.53***	3.14	***3.53***	***3.94***

Strategic Manufacturing Goals: Cost, Quality, Dependability, Flexibility, Innovation

TABLE 9

AEA
n=82

Tactical Program:	Cost	Quality	Dependability	Flexibility	Innovation
	Strategic Manufacturing Goals:				
Quality Awareness	***3.84***	***4.73***	***3.89***	2.51	2.52
Statistical Process Control	***3.68***	***4.30***	***3.58***	2.41	2.09
Just-in-Time	***3.96***	2.75	3.19	3.40	2.36
Design for Mfrability	***4.46***	***4.28***	***3.87***	3.24	3.23
Concurrent Engineering	***3.38***	***3.51***	3.29	3.22	***3.86***
Workcells	3.04	***3.69***	***3.43***	***3.49***	2.97
Self-directed Teams	3.09	***3.71***	***3.61***	3.29	***3.44***
Reducing Mfg Cycle Time	***4.20***	2.80	***3.73***	***3.84***	3.02
Time-based Competition	3.04	2.75	***3.56***	***3.56***	3.28
CAD/CAM	***3.44***	***3.67***	3.02	***3.48***	***4.02***

TABLE 10

KRP
n=25

Tactical Program:	Cost	Quality	Dependability	Flexibility	Innovation
	Strategic Manufacturing Goals:				
Quality Awareness	3.24	***4.84***	3.28	2.36	2.56
Statistical Process Control	***3.80***	***3.68***	***4.28***	2.72	1.68
Just-in-Time	***3.88***	2.36	***4.44***	3.28	1.92
Design for Mfrability	***4.64***	***3.64***	***3.60***	2.76	2.72
Concurrent Engineering	***4.04***	***4.16***	3.40	3.24	2.88
Workcells	3.08	***3.48***	3.36	3.12	2.40
Self-directed Teams	2.80	3.24	2.96	***3.40***	***3.56***
Reducing Mfg Cycle Time	***4.08***	2.40	***3.92***	3.16	2.40
Time-based Competition	3.08	2.20	3.52	3.04	2.64
CAD/CAM	***3.56***	***3.96***	3.28	3.12	3.08

TABLE 11

Program Correlations: Manufacturing Tactics - Strategy

	Cost	Quality	Dependability	Flexibility	Innovation
MSCA/EMPO	0.82	0.82	0.32	0.85	0.92
MSCA/AEA	0.70	0.87	0.66	0.8	0.91
MSCA/KRP	0.83	0.7	0.44	0.62	0.75
EMPO/AEA	0.87	0.92	0.57	0.91	0.98
EMPO/KRP	0.84	0.78	0.19	0.88	0.83
AEA/KRP	0.82	0.87	-0.10	0.72	0.82

CONCLUSIONS

This paper has used empirical data to explore issues related to the manufacturing/marketing interfaces. We first assessed the critical linkages between the competitive market environment faced by a firm and its elements of manufacturing strategy; we found strong agreement on the highest marketing/manufacturing linkages across the four executive program groups. Next, we observed agreement in average manufacturing priorities across groups of executives, but a substantial divergence in manufacturing priorities faced by individual firms in our sample. Finally, we observed relatively close agreement on the linkages between manufacturing tactics and manufacturing strategy priorities with the notable exception of Dependability.

This linkages framework can be used by a firm to translate the competitive market environment into relevant manufacturing strategy priorities, and these strategic priorities can be further used to assess which specific tactical programs will be most appropriate for the firm's competitive environment.

REFERENCES

Beckman, Sara L., Boller, William A., Hamilton, Stephen A., and Monroe, John W. (1989), "Using Manufacturing as a Competitive Weapon: The Development of a Manufacturing Strategy", Chapter 3 of Strategic Manufacturing: Dynamic New Directions for the 1990's, Patricia E. Moody, (Ed.), Dow Jones-Irwin, APICS Series in Production Management, November 1989.

Fisher and Yates, Statistical Tables for Biological, Agricultural and Medical Research, Fifth Edition, Hafner Publishing Company, New York, 1957, p. 59.

Hausman, Warren H. and Montgomery, David B. (1985)," Managing the Marketing/Manufacturing Interface," Issues (Autumn); reprinted in Gestion 2000 (Vol.2, No.5, 1986).

Hausman, Warren H. and Montgomery, David B. (1990)," Making Manufacturing Market Driven," Research Paper #1103, Graduate School of Business, Stanford University.

Hayes, Robert H., and Wheelwright, Stephen C.(1984), Restoring Our Competitive Edge: Competing Through Manufacturing, John Wiley & Sons, 1984.

LINKING TECHNOLOGY AND BUSINESS STRATEGIES: A METHODOLOGICAL APPROACH AND AN ILLUSTRATION

Arnoldo C. Hax and Manuel No
Sloan School of Management
Massachusetts Institute of Technology

ABSTRACT

Linking technology and business strategies is a demanding task that has central importance in strategy formation. Now that technology is a critical source to achieve and sustain competitive advantage, the ability to incorporate technology into a business strategy can make the difference between a winning or a losing strategic alternative. This paper discusses a methodology that can be used to explore systematically the way to link business and technology strategies. The authors illustrate the application of the suggested methodology in a real life setting.

INTRODUCTION

The role of technology has become so pervasive in the business world that is appropriate to say there is hardly any significant industry that can be classified as low-tech. In fact, technological forces are restructuring industries and defining new ways to compete. Managers are confronted with the demanding task of accelerating the speed at which innovations in new products and processes are translated into profitable commercial ventures.

Though there are many studies that analyze the process and sources of innovation (E.A. von Hippel, 1988), the disruptions introduced by new technologies (W.J. Abernathy and J.M. Utterback, 1978; J.M. Utterback and L. Kim, 1986; P. Anderson and M.L. Tushman, 1990), the concept of core competencies (C.K.Prahalad and G. Hamel, 1990), the strategic management of technology (L. Steele, 1989; B.L. White, 1988) or the human issues related with technology (U.E. Gattiker and L. Larwood, 1988; B. Twiss and M. Goodridge, 1989), there is less documentation on how to develop a strategic plan that integrates technology into the business strategy of a firm. This paper suggests a methodology to accomplish this task. The methodology is consistent with the strategic planning framework proposed by A. Hax and N.S. Majluf (1991), and is further illustrated by applying it to a start-up company in the massive parallel computer (MPC) industry. A brief description of the company, Masscalc, is provided in Appendix I. For a more detailed account of this case the reader is referred to No (1991).

A FRAMEWORK FOR THE DEVELOPMENT OF TECHNOLOGY STRATEGY

The formation of technology strategy takes place at all the key hierarchical levels of the firm: corporate, business, and functional levels. Figure 1 illustrates the primary tasks that we identify as relevant in the development of technology strategy. For a more detailed account of the strategic planning process, see Hax and Majluf (1991).

First, top managers have to decide, as part of the corporate strategy of the firm, what is the role to be played by technology in advancing the firm's competitive capabilities, the amount of resources to be allocated to technology, and the aggressiveness the firm will use in the innovative process and in imbedding technology into the firm's products and processes. Corporate attention is required since frequently a given technology is shared by several businesses and affects various managerial functions. Therefore, its strategic development cannot be totally decentralized at the business and functional levels. The elements of corporate strategy that communicate more pointedly to the technological requirements are the mission of the firm - particularly the statement of unique competencies - and the corporate strategic thrusts - an expression of the primary issues the firm has to address to establish a strong competitive position.

Next, technology strategies are formulated at the business level. During the process of business strategy formation we need to define the technological support required to create or reinforce the competitive advantage sustained by each business unit. This is supplied by the mission of the businesses and their respective strategic action programs. Obviously, a technology strategy cannot be created in isolation from the corporate objectives and the businesses it is intended to support.

Finally, at the technology level resides the task of interpreting all the requirements emerging from corporate and business levels, which will become the critical inputs for shaping the technology strategy of the firm. At this stage it is also necessary to identify the portfolio of specific technologies the firm will be using in supporting its business strategies. This leads to the definition of the strategic technology units (STUs), the central focus of attention in the development of technology strategy. The STU identifies the skills or disciplines that are applied to a particular product or process in order to gain technological advantage. The STUs should contain all the core technologies used now or needed in the future across the whole organization.

The STUs are critical to the execution of the technology environmental scan and internal scrutiny, the next tasks in the planning process. The environmental scan is aimed at obtaining an understanding of the key technology trends, assessing the attractiveness of each STU, and identifying technological opportunities and threats. This form of analysis we referred to as technology intelligence. Its purpose is to generate all the relevant information concerning the current and future state of development of the technology function. It is not only the existing managerial practice and state of technological progress that are important to detect. Even more critical is the recognition of future trends, state-of-the-art developments, and their embodiment in actions by competitors.

With regard to the internal scrutiny, besides the recognition of strengths and weaknesses associated with each STU, we need to determine the specific technological competencies we should build to gain competitive advantage. We conduct this analysis by examining the strategic categories of decisions linked to the technology function. A detailed listing of the decisions is included in Appendix II, taken from Hax and Majluf (1991).

Finally, we have the remaining tasks of defining broad and specific action programs, and budgets. These tasks represent the final output of the technology strategy formation process. They should respond to the corporate and business requirements as well as the challenges emerging from the environmental scan and internal scrutiny activities.

TECHNOLOGICAL REQUIREMENTS

As shown in Figure 1, the first step in developing the technology strategy is to derive a clear, unified statement of coherent strategic requirements that the company places over the technology function. The identification of these requirements helps to create a common understanding among top management of the horizontal nature of the technology function allowing them to identify and exploit potential synergies among distinct but related business units. More importantly, this step provides a mechanism that establishes an effective linkage between corporate, business and technology strategies.

Table 1 shows some of the technological requirements put onto Masscalc's technology function. Although we list those requirements by corporate and business strategies, this distinction is rather blurry in this case since Masscalc is at this time in only one business, the massive parallel computer (MPC) business.

THE DEFINITION OF STRATEGIC TECHNOLOGY UNITS (STUs)

The strategic technological units are a planning tool used to shape the strategic response to the aforementioned technological requirements. Thus, the proper selection of STUs is one of the most critical elements of the proposed methodology.

An STU refers to a discrete technology or group of technologies that are used by the company. The cluster of STUs should encompass any technology which has impact on the company's overall competitive position in the marketplace. To be effective, the authors have found that any STU should:

- be broad enough in order not to leave out potential innovations, yet specific enough so as to allow a clear understanding of the technological position of the company.
- have continuity, i.e. the STU will exist over a relatively long period of time in order to develop expertise and management control. This does not preclude the underlying product and process technologies included in a given STU to evolve through time.
- be critical to the product or service. It is recognized as a potential source of competitive advantage.
- require a set of distinctive technical capabilities. Each STU will represent a unique contribution.

Table 2 shows the STUs identified at Masscalc. Note that some of the STUs cut across organization units (STUs 1 and 2 encompass activities of the R&D and the Engineering Divisions). Also, some of them do not relate directly to the core business, but have a strategic relevance in supporting the corporate and business strategies. This is the case of "demonstration technologies" or "service", two supporting technologies with an important strategic role in Masscalc.

ENVIRONMENTAL SCAN

Only a deep knowledge of the intrinsic characteristics of the technologies used by a firm can generate the high-quality strategic thinking required for the healthy, long-term development of technology-based competitive advantages. The objective of the environmental scan is to gain this knowledge and to derive from it the degree of attractiveness of each technology as well as the opportunities and threats that technology presents to the firm. This analysis is done based on the STUs previously identified.

Developing strategic advantages from technology requires recognizing the trends followed by each specific technology, deciding which are the innovations the firm is going to incorporate, and setting up the internal means to take advantage of those innovations.

An important first step to facilitate this task is to identify the potential sources of innovation for each STU. Eric von Hippel (1988) has generated seminal work in this field. By conducting a large number of empirical studies, he has been able to pinpoint the sources of innovation in a large variety of industrial developments. The source of innovation varies greatly depending on whom is expected to receive the benefits from the innovational efforts. Primary sources are users, manufacturers and suppliers. Of special interest is the ability to identify lead users, if they are relevant to the innovation process. Lead users combine two characteristics: they have a need which is in advance of the general market, and they expect high benefits from a solution to that need. Whenever lead users do exist, it is of paramount importance to follow their innovation progress closely since they could be ahead of the market in their innovation capabilities. Table 3 shows the sources of innovation for each of the STU's of Masscalc.

The second task of the environmental scanning process is to assess the degree of attractiveness of each of the technologies the firm is using or is considering to use in its products and processes. A technology with a high degree of attractiveness is one that, when applied, will enhance significantly the competitive position of the businesses it supports. To assess the attractiveness of the technologies, we have to define those factors, normally external to the firm, that allow us to analyze the impact of each technology. The factors that we have used in our case are illustrated in Table 4, that shows the current and future profile for the STU number 3, board and systems design.

A similar assessment should be conducted for each of the STUs, leading toward a reflection of the overall opportunities and threats the portfolio of technologies presents to the firm.

INTERNAL SCRUTINY

The internal scrutiny process is a disciplined approach to identify the technological strengths and weaknesses of the firm against its most relevant competitors. The process starts with the identification of the critical success factors associated with each STU. Those factors represent capabilities controllable by the firm in which it has to excel to achieve a competitive superiority in each STU. Once the factors are identified, we conduct a competitive profile, measuring the position of the firm now and in the future.

In the internal scrutiny phase the future profile does not represent a trend forecast, as was done in the environmental scan, but a desirable position the firm would like to achieve against its leading competitors.

Table 5 shows the analysis as performed on the STU 3 of Masscalc, the board and system design. Notice that there are two categories of evaluation factors : those that measure the technological capabilities of the firm (knowledge, equipment, patents, etc.) and those that measure its efficiency in embodying this knowledge into products and processes.

A second element of the technology internal scrutiny that we have found particularly useful is to analyze the strengths and weaknesses of the existing policies the firm follows in each of the critical categories of decision making. The taxonomy of decisions that are relevant to technology strategy is taken from Hax and Majluf (1991). They propose to consider seven key categories of decisions: technology intelligence; technology selection; timing of new technology introduction; modes of technology acquisition; technology horizontal strategy; project selection, evaluation, and resource allocation; and technology organization and managerial infrastructure. Brief definitions of these categories of decisions are presented in Appendix II.

Table 6 shows the description of the policies adopted by Messcalc in each technology decision making category and their corresponding strengths and weaknesses. The high degree of informality the firm has in its technology strategy emerges from these descriptions. This behavior is quite common in start-up entrepreneurial firms.

THE TECHNOLOGY ATTRACTIVENESS-TECHNOLOGY STRENGTH PORTFOLIO MATRIX

Portfolio business matrices have been in use for over twenty years in American industry. They represent useful tools to reflect on the overall strength of the business portfolio of a firm (see, for example, Hax and Majluf, 1984, Chapters 7, 8, 9, and 10). The concept of business portfolio matrices can be extended easily to address the strength of the overall portfolio of technologies available to the firm. the technology portfolio matrix graphically displays all of the STUs of the firm according to two dimensions: technology attractiveness and technology strength. These two dimensions were already assessed in our study on the environmental scan and internal scrutiny processes, respectively. Recall that Table 4 illustrated how to evaluate current and future attractiveness of STU 3, board and system design; while Table 5 showed how to specify the current and future strengths of that same STU. Those tables evaluated attractiveness and strength through the use of several relevant factors. It is now required, either subjectively or by assigning different weights to each factor, to translate these multifactor profiles into a single measure of technology attractiveness and strength.

Figure 2 shows the technology portfolio matrix for Masscalc. The circles identify the existing position of each SBU, the dots the future position. Ideally we would like to have all the STUs in the high-attractiveness, high-strength cell of the matrix, such as STU 1. What is critical is to reflect on the competitive moves that have to be made in order to gain competitive strength in highly attractive STUs such as 2, 11, 3, 4, 5, and 10. The amount of effort and resources to be allocated to each STU depends both on our ability to gain competitive advantage, and the projection of future attractiveness of a given STU. In our example, STU 6 seems to be in a very precarious current position, but since its attractiveness is projected to be improved significantly, it makes sense to intend to raise the firm's competitive strength in it.

It is important to separate the current portfolio representation from its future projection. The current position should be the result of an objective, factual diagnosis of existing technology attractiveness and the firm's technological competencies. The future is more speculative, and needs to be critically examined in terms of the degree of confidence in the future technological trends, and the firm's capacity to improve its competitive standing. We have found the technology portfolio matrix to be a powerful diagnostic tool.

FORMULATION OF STRATEGIC ACTION PROGRAMS AND BUDGETS

Having identified the technological requirements generated from corporate and business strategies, and completed the environmental scan and internal scrutiny processes, we are ready to address the last stage of technology strategy formation: the development of broad and specific action programs, as well as the budgets which translate into financial terms the strategic and operational commitments implicit in the technology decisions.

To be consistent with the framework we propose, the strategic action programs should:

- Respond to the technological requirements emanating from corporate and business strategies.
- Seize the opportunities and neutralize the threats identified in the environmental scanning process.
- Reinforce the strengths and eliminate the weaknesses detected in the internal scrutiny process.
- Address all the issues linked to the strengthening of the portfolio of technologies of the firm.

Table 7 provides a rather incomplete representation of the technology broad action programs of Masscalc. In a real-life situation a higher degree of specification and comprehensiveness will be required.

TECHNOLOGY POLICIES

When the strategic analysis uncovers some serious deficiencies in the technology policies of the firm, it might be necessary to reevaluate them. Technology policies tend to be broad guidelines that define the scope in which technology decisions are to be made. These policies have some inherent stability and, therefore, are not supposed to be redefined at the end of every planning cycle. In the case of Masscalc, managers opted for issuing a simple statement of technology policies that group the seven categories of decision making we used for policy evaluation under three headings: innovation policies (including technology intelligence, technology selection, timing of new technology introduction, and modes of technology acquisition), technology dissemination and resource allocation (including technology horizontal strategy, and project selection, evaluation, resource allocation, and control), and technology organization and managerial infrastructure. Table 8 describes Masscalc's technology policy statement.

CONCLUSION

The paper presents an orderly process for the development of the technology strategy of a firm. The methodology has proven to be effective in providing an appropriate diagnosis of the existing state of technology utilization and to reflect on the necessary changes to be made in order to make technology more effective in supporting the competitive position of the firm's businesses. The methodology should be regarded as a process that facilitates communication and discussion among the key firm's managers rather than a rigid step-by-step procedure.

The Masscalc case discussed throughout the paper can only be regarded, for reasons of space and confidentiality, as a highly simplified illustration as opposed to a comprehensive realistic example.

APPENDIX I
Masscalc - A High-Performance Computer Company

The company that is used to illustrate the application of the methodology recommended to link technology and strategy is briefly described in this appendix. In order to protect the confidential nature of the strategic audit we conducted, we have slightly modified its content and used the fictitious name Masscalc. High-performance computers are devoted to applications that require very large computing capabilities. The whole size of the high performance computers' market accounted in 1990 for $10 billion and is growing at 20%. There are about 20 companies that serve most of the world market. The technological environment of the industry is very volatile, with incremental innovations being announced every few months. Also, the cycle of major innovations in the technologies used is short, requiring a sustained effort to improve actual models while working in the development of the technology of the next series of products.

Massive parallel computers (MPC) are a new family of high performance computers representing one of the most recent and significant technological breakthroughs of the industry. Most of the companies manufacturing MPC are new companies, and none of them sell any other type of computers. Nowadays, MPC competes in the market arena with two much more mature technologies (supercomputers and array processors), and though MPC still accounts for a very small share of the market, most experts predict a very impressive market share increase.

Masscalc is a start up company committed to designing, producing and selling MPCs. Its primary target is to provide computers to support the production rather than the R&D function of its customers, as most high-performance computer companies do. Masscalc'c challenge is to be able to produce large volumes of MPCs that are affordable, user friendly, and able to operate with commercially available software. This will transform the company from a design-focused organization into a massive producer of MPCs.

At the time of this study, Masscalc had successfully introduced its first model, which received an excellent review by the industry experts. All the computer parts were subcontracted and manually assembled in-house. A well-conceived and implemented technology strategy is critical to Masscalc's success.

APPENDIX II
Major Categories of Strategic Decisions Linked to Technology(**)

1. TECHNOLOGY INTELLIGENCE

An effort oriented at gathering information concerning the current and future state of technology development. Some of the tasks associated with it are: Identification of strategic technical units (STUs), evaluation of competitive technical strengths by STU, detection of the focus of innovation by key product areas (users, manufacturers, suppliers, others), collection and comparison of expenditures in technology by key competitive firms.

2. TECHNOLOGY SELECTION

It addresses the issue of selecting the technologies in which the firm will specialize, and the ways in which they will be embodied in the firm's products and processes. Some of the issues to be recognized are: selection of the technologies needed for product and process innovation, assuring the congruency of technology development with the business life cycle and with the desired business strategy, and assigning the appropriate priorities to resulting technological efforts.

3. TIMING OF NEW TECHNOLOGY INTRODUCTION

It involves the decision as to whether to lead or to lag behind competitors in process and product innovations. Issues to be addressed are : identifying the benefits and risks associated with a leadership and followership strategy, and assuring the congruency of the selected technology strategy with the generic business strategy.

4. MODES OF TECHNOLOGY ACQUISITION

The extent to which the firm will rely on its own internal efforts in developing internal capabilities, versus resorting to external sources. The options available for the modes of technology acquisition of products and processes are: internal development, acquisition, licensing, internal ventures, joint ventures or alliances, venture capital, and education acquisition.

5. HORIZONTAL STRATEGY OF TECHNOLOGY

It consists of identifying and exploiting technological interrelationships that exist across distinct but related businesses. It is a mechanism by which a diversified firm enhances the competitive advantage of its business units. Sources of technological interrelationships are: common product technologies, common process technologies, common technologies in other value-added activities, one product incorporated into another, and interface among products.

6. PROJECT SELECTION, EVALUATION, RESOURCE ALLOCATION, AND CONTROL

The principal concern in this case is the appropriate allocation of resources to support the desired technological strategy. Issues to be addressed are: criteria for resource allocation, project-oriented resources versus loosely controlled funds to support and plan projects, the degree of fluctuation in technology funding, and the magnitude in the profit gap to be filled by new products.

7. TECHNOLOGY ORGANIZATION AND MANAGERIAL INFRASTRUCTURE

It is oriented toward the definition of the organizational structure of the technology function. It includes the identification of the horizontal coordinating mechanisms needed to exploit the technological interrelationships existing among the various business units and the activities of the value chain. Issues to be considered are: centralization versus decentralization of the technology function, development of career paths for scientists and technical professionals, use of project team, use of lateral mechanisms to facilitate sharing technological resources, design of motivational and reward systems for scientists and technical professionals, degree of involvement of top managers in technological decisions, decision making process for resource allocation to technological projects, protection of technological know-how, patents policies, and publication policies.

(**) Extracted from: A. C. Hax and N. S. Majluf (1991). For useful references in technology see R. A. Burgelman and M. A. Maidique, *Strategic Management of Technology Innovation* (Homewood, IL: Richard D. Irwin Press, 1988); M. Horwitch (ed.), *Technology in the Modern Corporation: A Strategic Perspective* (New York: Pergamon Press, 1986); M. E. Porter, *Competitive Advantage: Creating and Sustaining Superior Performance* (New York: The Free Press, 1985), Chapter 5; E. B. Roberts (ed.), *Generating Technological Innovation* (New York: Oxford University Press, 1987); D. J. Teece (ed.), *The Competitive Challenges: Strategies for Industrial Innovation and Renewal* (Cambridge, MA: Ballinger Publishing Co., 1987); B. Twiss, *Managing Technological Innovation* (London: Longman Group, 1982), and E. von Hippel, *The Sources of Innovation* (New York: Oxford University Press, 1988).

REFERENCES

Abernathy, W.J., and J.M. Utterback, "Patterns of Industrial Innovation", *Technology Review*, Vol. 80, no. 7, pp. 40-47 (1978).

Anderson, P., and M. L. Tushman, "Technological Discontinuities and Dominant Designs: A Cyclical Model of Technology Change", *Administrative Science Quarterly*, Vol. 35, pp. 604-633 (1990).

Burgelman, R. A., and M. A. Maidique, *Strategic Management of Technology Innovation* (Homewood, IL: Richard D. Irwin Press, 1988).

Gattiker, U. E., and L. Larwood, *Managing Technological Development: Strategic and Human Resources Issues* (Hawthorne, NY: Walter de Gruyter, 1988).

Hax, A. C., and N. S. Majluf, *Strategic Management: An Integrative Perspective* (Englewood Cliffs, NJ: Prentice Hall, 1984).

Hax, A. C., and N. S. Majluf, *The Strategy Concept and Process: A Pragmatic Approach* (Englewood Cliffs, NJ: Prentice Hall, 1991).

Horwitch, M. (ed.), *Technology in the Modern Corporation: A Strategic Perspective* (New York: Pergamon Press, 1986).

No, M., "Developing a Methodology for Technology Strategy. An Application in the High Technology Industry", M.S. Thesis, Sloan School of Management, MIT (1991).

Porter, M. E., *Competitive Advantage: Creating and Sustaining Superior Performance* (New York: The Free Press, 1985).

Prahalad, C. K., and G. Hamel, "The Core Competence of the Corporation", *Harvard Business Review*, Vol. 90, no. 3, pp. 79-91 (1990).

Roberts, E. B. (ed.), *Generating Technological Innovation* (New York: Oxford University Press. 1987).

Steele, L., *Managing Technology* (New York: McGraw Hill, 1989).

Teece, D. J. (ed.), *The Competitive Challenges: Strategies for Industrial Innovation and Renewal* (Cambridge, MA: Ballinger Publishing Co., 1987).

Teece, D. J., "Profiting from Technological Innovation: Implications for the Integration, Collaboration, Licensing and Public Policy", *Research Policy*, Vol. 15, no. 6 (1986).

Twiss, B., *Managing Technological Innovation* (London: Longman Group, 1982).

Twiss, B., and M. Goodridge, *Managing Technology for Competitive Advantage. Integrating Technological and Organizational Development: From Strategy to Action*, (Aulander, NC: Pitman, 1989).

Utterback, J. M. and L. Kim, "Invasion of an Established Business by Radical Innovation", Proceedings from *The Management of Productivity and Technology in Manufacturing*, pp. 113-151, edited by P. R. Kleindorfer (New York: Plenum Press, 1986).

von Hippel, E. , *The Sources of Innovation* (New York: Oxford University Press, 1988).

White, B. L., *The Technology Assessment Process. A Strategic Framework for Managing Technical Innovation* (Westport, CT: Quorum Books, 1988).

Figure 1. A framework for the development of technology strategy

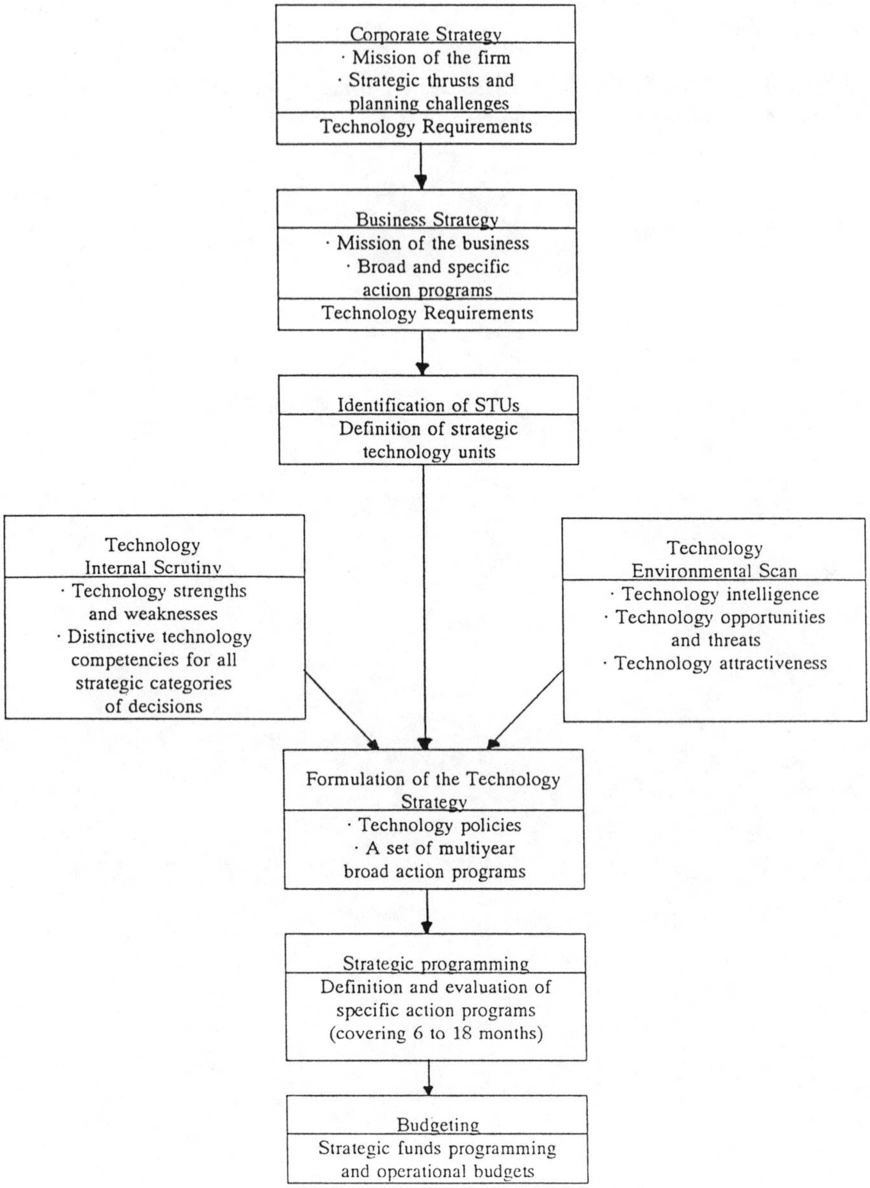

Table 1. Technological requirements(*)

Implied by	Technological requirements
Corporate Strategy	• Acquire and develop those technologies and procedures required to ensure high quality and reliability in large scale production of existing Massive Parallel Computer (MPC) product line. (period of accomplishment: 12 months). • Develop the technological capabilities needed to design and bring to the market a new generation of MPC. Time period: 3 years. • Enhance existing product line with minor innovations every six months.
Business Strategy	• Reduce board and system manufacturing cost by 10% every six months by better use of available technologies. • Acquire technical capabilities (human and equipment) in the area of demonstration technologies to serve actual marketing needs. • Bring to the market high speed input/output and video devices in 9 months

(*) This is a selected subset of Technological Requirements. In real applications the list should be collectively exhaustive.

Table 2.
Identification of all relevant STUs to support competitive advantage

1. *System architecture:* Technologies related to the definition of the basic architecture of the computer.
2. *Chip design and engineering:* Technologies related with chip design and manufacturing. It includes alternative technologies to the one used right now.
3. *Board and system design and engineering:* Board and system design and manufacturing.
4. *Support software:* Includes microcodes, compilers, and basic libraries.
5. *Application software:* Technologies to support companies that develop software to run in Masscalc machines.
6. *Management of Information Systems:* Information systems to support all activities of the company, including marketing, sales and service.
7. *Process Technologies:* Procurement and control of suppliers' production processes as well as in-house assembly.
8. *Testing Technology*: Technologies used to test subassemblies and the whole system.
9. *Demonstration technologies:* Includes video and communications vehicles to help preparing and delivering shows, demonstrations, etc.
10. *Peripherals*: Technologies required to design or subcontract the design of high speed peripherals for visualization and image processing.
11. *Service*: Technologies and methodologies for delivering service to the computer industry (e.g.: remote diagnosis, education of technicians, etc.)

Table 3. Sources of innovation by STU

STU	POTENTIAL SOURCE OF INNOVATION
1. Systems architecture	Competitors, universities
2. Chip design and engineering	Suppliers, competitors, and other computer companies
3. Board and system design and engineering	Suppliers, and electronic and computer companies
4. Support software	Lead users, suppliers, competitors
5. Application software	Lead users, suppliers, competitors
6. Management of information systems	Suppliers, industry in general
7. Process technologies	Suppliers and companies with analogous production processes
8. Testing technologies	Electronic companies and suppliers
9. Demonstration technologies	Lead users, competitors and other computer companies
10. Peripherals	Lead users, and electronic and peripherals companies
11. Service	Lead users, and competitors

Table 4. Technology attractiveness

X: 1992
O: 1994-95

STU 3: Board and system design

FACTORS	=	–	E	+	++	COMMENTS
Potential for enhancing competitive advantage in						Most of the innovations are incremental, so a small group of people can keep up with them.
– products				XO		
– process				XO		
Rate of technological change			O	X		
Potential for long term value added			XO			
Impact on:						It is a key technology in terms of cost, performance, and quality. It drives most of the manufacturing and assembling processes and has strong implications in procurement
– cost					XO	
– performance					XO	
– quality					XO	
– differentiation				XO		
Impact on entry barriers			XO			The technology has a moderate impact on changing the industry structure, the barriers to entry, and the industry standards.
Impact on setting industry standards.			XO			
Impact on improving industry			XO			

Key:

=: The STU is not relevant as a source of competitive advantage.
- : Potential for minor support.
E : Even - The STU supports average performance.
+ : Potential for mild competitive advantage.
++ : Potential for strong competitive advantage.

Table 5. Technology strength.

X: 1992
O: 1994-95

STU 3: Board and system design

Major competitor: ABC

FACTORS	=	-	E	+	++	COMMENTS
Technology capabilities						Masscalc has low level of human resources in this area. Its intentions are to change that due to the technology importance in cost reduction and reliability of the final product.
Human resources		X		O		
Equipment and laboratories		X		O		
Access to external sources			XO			
Recent patents			O	X		
Technology embodiment:						The company is selecting technologies that do not optimize cost and that increase risk, focusing only on higher performance, contrary to the business strategic focus.
Cost reduction in design		X		O		
Cost reduction in manufacturing		X			O	
Effective use of manufacturing standards			X	O		
Procurement eagerness		X				
Quality of product			XO			

Key:

=: The STU is not relevant as a source of competitive advantage.
-: Potential for minor support.
E: Even - The STU supports average performance.
+: Potential for mild competitive advantage.
++: Potential for strong competitive advantage.

Table 6. Characterization of present technology policies regarding the major categories of decision making

Decision Category	DESCRIPTION OF POLICY	STRENGTHS	WEAKNESSES
Technology intelligence	There is no action plan to capture external innovations nor any measure of how much information is coming in and how it is disseminated through the company	The informal policy may be very adequate now that the company is small.	It is very dangerous because many important innovations, requirements, or problems with the product, will probably not be captured on time.
Technology selection	With the exception of STUs 1 and 2, technologies are selected without consideration of the business mission.	None	Cost ineffective. May endanger some of the key business objectives.
Timing of new technology introduction	Incorporate advanced versions of the technology once less sophisticated, safer ones are already in the product.	Low technological risk. Leverage of experience	By not defining the policy in terms of STUs and its support to the business, there is a risk of missing the right time for technology introduction.
Modes of technology acquisition	Basically internal, with very few exceptions.	Easier coordination of proper development and use of technologies	Sub-optimization of resources. Very unlikely to be able to excel in all areas.
Technology horizontal strategy	Relies basically on informal communications.	Better disposition to share technology	Requires a strong culture to maintain this approach as the company grows.
Project selection, evaluation and resource allocation	Projects selected based on market inputs	Supports the company market driven approach.	Potential for losing long term innovations.
Technology organizational & managerial infrastructure	Most of the responsibilities for long and short term rely on the same people.	Gives control over what is going on, and facilitates coordination of previous policies.	May generate large decision problems as the company and the breadth of products grow.

Figure 2. Technology portfolio matrix

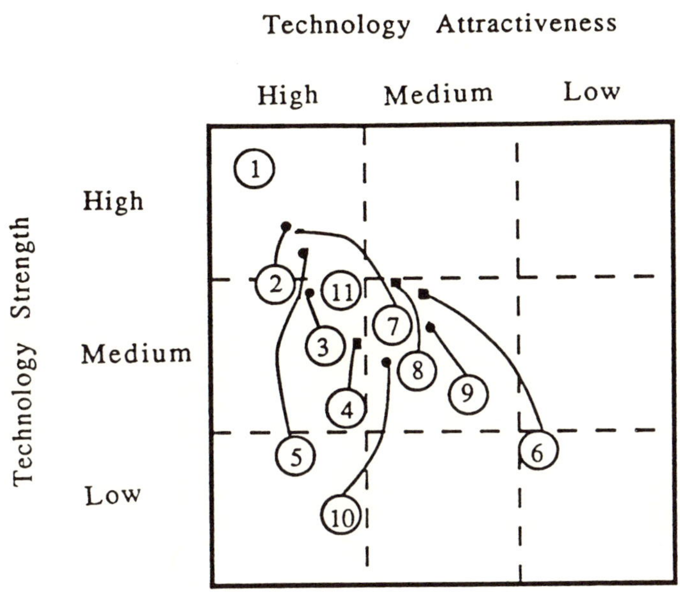

STU representation:

1. Systems architecture
2. Chip design and engineering
3. Board and system design and engineering
4. Support software
5. Application software
6. Management of information systems
7. Process technologies
8. Testing technologies
9. Demonstration technologies
10. Peripherals
11. Service

Table 7. Technology broad action programs

ACTION PROGRAM	RESPONDS TO(*)	1st MILESTONE
Define the specific needs of people in engineering and procurement for complying with short product cycles and incremental innovations	TR, IS	June 1992: the specification of needs
Define the specific technologies to be used in board design according to the established innovation policies.	TR, ES	September 1992: critical analysis of available technologies rated by cost, performance and riskiness.
Develop and launch a program to acquire the next generation of board design and manufacturing techniques with special focus on cost reduction and quality.	ES, TR	September 1992: critical analysis of new developing technologies.
Create a unit fully responsible for board design and manufacturing.	IS	December 1992: description of needs to be covered by this unit and means needed to achieve it.
Set up review meetings for chip and architectural design each six months. The purpose of these review meetings should be to track and evaluate external upcoming innovations.	IS	First review meeting in September, 1992.
Introduce demonstration technologies according with technology policies and marketing requirements.	TR, IS	In two months there should be a specification of what the company needs in the near future and which companies can provide the service.
Establish a program to develop a new generation of chip and architecture within three years.	IS	Define project by June 1992.

(*) This column identifies the appropriate state of the technology planning process process a particular action program is responding to, according to the following convention:
TR = technology requirements generated by corporate or business strategies
ES = environmental scanning process
IS = internal scrutiny process

Table 8. Technology policies

Innovation policies

- Be leaders in introducing incremental concepts in system architecture and chip design.
- Be followers in major innovations in chip manufacturing and board design and manufacturing technologies.
- Select those technologies that lead the company toward standards, mainly in those technologies that are not the core of the business.
- Select standard hardware and software available in the market or that can be designed and manufactured outside without interfering with company's proprietary knowledge.
- Acquire from outside all the support software that is not crucial for the proprietary technology or expertise of the company.
- With regard to demonstration technologies and other non-crucial activities, look for an agreement with some external company and an internal coordinator.

Technology dissemination and resource allocation

- Set a program for temporal rotation of people. It should include:
  - Interchange between people at R&D center and engineering.
  - People working at the R&D center should be allocated one month every three years as marketing support personal, in a rotational basis.
- Maintain policies regarding resource allocation.

Technology organization and managerial infrastructure

- Establish a program for evaluation of new changes in current designs, to ensure that each new innovation included is appropriate in terms of factors such as market needs, cost reduction, better service to either final customers or software companies, etc. Establish priorities among these factors.
- The R&D organization will be under the Engineering and Manufacturing department. But, R&D will be seated at all the top-management committees to ensure that its long term objectives are pursued.
- Lower down the organizational level at which technological decisions are made
- Set up, within the evaluation program, an analysis of technological decisions made and its agreement with satisfaction of technological requirements and technology policies.

STANDARDIZATION AND THE STRATEGIC MANAGEMENT OF TECHNOLOGY

Richard S. Rosenbloom
Harvard Business School
Boston, Massachusetts

Effective strategies reflect a high degree of "fit" with the competitive environment (Andrews 1987). Product standardization has long been a visible element of the environment for competitors in many industries, but its incidence and significance have increased greatly in recent years. In some industries (e.g. microprocessors), the fortunes of rival firms have been profoundly shaped by the consequences of standardization. As industrial managers cope with the dynamic forces of an era dominated by information technologies, standardization will be an increasingly common element of the strategic mix.

My aim in this essay is to suggest an approach to research that would explore the relationship between standardization, broadly defined, and the strategic management of technology. There is an extensive literature on each topic. For example, a recent review article on The Economics of Compatibility Standards (David and Greenstein 1990) lists more than 150 items, of which roughly 75% were published after 1985. The principal matters addressed in this literature have to do with the mechanisms by which standards are established and the extent to which they produce efficient, or socially optimal, outcomes. Economic models, so far, shed little light on the differential consequences that standardization can have on competing firms. Hence there are few connections between that body of work and the literature of strategic management, especially technology strategy.

Creating those connections is complicated by the fact that there is no generally accepted paradigm for the discussion of technology strategy. As noted in a recent comprehensive review of the relevant literatures (Adler 1989, p. 26):

> The development of a technology strategy framework has become increasingly urgent as firms become more conscious of the potential of technology as a competitive weapon ... The result is an increasing number of firms attempting to manage technology strategically and a burgeoning case literature, but as yet no consensus on the substance of a technology strategy framework.

For a given firm to gain (or avoid losing) competitive advantage in the wake of standardization, its strategic choices in the management of technology must be aimed, at least in part, at influencing the relevant standardization processes and exploiting their consequences. The task of the technology strategist therefore must include anticipation of the likely character and magnitude of the strategic consequences of new standards as they can be expected to emerge following significant innovation.

The most significant competitive consequences in this context are the potential to enhance or diminish the salience of firm-specific assets,[1] and to create opportunities for market entry, for innovation, or for the attainment or retention of efficient scale (Teece 1986).

The remainder of this paper is organized in four sections. The first discusses the nature of standards and the process of standardization. The next section sets standards in the context of common patterns of innovation and technological advance. The third section addresses the principal competitive consequences of standardization. A brief final section illustrates the concepts with two examples.

Standards and Standardization

A "standard" may be defined as a set of design specifications which constrain the actions of incumbent and prospective producers of the standardized

---

[1]This is similar to the notion of competence-enhancing and competence-destroying innovations (Tushman and Anderson 1986).

good.[2] Standards vary in detail across a broad spectrum of possibilities. At one extreme, a standard may be a broad categorical definition ("standard" automobiles are powered by Otto engines that burn gasoline). At the other extreme, a standard may provide a detailed and precise specification of materials, dimensions, or configurations (e.g. the standard design for lightbulb bases and mating lamp sockets). A standard may be voluntary or compulsory; implicit or explicit; narrow in scope and detailed in content, or general in scope and broad in content. Standards are ubiquitous, often even "invisible," because taken for granted.

In any class of products, standards may constrain innovation at various levels in the hierarchy of design (Clark 1985). At the peak of the hierarchy is a "core technology." Reference to "jet transports," "silicon logic," or "diesel engines" implicitly identifies a principal underlying technical capability required to produce them. Specific products are then designed with a certain "architecture" (Henderson and Clark 1990) which establishes the relationships among components, as in "supersonic transports," "notebook computers," or "front-wheel drive cars." Within an established architecture, innovation may alter the particular materials and components employed, or the detailed design thereof. These concepts constitute an ordered set of levels of technological change (Table 1).[3]

Table 1

Levels in the Hierarchy of Design

- Core Technology
- Product Architecture
- Materials and Components
- Details of Design

The strategic implications of a standard are, in part, a function of the level in this hierarchy that the standard addresses. If the only "standard" is the use

---

[2]Coles (1949, page 107) defines a standard as "... a physical, written, graphic or other representation of a product or a procedure established by authority, custom, or general consent with which other products or procedures of a like nature are compared for identification or measurement or to which they are made to conform.

[3]This is an adaptation of the structure suggested by Clark (1985) and later modified by Henderson and Clark.

of a common core technology, e.g. gasoline-fueled internal combustion engines, the consequences are different from those flowing from detailed specification of an interface standard defined at the level of an individual component. The standard implied by a dominant design (Abernathy and Utterback 1978) falls in between the preceding cases, as it identifies a particular synthesis of components and a defined architecture, within a certain core technology. The prevailing VHS standard for home video, for example, combines both a dominant design (1/2" magnetic cassette recorder) and an interface standard specifying the cassette dimensions and tape loading mechanism.

A specification may become a de facto standard as a result of the independent choices of a predominant number of producers (e.g. "IBM compatible PCs"), reflecting their perception of buyer preferences and their strategic aspirations. Alternatively, a de jure standard may be imposed on all producers by a body empowered by law (e.g., television receivers produced for sale in the United States must conform to FCC specifications for broadcasting) or by mutual agreement (e.g., NEDA standards for electrical motors).

In many cases (e.g. the development of television broadcasting, introduction of the Universal Product Code for supermarkets, compact disc audio), standardization of critical interfaces preceded the growth of the market. Of equal interest are those situations in which the dynamic workings of demand drive the process of de facto standardization. The VHS standard for video, and the IBM-compatible PC design are oft-cited examples. When producers or users have a direct stake in the outcome of the standardization process, "sponsors" (David & Greenstein 1990) play an active role, and "strategic maneuvering" may be decisive (Cusumano, Mylonadis, and Rosenbloom 1992).

Patterns of Technical Advance

While standards continue to emerge throughout the life cycle of a product or industry, the most decisive choices usually are made in the first phases of an emerging industry (e.g. VCRs in the 1970s, personal computers in the early 1980s). Managers, therefore, make their critical strategic choices in anticipation of or against the background of the characteristic events of that sort of confusing and chaotic context. We shall use a simple evolutionary framework—variation, selection, and retention—as a metaphor characterizing a common pattern in the emergence of an industry based on a new technological paradigm (Dosi 1982).[4] Clearly, other models might also serve well, but this one will help us to illustrate pertinent choices, as they developed in the history of television. Figure 1 is a schematic representation.

Figure 1

A Pattern of Technical Advance

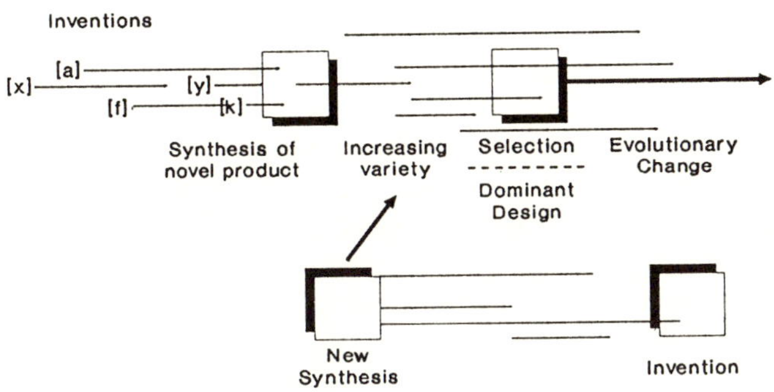

As in many cases, this story begins with a period of "prehistory," decades long, during which a number of requisite constituent inventions occurred (in Figure 1, the [x] etc.). The development of television traces to inventions by Nipkow in 1884, but it took nearly a half-century before all the ingredients were ready for synthesis of a workable system. Demonstrations of television began in London in the 1920s, and broadcast tests in England, Germany, and the U.S.A in the 30s.

A first synthesis is often followed by an explosion of diversity, as rival technologies and architecture compete for acceptance (Arthur 1988). For television, RCA, the American pioneer, fought in the 1930s to establish a system based on the iconoscope, against a rival technique invented by Farnsworth. In England, RCA's licensees competed with sponsors of Baird's

---

[4]Sahal (1982) uses the term "guidepost" for a similar concept.

mechanical disc. During this period of increasing variety, prospective producers developed distinctive capabilities (firm-specific assets) which were differentially useful for alternative product formulations.

A some point in every such story, choices made by producers, users, or regulators begin to reduce the diversity. A dominant design may emerge, or interface standards may be agreed or imposed, excluding alternative architectures or technologies. By 1940, the FCC in the U.S.A. and the BBC in England chose the RCA technology, and defined standards for picture resolution and broadcast frequencies to which TV equipment producers were forced to conform for decades after. Alert producers pursued strategies designed to influence these choices in directions favorable to their interests.

The de jure definition of broadcast standards defined the successful technical paradigm (defined by a core technology and specific product architecture) for television, punctuating a process of extraordinary innovation.[5] (Rosenbloom 1985) Implicit in the paradigm was an agenda of techno-economic problems to be resolved by subsequent, evolutionary, innovation. For example, increasing screen size was an obvious goal for early technical efforts in television receiver development. The acceptance of a new technical paradigm thus defined the likely trajectory of technical advance.

The establishment of a new core technology or architecture is an extraordinary event, relatively rare in occurrence but potentially sweeping in its consequences. Within a given paradigm or dominant design, as producers work out the implied technological trajectory, a period of "normal change" ensues. This more common pattern of innovation produces a sequence of product generations or minor modifications of elements of design. Although any one step in this process usually has modest economic consequences, the cumulation of these frequent events may have very substantial effects (Hollander 1965).

As part of the process of "normal change," significant innovation may still occur within the constituent components and materials (e.g. wireless remote control, electronic tuning). By grouping related changes at this level,

---

[5]The de facto emergence of a dominant design will have the same effect. A "dominant design" can be seen as a paradigm that has been adopted by a preponderant share of producers (Anderson and Tushman 1990).

producers create new product "generations." Minor, "evolutionary" change takes place continually in the details of design, subject to constraints imposed by standards applicable to specific components. Eventually, another radical innovation may emerge to introduce a new period of extraordinary change and rivalry over standards (e.g. color TV in the 1950s, digital transmission in the 1990s). This notion of technological advance as a process of punctuated equilibrium fits the histories of many important technologies and industries (Tushman and Anderson 1986).[6]

Strategic Consequences

A wide range of strategic issues in the management of technology are potentially influenced by standardization. I suggested earlier that what matters most are the ways that existing or prospective standards influence entry (and exit) decisions and the ability of a producer to create proprietary competitive advantage. For example, in an emerging industry, the widespread acceptance of a standard for critical aspects of design may accelerate growth in demand, creating opportunities for pioneering incumbents and imitative entrants. In an established industry, entrenched standards may constitute formidable barriers to innovation or to entry with superior non-conforming designs (e.g. radial tires in the 1960s or HDTV and flat panel TV screens today). On the other hand, the emergence of a new standard may extinguish demand for products shaped by earlier standards (e.g. bias-ply tires in the 1970).

To understand the likely consequences of standardization in a specific case, we need to set it in some strategic context. In general, the consequences will vary with factors at three levels: the technology, the industry, and the firm. In the technological context, the ease of appropriability is clearly relevant, as is the level in the design hierarchy at which change might take place. Standardization will have different strategic consequences at each level in the hierarchy of types of innovation. The strategic consequences of standardization also depend upon the competitive salience of the aspect of design that is affected. Industry factors include the nature of existing standards, and the degree of maturity of the industry—is it newly emerging,

[6]The generic characteristics of this theoretical construct as applied in various fields of scholarship are ably explored by Gersick (1991), although without specific reference to technology.

rapidly growing, or very mature and perhaps declining? Finally, standards will influence specific firms differently, according to whether they are incumbents or prospective entrants, and whether or not they have access to those complementary assets necessary to commercialize innovations in a given setting.

In general terms, competitive advantage flows from the attainment of product differentiation or low-cost production. The ability to exploit available technological trajectories through a stream of improvement innovations is a function of the strength and character of existing standards, potentially restricting opportunities to create or sustain product differentiation. Cost positions may be shaped by the ability to gain and hold efficient scale. To the extent that scale economies are available and standardization stimulates primary demand, it will create opportunities to realize lower costs. Both elements—innovation and scale—and thus both strategies—differentiation and cost—are therefore powerfully shaped by standardization outcomes.

Two Illustrative Cases

To illustrate these concepts, we can consider two examples from consumer electronics, the Walkman innovation and current rivalry over the next stage of audio recording technology.

The Walkman, which created a wholly new and important market segment, was the result of a novel synthesis of existing technical elements, arranged in an innovative product architecture. Its design conformed to the existing interface standard for 1/4" magnetic tape players, thus opening the door to rapid achievement of large scale. Despite weak protection against imitation, entry was limited by the need for important complementary assets possessed only by firms already established in the consumer entertainment electronics businesses. The dominant design established by the innovator, Sony, left room for substantial "normal change" leading to successive product generations enabling continued product differentiation and a further segmentation of demand (Sanderson and Uzumeri 1990).

The consumer audio market in 1992 will see the introduction of two new products, the Digital Compact Cassettes (DCC) by Philips and the Mini Disc (MD) by Sony, aiming to tap the large existing demand for audio recorder/players currently based on the 1/4" cassette. (Riggs 1991) The new technologies of digital magnetic recording and erasable optical discs make it possible to offer CD-quality sound in compact devices also capable of recording. A third contender, already on the market, is the Digital Audio Tape (DAT). DAT blends the architectures of audio and video cassette recorder, without compatibility with either. Because the DCC retains the core technology and architecture of the 1/4" cassette recorder, it is backward compatible with existing software. Sony's MD exploits a new core technology, with all that that implies. Among other consequences, Sony faces conflicting pressures to recruit other firms into an alliance in order to establish the credibility of the MD standard, although this will limit its ability to gain scale rapidly in order to attain some cost advantage. All three of the new technologies are protected from ready imitation by the novelty and complexity of the core technologies and the likely importance of know-how and experience.

As these two examples suggest, a winning strategy can depend significantly on the consequences of standardization. In an information age, standardization will be an increasingly common element of the strategic mix. In these brief remarks I have tried to suggest the value of research that would begin to link the extensive available literatures on the economics of standardization and on the strategic management of technology. As a first step, I have suggested a set of categories around which such research could be organized.

REFERENCES

Abernathy, William J. and James M. Utterback. 1978. "Patterns of Innovation in Technology." Technology Review, 80:7, 40-7.

Adler, Paul. S. 1989. "Technology Strategy: A Guide to the Literatures." In Richard S. Rosenbloom and Robert A. Burgelman (eds.), Research on Technological Innovation, Management and Policy, 4, Greenwich, CT., JAI Press.

Anderson, Phillip, and Michael L. Tushman. 1990. "Technological Discontinuities and Dominant Designs: A Cyclical Model of Technological Change." Administrative Science Quarterly, 35, 604-633.

Andrews, Kenneth R. 1987. The Concept of Corporate Strategy. Irwin, Homewood, Illinois. (Third Edition)

Arthur, W. Brian. 1988. "Competing technologies: an overview." in Giovanni Dosi, Christopher Freeman, Richard Nelson, Gerald Silverberg, & Luc Soete, (eds.) Technical Change and Economic Theory, Pinter Publishers Limited, London, 590-607.

Clark, Kim B. 1985. "The Interaction of Design Hierarchies and Market Concepts in Technological Evolution." Research Policy, 14, 235-251.

Coles, Jessie V. 1949. Standardization of Consumer Goods. Ronald Press, New York.

Cusumano, Michael, Yorgis Mylonadis, and Richard S. Rosenbloom. 1992. "Strategic Maneuvering and Mass-Market Dynamics: The Triumph of VHS over Beta. "Business History Review (forthcoming).

David, Paul A. and Shane Greenstein. 1990. "The Economics of Compatibility Standards: An Introduction to Recent Research." Economics of Innovation and New Technology, Harwood Academic Publishers, 1, 3-41.

Dosi, Giovanni. 1982. "Technological paradigms and technological trajectories." Research Policy, 11, 147-162.

Gersick, Connie J. G. 1991. "Revolutionary Change Theories: A Multilevel Exploration of the Punctuated Equilibrium Paradigm." Academy of Management Review, Vol. 16, No. 1, 10-36.

Henderson, Rebecca M. and Kim B. Clark. 1990. "Architectural Innovation: The Reconfiguration of Existing Product Technologies and the Failure of Established Firms." Administrative Science Quarterly, 35, 3-30.

Hollander, Samuel. 1965. The Sources of Increased Efficiency: A Study of DuPont Rayon Plants. MIT Press, Cambridge, Mass.

Riggs, Michael. 1991. "Digital Decisions." Stereo Review, October, 54-58.

Rosenbloom, Richard S. 1985. "Managing Technology for the Longer Term: A Managerial Perspective." In Kim B. Clark, Robert H. Hayes, and Christopher Lorenz, eds., The Uneasy Alliance: Managing the Productivity-Technology Dilemma. Harvard Business School Press, Boston, Mass. 297-327.

Sahal, Devendra. 1981. Patterns of Technological Innovation. Addison-Wesley, Reading, Mass.

Sanderson, Susan Walsh, and Vic Uzumeri. 1990. "Strategies for New Product Development and Renewal: Design-Based Incrementalism." Center for Science and Technology Policy, Rensselaer Polytechnic Institute, Troy, NY.

Teece, David J. 1986. "Profiting from Technological Innovation: Implications for Integration, Collaboration, Licensing and Public Policy." Research Policy, 15, 285-305.

Tushman, Michael L. and Phillip Anderson. 1986. "Technological Discontinuities and Organizational Environments." Administrative Science Quarterly, 31, 439-465.

RESEARCH IN MANUFACTURING STRATEGY: A CROSS-FUNCTIONAL PERSPECTIVE

Uday S. Karmarkar

William E. Simon Graduate School
of Business Administration
University of Rochester
Rochester, New York 14627

ABSTRACT

One of the ubiquitous features of research in manufacturing strategy, is that there is invariably the need to consider the interfaces between the operations function and other functions in the firm. This is also true of practice in devising and implementing manufacturing strategy. Addressing these interfaces is thus valuable for research in strategy implementation in addition to strategy analysis and generation.

This paper discusses research topics in manufacturing strategy that emphasize interfaces and interactions with other functions. Future research directions which appear to have a high potential, are also discussed. Often the insights and contributions of the research arise from transferring a paradigm or methodology from one functional area to another. In other cases, the contribution is in correcting the biases that arise in practice and in research, from excessive functional specialization.

INTRODUCTION

Research in manufacturing strategy has become more popular in schools of management over the last decade and it has become acceptable to treat problems in manufacturing strategy as the subject of academic research. Perhaps this is also motivated by the seeming inability of U.S. firms to compete successfully in many industry sectors, and the apparent irrelevance of much of traditional POM research even at the plant level. Whatever the reason, research in manufacturing strategy is burgeoning in academia and seemingly finding an audience in industry as well. In addition to research articles, recent books on manufacturing strategy include those by Buffa (1984), Clark et. al. (1988) and Hax and Majluf (1991).

There are several alternative paradigms for research in the strategic aspects of manufacturing operations. For example, a traditional perspective is to consider the range of operations decisions that involve resource acquisition and configuration. Thus facility location, system design, process choice and technology choice qualify as strategic issues. Another view is to define strategy as those issues that affect the competitive success of the firm. In some sense this is too broad a definition, but it draws attention to the impact of operations on the marketplace, in contrast to the traditional cost based view. Hence, market and revenue related issues such as quality, customer service, process flexibility, product introduction and product design, can begin to receive appropriate attention. A third perspective, and the one dealt with below, recognizes that many strategic issues involve interactions between the traditional operations functions and other functions in the firm.

Such a functional viewpoint may seem limiting. After all, "breaking down functional silos" is still high on the list of current consulting slogans. However, functional divisions exist in firms because of certain organizational efficiencies in hiring, training, evaluating and promoting personnel. It is likely that they will remain. A realistic view of strategy generation and implementation requires a close understanding of which functions are affected, what new relationships or interactions are created, and perhaps even how they might be reorganized. Functional divisions are also very much present in research, possibly for similar reasons. There is a great deal of knowledge and methodology that has been developed within specific areas, and there appear to be potential gains in simply bringing these together. The direction of this paper then, is to describe certain areas of interaction between functions, that are encountered in studying strategic issues in manufacturing. This might be termed a bottom up approach to strategy research that has the advantage of being able to draw on existing work as a basis. I do not possess a grand plan for this view. The following is simply a collection of research topics that link operations management with other functional areas. Nor is this intended to be a survey; I make no claims of completeness. Rather this is a description of the research direction being taken by a few colleagues, and the references are primarily there to serve as examples. I hope to describe an approach to strategy research that seems to have considerable potential.

OPERATIONS AND FINANCE

Any text on operations management, and indeed much traditional research in this field, is full of models of decisions. Almost without exception, these decisions are cast in terms of minimizing costs or expected costs. Yet, the modern theory of finance identifies risk and the pricing of risk as a central issue in the valuation of firms. Given the objective of maximizing the value of the firm, risk must also play a part in any decision made in operations, at least when outcomes can be random. Conversely, operations decisions can affect the riskiness of a firm simply by affecting the sensitivity of costs and revenues to stochastic variations in the environment or in the firm itself. There are many research issues raised here. A very simple one is: how should traditional models in operations management be modified to include risk considerations? Going further, what is the process by which the objective of the firm can be decomposed into the appropriate sub-objective for operations decisions? To cast the problem in a somewhat different way, operations decisions directly affect cost structure and hence, "operating leverage". That is to say, the fixed to variable cost relationships in the firms can be altered by decisions in operations. It is well known that operating leverage affects the riskiness of firms as measured by their "betas" (Lev, 1974). It is fair to say that there is little recognition of this relationship in formulations of decisions in operations.

There is other important subtext here. Casual observation suggests strongly that there is a wide communication gap between manufacturing and financial control functions in firms. A common symptom of this gap is the adversarial nature of the interactions between these functions in dealing with decisions such as capital asset acquisition. A typical argument might revolve around the degree of risk involved in acquisition of a new technology. There are many papers in the research and general literature inveighing against the myopia of financial evaluation methods. Suppose we accept for the moment that there are shortcomings if not in the theory, then in the way these methods are implemented. What is the alternative process by which these critical decisions are to be made? Again, casual observation suggests that when standard means are insufficient, decisions are made by a complex ad hoc process, often adversarial, involving many managers from many functions in many meetings. This can work on a few occasions, but can be too costly a way of doing business on a continuing basis. The existence of a standard evaluation process reduces the costs of making decisions. I would propose that a reasonable aim for research in this area is to develop the analysis that permits such standard methods.

As a leading example, Singhal (1988) addresses the issue of safety stock decisions, taking risk into account. The capital asset pricing model (CAPM) is a basic tool in his analysis. He points out that from a cost structure viewpoint, buffer inventories are a fixed cost and thereby increase risk. He analyzes data from several companies and finds an association between inventories and riskiness, although the evidence is weak. He also analyzes a newsboy problem with a risk adjusted objective function. One derived implication is that the inclusion of risk considerations results in an unambiguously lower choice of stock level. A similar approach is followed in the papers by Singhal and Raturi (1990) and Raturi and Singhal (1990) which examine (Q,R) and basestock models with risk. A broad general conclusion is that inventories often constitute a fixed cost, and thus add to risk, so that the term "safety stock" might be considered a bit of a misnomer.

Capital equipment purchases are an obvious area for investigating the impact of risk. Manufacturing systems are widely recognized to have increasingly larger investment requirements, due to new systems technologies. Incremental equipment purchases are not possible in highly integrated and automated environments. The resulting shift to a high fixed to variable cost ratio in cost structure has been examined by Lederer and Singhal (1988). While the high fixed costs of systems might be expected to increase risk, they note that *future* cash flows might actually be less risky, since all the fixed costs are incurred at the front end. Equally significant is the issue of the specific nature of the application, and the source of the variation that creates risk. They illustrate that in some cases, it can happen that the products to be produced on a specific technology can have low levels of demand risk as measured in terms of co-variance of forecast demand with the market portfolio. Both these issues are important, since typical practice in financial control attempts to compensate for risk by raising the hurdle rate used to discount future cash flows. Erroneous computation of the hurdle rate can seriously bias the evaluation of new manufacturing systems in a conservative direction. Lederer and Singhal use prototypical data from case studies to illustrate that the correct hurdle rates can actually be quite low.

International manufacturing and distribution operations lead to certain problems closely related to financial policies. One such issue is that of exchange rate risks. Traditional models of facility location, production allocation, and transportation must be augmented to consider the implications of stochastic variations in exchange rates. International location-allocation decisions also create an opportunity for hedging against exchange and demand risks. Investing in the ability to manufacture in a certain country can provide a way of reducing risk due to these causes. While it is possible to hedge these risks by purely financial means, manufacturing, transfer of product, transfer pricing manipulations, and profit reporting all provide alternative routes to reducing the impact of unforeseen variations. The question of location and capacity choice in different economic regions and markets, also necessitates risk management due to demand variations in different markets. Hodder and Jucker (1985) and Hodder and Dincer (1986) describe models that incorporate many of these issues. Cohen and Huchzermeier (1991) address the inclusion of exchange rate variations in international location and logistics formulations. Gray (1991) analyzes the effect of variability in demand forecasts on location decisions. The results cannot be characterized in a simple way, since both risk and return determine location policies.

Some recent work by Lederer and Singhal (1992) takes a different tack. Since technology decisions affect the cost structure of firms, the choice of technology has an effect on riskiness. Hence, financing decisions for firms are likely to be affected since the attractiveness of the firm to bond (debt) holders and equity holders is affected by the degree of risk. Lederer and Singhal argue that this is the case, and that furthermore, the financing and technology choice decisions should be made jointly, since they are interrelated. Potential topics for further research include any operations issues with substantial stochastic variation. Much of the research mentioned above considers risk due to demand variation or exogenous factors such as exchange rate variations. Supply side variations due to poor system performance, failure, poor yield or delivery, all also have an impact on riskiness. These have received less attention to date.

OPERATIONS AND COST ACCOUNTING

There has been a burst of interest in cost accounting recently, primarily sparked by the efforts of Kaplan (1983a, b) and some of his colleagues. Kaplan's own work has been directed towards improvement of cost accounting methods to better reflect the realities of modern manufacturing. Limiting the discussion to manufacturing for the most part, it would seem that the entire question of costing is inextricably intertwined with operations management. Most obviously, operations decisions cast as cost minimization formulations, require knowledge of cost coefficients. If costs are used in performance measurement, these costs also *de facto* become part of the objective function for the manager being measured. However, there is a more fundamental relationship. If one takes the view that costs should reflect opportunity costs, then the characteristics of manufacturing systems must be reflected in costing methods.

At this juncture, it is useful to distinguish between accounting costs and opportunity cost estimates for decision making purposes. Accounting systems record actual expenses and expenditures, and accounting costs are developed by transformations on these streams. On the other hand, opportunity costs can very well include revenue inflows as well as cost outflows. For example, investing in process control may lead to quality improvements, that ultimately lead to increased market share in a competitive marketplace. Accounting costs can only address possible operating cost reductions, but cannot provide information on future revenue related benefits. To take another example, past expenditures are irrelevant from an opportunity cost viewpoint, because they are sunk. Yet costing systems primarily address such expenditures, with some consideration of accrued expenses. Standards developed from such historical costs can have predictive value only in a completely stable environment.

While opportunity costs related to future revenue streams might be easily distinguished from accounting costs, there are other categories of opportunity costs or imputed costs where the distinction is less obvious. As an example consider the shadow price of a scarce resource. At a particular decision level, say that of aggregate capacity planning, resource levels may be fixed and the expenditures on those resources sunk. The limited capacities of those resources (machines, labor, storage space) act as constraints with implicit costs (e.g., Jaikumar and Cooper, 1984). It is hard to see how accounting costs can reflect these implicit costs in any systematic way. In particular, these implicit costs are very dependent on the set of actions or decision variables available to the decision maker, and may also depend on conditions such as the level of utilization, congestion, or the degree of uncertainty in forecasts.

Apart from these fundamental problems in capturing opportunity costs in costing systems, traditional costing methods have some built in biases. There is a tendency to define costs in terms of production quantities. Furthermore, cost functions are typically assumed to be separable, linear (or affine) and constant over time. Perhaps in the production of tangible objects, it is tempting to think that if one unit costs $x, then 5 units must surely cost $5x. There appears to be less of a tendency towards such approximations in services, perhaps because of the complexity of defining or measuring outputs. It is however, clear from many studies of the characteristics of manufacturing and service systems (e.g., Rummel, 1989; Noreen, 1990), that production cost functions can be highly non-linear. Linear or affine models will tend to lead to extreme point solutions in many decisions where an interior solution might well be more appropriate. Product mix decisions are an example. Furthermore, costs can depend on many decision variables other than volume, such as batch sizes, sequences, transaction volume, number of products and so on.

The practical issue here is that costs obtained from accounting systems are in fact widely used to support and evaluate operations decisions ranging from dispatching at machines, to order release, make-buy and equipment acquisition. Yet there is a broad opinion in industry that standard costing methods are not adequate for operational decision making (Karmarkar, Lederer and Zimmerman, 1990). Nor are these problems new; the eponymous Mr. Gantt (1990) describes a case of allocation of overhead costs leading to poor make-buy decisions, to the extent of shutting down production. The story has a familiar ring to it.

What then are the important research questions? One of the most basic is that of the determination of costs and cost coefficients in objective functions for operations decisions. Grubbstrom (1980), Porteous (1985), Rummel (1989) and others have discussed the derivation of setup and holding cost approximations from underlying discounted cash flow models. If the presumed cost structure is valid, then the approximation is indeed good. However, Rummel argues that in many cases, the relationship between actions and cash flows is tenuous, and the conventional models can be in error. Aucamp (1984) and Karmarkar and Rummel (1990) point out that there may in fact be no cash flows involved in transactions such as setups. In that case, "costs" may arise due to capacity constraints or the opportunity costs of lost production, and the implicit cost of a setup may have little to do with historical cash flows. Carrying the argument in a different direction , they suggest that in some cases, defining costs for batching on the domain of transaction costs (setups) and inventory levels, need not be the best form for the objective for batching decisions. There is evidence to suggest that defining cost as a function of lead time is a superior approximation to the underlying cash flows (Rummel, 1989). Morton et al. (1988) study the appropriate objective for sequencing and scheduling decisions, and justify lead time based objectives among other issues. There is some convergence in approaches such as Morton and Singh (1988), Dobson et al. (1987), Morton et al.(1988), Chase and Rajagopalan (1991) and Karmarkar, Kekre and Kekre (1990), all of which suggest that for manufacturing shops with queues, "setup costs" for batching decisions are related to queue length or busy periods. This makes intuitive sense in that the opportunity cost of a setup is primarily to the delay created for waiting work and the resulting costs of WIP and tardiness.

Certain costing practices are of particular significance for operations. The allocation of fixed costs is one that has received a great deal of attention. Not only is fixed cost allocation used to assign overhead costs to products and activities, but allocation is also implicit in depreciation and amortization calculations, where the costs are spread over time rather than products or units. The allocation of fixed costs has long been academically perplexing, because it seems clear that in many cases, fixed costs are sunk, and should have little bearing on decisions after they are incurred. Yet allocated fixed costs are not only routinely used in performance measurement, product costing and setup costing, but also for pricing and profitability analyses. Karmarkar and Pitbladdo (1990a) resolve this apparent difference between practice and theory by showing in the context of a model of competition with fixed costs, that prices approximate average costs rather than fixed costs. However, this is not because of allocation and "pricing to recover fixed costs". Rather it is because fixed costs restrict entry, and firms can then restrict output and raise prices till revenue exceeds variable and fixed costs. They suggest that cost allocations might be interpreted as heuristics that approximate complex competitive environments; nevertheless it appears that most such allocation methods are likely to distort short term incentives, with the exception of allocation proportional to gross contribution.

Zimmerman (1979) conjectures that fixed cost allocation serves as a proxy for utilization of scarce capacity (shadow prices) or for externalities (congestion and WIP). When the structure of shadow prices in such cases is examined (Karmarkar and Pitbladdo, 1990b), there is appears to be little connection between fixed costs and either shadow prices for capacity or the costs of externalities. Incidentally, they also show that shadow prices or imputed costs of capacity in competitive environments do not depend on the capacity of one firm, but on the the total capacity in the market place. Entry of a new firm, can reduce the value of capacity in every other firm. Furthermore, the imputed cost (or value) of the marginal capacity unit also depends on the elasticity of demand curves for the set of products produced. In retrospect these observations seem obvious; yet they are not part of conventional thinking in academia or practice.

My own conclusions are that from the point of view of industry practice, the use of "costs" from internal company costing systems will survive for a long time partly due to inertia, and partly because in many applications, direct material and labor are still fairly large portions of controllable costs. However, there is already general recognition of the fact that revenue and market impact are central issues in manufacturing performance that cannot be captured by historical costs. As yet, this direction has found expression mostly in top down programs such as JIT and Total Quality Control, which substitute alternative paradigms for performance and create different incentive systems. Unfortunately, while they can serve to shake loose the grip of traditional costing measures, they often create the impression that costs can somehow be ignored and that cost tradeoffs are irrelevant (Zipkin, 1991). Creating a consistent approach to setting decision objectives, that combines both cost and revenue implications is an important research task.

JIT AND TQC AS CROSS-FUNCTIONAL PROGRAMS

I have suggested that there is a basic issue underlying the discussion above: traditional financial methods of objective setting and performance measurement are proving to be inadequate under current conditions in manufacturing. This may be partly due to changing relative prices and capabilities in factor markets (labor, delivered material, capital equipment, technology), and partly due to changing competitive conditions in product markets. At any rate, new approaches to performance measurement have appeared which are perhaps most clearly seen in programs such as Just-in-time (JIT) and Total Quality control (TQC).

While JIT is sometimes described as a philosophy, I find it more useful to think of it as a recognition of and emphasis on, lead time related costs and benefits. In other words, instead of focussing on specific techniques, it is more revealing to characterize JIT as a performance measurement, and therefore incentives system. Specific techniques and methods are generated naturally as a response to these incentives or objectives. Seen from this perspective, JIT has two results. One is that within functions, certain "costs" which are not picked up by traditional costing systems, are easily identified and measured. An example of such a cost is manufacturing lead time due to queues and congestion. Another is the category of overhead costs related to supporting activities or to transactions or to WIP. These costs also happen to correlate with lead times. In addition, the JIT approach crosses functional boundaries. In one sense, lead time measures are just as applicable to transactions flow and service operations in marketing or financial control, as to material flow and inventories. A second aspect of lead time measurement, is that it naturally highlights interfaces between functions where delays often occur. For example, reducing finished goods inventories couples manufacturing more tightly with distribution and sales. To take an extreme viewpoint, inventory reduction by definition makes manufacturing more like a service operation, with the side effect that as in services, the normally isolated production function is forced to become more aware of the customer and the marketplace. While JIT methods by and large deal with lead times in material and transaction flows, "time-based competition" (e.g., Stalk, 1988) takes a broader view of the domain of application for lead time measures. Hence, design, product development, and many other activities of strategic importance can also be addressed by this approach.

It seems clear that lead times in manufacturing (and service for that matter) are strongly correlated with performance at the operational level (Karmarkar, 1991). The connection between lead times and overall company performance has been empirically supported (e.g., Schmenner, 1988). Other recent work has begun to look at some of the institutional causes for lead times, having to do with transactions and procedures (Hegde et al., 1991). There is still ample room for research into the relationships between lead times and firm performance. For example, while there is anecdotal and theoretical support for the benefits of rapid product introduction, I am not aware of empirical studies on the subject. There are some areas where theoretical models are missing as well; an example is the relationship between manufacturing lead times and quality management.

Total Quality Control (TQC) can also be interpreted as a cross-functional technique for creating incentives and setting objectives. Consider the tools available for performance measurement and objective setting at the strategic level. Financial valuation methods typically use cash flow projections for evaluation, and market value as a basis for incentives. The former is often very difficult to apply in practice, since it is extremely difficult to estimate all the cash flow consequences of say, an improvement in product design, or an investment in process equipment. This is particularly true of market related or revenue consequences of manufacturing decisions. Market valuation (share price) on the other hand, is the end result of estimates made by the many agents in the marketplace, in the course of supplying capital to the firm. In some sense, it is intriguing that valuation of the firm is based on a factor market rather than a product market — on reflection, the reasons are apparent. However, what is clear is that this valuation is not a very useful guide to action. To stretch the analogy a bit, using share price as a measure is a little like a football coach being paid on the basis of the Las Vegas line on the score in the next game. The betting odds might be an accurate summary valuation of the team's chances, and they could indeed provide incentives in the right direction. However they give little guidance to the coach in terms of coaching strategy, or team development. In my opinion, the enthusiasm for TQC in industry is at least partly due to its potential as a strategic initiative that allows objective setting across functional areas. It has the advantage that it is action oriented so that strategy implementation is built in. Furthermore, it is a top-down approach as indeed any cross-functional method must be. Finally, TQC (and JIT) are approaches which take market competition into account. While techniques like cash flow analysis have the ability to do so, they are often applied as though the firm exists in a market vacuum. In other words, there is often no active recognition of competition. Van Wassenhove (1991) has commented that policy directives in firms are increasingly taking a programmatic guise, and that firms address issues of both costs and competition through cross functional programs.

From a research viewpoint, there are several issues raised by such programmatic approaches. As I have interpreted them here, they are an alternative to traditional means of decomposing the overall value of the firm objective into local objectives. In other words, instead of breaking down profitability into revenue (for sales) and costs (for say production), a program such as JIT sets objectives for the firm, that while imprecise, have implications for all functions, and provide a common target. Is is possible to formalize the decomposition of company objectives into a set of such programs? Do the programs collectively span the complete action space and outcome space? How are competitive issues reflected in these programs — i.e., what happens when all competitors are engaged in TQC, JIT, CIM or other types of program implementation? It would seem to be worthwhile to delineate the proper domains of applicability of these programs. It is quite possible that uncritically applied, they can create incentive distortions, strategic misdirection, and misallocation of resources.

OPERATIONS AND INFORMATION SYSTEMS

The close tie between operations and information systems is obvious at the operational level. In issues of strategy, the connections are less clear and less direct. Indirectly, automation and CIM constitute a very significant strategic issue for manufacturing, that has been amply discussed by sources such as Ayres and Miller (1981) and Jaikumar (1986b). Yet, a recent survey by Johansen et al. (1991) shows that CIM technology adoption is typically driven by bottom up initiatives, using traditional cost minimization means of justification, and with little integration across departments or functions. The picture that emerges is that CIM, unlike JIT or TQC, has not become a top-down program used tactically to implement a strategic directive. One could cynically conjecture that unlike say JIT, CIM issues are more difficult for senior managers to comprehend, and that as a result CIM has not had the same program visibility except in a few technically oriented firms (e.g., Allen-Bradley).

A very significant issue in information systems for manufacturing in coming years, is their use in leveraging the productivity of white collar management. Certainly aspects of CIM such as CAD/CAM and CAE are related to this question. However, I am referring to the broad topic of improving the productivity of all levels of management. Just as automation and CIM enhance shop floor productivity, information systems are the only means of leveraging or automating managerial or supervisory productivity. There are similar issues present in factory communication, and EDI applications in purchasing and sourcing. To cite a couple of examples, Hegde et al. (1991) have empirically studied some of the interactions between information technologies and manufacturing performance. Groenevelt and Saxena (1991) model the interaction between speed of communication and JIT program performance.

These issues are not novel. Yet, I would argue that we do not know very much about capital for labor substitution in terms of the automation of management functions. We do not know what the rates of substitution are, or even how to measure quantities of information or of management activity. Yet there is a visible trend towards eliminating layers of management from manufacturing, and it is apparent that the remaining people are severely pressed to keep up with work loads. There is a far greater demand for computer based methods of planning and analysis than in the past, but I do not perceive a systematic process underway. In other words, the substitution is occurring rather haphazardly when it does occur, and is being driven by cost reduction pressures. The current interest in "process reengineering" is the closest to a systematic approach, but it lacks a strategic direction. In many cases, there is a tendency to outsource activities ranging from production of parts to running information systems themselves. I do not have any neat paradigm that captures these issues, but I am persuaded that automation of management activity through intelligent information systems will emerge as a crucial issue for research in the next decade.

OPERATIONS AND MARKETING

The interaction between production operations and marketing constitutes a substantial topic. At one level there are interactions between traditional views of operations decisions, and marketing issues. Examples of these are decisions on capacity, location, service and inventory tradeoffs, stock or make-to-order and so on. In addition, at the operational level there are interactions between promotion activity and production planning, manufacturing lead times and inventory or service levels, and sales and production planning with respect to volume and product mix. Decisions such as entry, product mix, product line, pricing and positioning are all affected by the nature of the production function. Dobson and Kalish (1990) study the product mix problem from this perspective. Dobson and Lederer (1990) undertake an ambitious analysis that integrates pricing, route structure and flight scheduling in airline network competition. While this is a service problem, it shows the close relationship between operations decisions at both operational and strategic levels, and marketing issues such as positioning and pricing. I will here focus more on some areas that are still emerging as research issues, and which go somewhat outside traditional manufacturing topics.

Class and Quality: I have already discussed quality management as a cross functional program. However, formal views of quality in marketing and manufacturing differ to some extent. In marketing, multi-attribute descriptions of products are becoming the standard. They permit marketing models to be cast in a manner consistent with economic models of competition, and are especially valuable in product positioning and pricing analysis. However, in many of the basic papers on this topic (e.g., Rosen, 1974; Mussa and Rosen, 1978), the term "quality" is taken to mean the level of *performance* of the product as measured by levels of attributes. This is inconsistent with the manufacturing or process view of quality, which is closer to *conformance* with specifications or expectations. It is however possible to develop a concept of quality that brings the two views closer (Karmarkar,1991; Karmarkar and Pitbladdo, 1990c).

Rather than "quality", suppose that we use the term "class" to mean the position of a product in terms of its defining attributes. Now the actual attribute levels achieved in a particular product unit may not be exactly equal to the desired class of the product. In other words, the population of all units of the product has a probability distribution around the desired class. If we nominally think of the class of the product as the location of this distribution, we can think of quality as some measure of the lack of dispersion in the distribution. In the simplest case of a one-attribute model with a Normal distribution of the attribute in the population, we could define μ as the class or position of the product and say,

$1/\sigma^2$ as the quality. This approach contrasts with that of treating quality as a separate attribute (e.g., Fine, 1986). Now, since the customer is facing uncertainty in the attribute levels of a purchased unit, customer preferences must be represented by a cardinal utility function rather than a value function. However, customer preferences for class, quality and price can be explicitly addressed. Karmarkar and Pitbladdo use a framework in which customer preferences are estimated in terms of the parameters of a utility model, a positioning decision is made, and the product is then manufactured. The manufacturing process results in output with a distribution of values; production costs are taken to be a function of the parameters of the distribution. Models of monopoly and perfect competition are developed. It is shown that for the case where price is separable from other attributes, both scenarios lead to the same positioning and quality level decisions for the product, differing only on price. The model is developed further to address uncertainties in customer preference estimation, and heterogeneous customers. It is shown that *product* quality may suffer from estimation and positioning errors in addition to poor *process* quality.

Such an approach has the potential for analyzing resource allocation across functions. For example, should more effort be spent in estimation of customer preferences or in improving process quality? This model can also address the selection of process quality control methods, in the context of a competitive environment, with quality niches in addition to performance positioning. It is also possible to include concepts like Quality Function Deployment (QFD). Horsky and Nelson (1992) point out that there is a gap between a multi-attribute model of a product based on attributes representing customer perceptions and one with attributes representing physical properties. In the marketing literature, this issue is partially captured by the dichotomy between search attributes and experience attributes. However, the more difficult issue is that of the relation between preference attributes which define product quality and class in the market place, and physical attributes as in engineering specifications which underlie process quality (Karmarkar, 1991). Recent work by Moskowitz et al. (1990, 1991) formalizes the QFD model as the development of the mapping from production specifications to customer preference attributes. This model makes it clear that QFD is more closely related to performance and positioning (i.e., class) rather than consistency and conformance.

Industrial or Business Marketing: Most of the research and teaching activity in marketing departments of business schools is aimed at consumer product marketing. In my estimation, one reason for this is that when the buyer is a firm, modeling demand requires very different approaches from those used in consumer markets. In the latter case, the large number of individual buyers dictates aggregate modelling of preferences, typically at the level of multi-attribute value or utility functions. Classes of customers are distinguished by parameter values which might be tied to say, demographic characteristics.

Conjoint analysis, utility function modeling and statistical tools such as regression and factor analysis are the technical methods that apply. With industrial markets, the issues are very different. First, the number of customers in a market place is often quite small — consider the market for an automotive part as an example. More crucially, demand is often determined by the nature of the production function of the buyer. If the product in question is a machine tool, it is essential to understand how it will be integrated into the buying firm's manufacturing system. Alternatively, if the product is a part that will be used in the buyer's eventual product, the specification of the part depends in a complex way on the design of the end product. Furthermore, the production process in which the part will be used, determines the service and delivery requirements for the part, in addition to quality.

Post sales support provides an example of demand characterization and segmentation based on customer usage characteristics in an industrial framework. Modeling the demand for product support involves the understanding of customer costs based on usage and production patterns. On the supply side, modeling provision of support requires models related to service and inventory but also product design, product reliability, modularity etc. (e.g., Lele and Karmarkar, 1983; Hegde and Karmarkar, 1989)

To make a long story short, demand functions in industrial marketing can be imputed from buyers' production functions. Hence industrial marketing analyses can require a large component of operations management. From another perspective, analysis of the manufacturing process of a firm is basic marketing information for any supplier of machines, equipment, parts, or services to the firm. To give a few examples, research on capital justification and technology acquisition in manufacturing is an analysis of buyer behavior for suppliers of capital equipment and technology. Programs such as JIT, TQC, and concurrent engineering have obvious implications for suppliers or parts and components. They have also had an impact on suppliers of manufacturing equipment, due to the movement towards cellular manufacturing. CIM and automation are creating a market for integrated manufacturing systems rather than stand alone equipment. This implies large changes for not only product design but also for marketing and support. The Okuma case study by Jaikumar (1986a) is a good example of the close relationship between manufacturing analysis, product design and market strategy in the advanced machine tool sector.

It seems that there is a vast research opportunity in industrial or business marketing, based on the use of models of manufacturing operations. Research in industrial marketing used to address issues like purchasing behavior and identifying the organizational location of the buying decision. Today it is shifting rapidly towards the kinds of concerns I have described above (for example see ISBM, 1991).

SUMMARY

I have attempted to describe a cross-functional approach to manufacturing strategy research, arguing that there are important research questions here, and also that strategy implementation requires such a perspective. Most companies are still organized along functional lines, and managers still trained in functional specialties. Like the general practitioner in medicine, the generalist manager runs the risk of being undervalued. I have suggested that at this time, our knowledge about the details of firm behavior is primarily organized along functional lines. There are advantages to assembling the available pieces together incrementally, rather than trying to start from scratch. In my examples, I have not tried to be encyclopedic. More complete accounts of manufacturing strategy issues can be found in other papers in this volume, and in surveys such as Fine (1989). I have also excluded many areas (human resource management for example). I do not mean to imply that they are not important.

REFERENCES

Aucamp, D. C. (1984), "A Caveat on Using Setup Costs in Lot-Sizing", *Production and Inventory Management*, First Quarter, 11-14.

Ayres, Robert U. and Steven Miller (1981), "Robotics, CAM, and Industrial Productivity", *National Productivity Review*, 42-60.

Bogue, M. C. and E. S. Buffa, (1986), *Corporate Strategic Analysis*, The Free Press, New York.

Buffa, E. S. (1985), "Meeting the Competitive Challenge with Manufacturing Strategy", *National Productivity Review*.

Buffa, Elwood (1984), *Meeting the Competitive Challenge: Manufacturing Strategy for U.S. Companies*, Dow Jones-Irwin, Homewood, IL.

Chase, Richard B. and S. Rajagopalan (1991), "A Production Planning and Scheduling System at a Chemical Testing Laboratory", Working Paper, Center for Operations Management Education and Research, University of Southern California, Los Angeles, CA 90089-1421.

Clark, Kim, Robert Hayes and Steven Wheelwright (1988), *Dynamic Manufacturing*, The Free Press, New York, New York.

Cohen, Morris and Arnd Huchzermeier (1991), "The Valuation of Longterm Supply Contracts Under Exchange Rate Uncertainty", presented at the Center for Manufacturing and Operations Management Annual Conference, *Manufacturing in Europe '92*.

Cohen, Morris A. and Hau L. Lee (1989), "Resource Deployment Analysis of Global Manufacturing and Distribution Networks", *Journal of Manufacturing and Operations Management*, **2**, 81-104.

Dobson, Greg and Shlomo Kalish (1990), "Bundle Pricing: Formulation and Heuristics", Working Paper, William E. Simon Graduate School of

Business Administration, University of Rochester, Rochester, New York 14627.

Dobson, Greg and Phillip Lederer (1990), "Airline Scheduling and Routing in a Hub and Spoke System", Working Paper No. CMOM 89-08, Center for Manufacturing and Operations Management, William E. Simon Graduate School of Business Administration, University of Rochester, Rochester, New York 14627.

Dobson, Greg, Uday S. Karmarkar and Jeffrey Rummel (1989), "Batching to Minimize Flow Times on One Machine", *Management Science*, **33**:6, 784-799.

Fine, C. H. (1989), "Developments in Manufacturing Technology and Economic Evaluation Models", prepared for *Logistics of Production and Inventory*, S. C. Graves, A. Rinnooy Kan and P. Zipkin (eds.), in the North Holland Series of Handbooks in Operations Research and Management Science.

Fine, C. H. (1986), "Quality Improvements and Learning in Productive Systems", *Management Science*, **32**:10, 1301-1315.

Fine, C. H. (1988), "A Quality Control Model with Learning Effects", *Operations Research*, **36**:3, 437-444.

Fine, Charles H. and Arnoldo Hax (1985), "Manufacturing Strategy: A Methodology and an Illustration", *Interfaces*, **15**, 28-46.

Gantt, H. L. 1990, "The Relation Between Production and Costs", *Journal of Manufacturing and Operations Management*, **3**:3, 251-267.

Gray, Ann E. (1991), "The Location of Production Facilities: Issues of Dynamics and Uncertainty", Unpublished Dissertation, William E. Simon Graduate School of Business Administration, University of Rochester, Rochester, New York 14627.

Groenevelt, H. and R. Saxena (1991), "Base-Stock Levels for Given Service Profiles", Working Paper, William E. Simon Graduate School of Business Administration, University of Rochester, Rochester, New York 14627.

Grubbstrom, R. W. (1980), "A Principle for Determining the Correct Capital Costs of Work-in-Progress and Inventory", *International Journal of Production Research*, **18**:2, 259-271.

Hax, Arnoldo and N. Majluf (1991), *The Strategy Concept and Process: A Pragmatic Approach*, Prentice Hall, Englewood Cliffs, New Jersey.

Hegde G. G. and Uday S. Karmarkar (1989), "Market for Product Support: An Equilibrium Analysis", Working Paper QM 86-24, William E. Simon Graduate School of Business Administration, University of Rochester, Rochester, New York 14627.

Hegde, G. G., S. Kekre and S. Kekre (1991), "Engineering Changes and Time Delays: A Field Investigation", *Journal of Manufacturing and Operations Management*, forthcoming.

Hendricks, Kevin B. and Vinod R. Singhal (1992), "Quality Awards and the Market Value of the Firm: An Empirical Investigation", Working Paper, School of Management, Georgia Institute of Technology, Atlanta, Georgia.

Hodder, James E. and M. Cemal Dincer (1986), "A Multifactor Model for International Plant Location and Financing Under Uncertainty", *Computers and Operations Research*, **13**:5, 601-609.

Hodder, James E. and James J. Jucker (1985) "A Simple Plant-Location Model for Quantity Setting Firms Subject to Price Uncertainty", *Computers and Operations Research*, **13**:5, 601-609.

Horsky, Dan and Paul Nelson (1992), "New Brand Positioning and Pricing in an Oligopolistic Market, *Marketing Science*, forthcoming.

Institute for the Study of Business Markets (1991), "Research Program Overview", The Mary Jean and Frank P. Smeal College of Business Administration, Penn State.

Jaikumar, R. (1986a), "Okuma Machinery Works Ltd.", Harvard Business Case # 9-687-038, Harvard Business School, Publishing Division, Boston, MA.

Jaikumar, R. (1986b), "Postindustrial Manufacturing", *Harvard Business Review*, (November-December), 69-76.

Jaikumar, R. and R. Cooper (1984), "Management Portfolio Analysis and Flexible Machining Systems", Working Paper, Harvard Business School.

Johansen, John, Uday S. Karmarkar and A. Seidmann (1991), "CIM Adoption and Implementation Strategies: A Study of Current Practice and Future Plans", Working Paper, William E. Simon Graduate School of Business Administration, University of Rochester, Rochester, New York 14627.

Jucker, James V. and Robert C. Carlson (1976), "The Simple Plant-Location Problem under Uncertainty", *Operations Research*, **24**:6, 1045-1055.

Kaplan, R. S. (1983a), "Measuring Manufacturing Performance: A New Challenge for Managerial Accounting Research", *The Accounting Review*, **68**, 686-705.

Kaplan, R. S. (1983b), "Yesterday's Accounting Undermines Production", *Harvard Business Review*, July-August, 95-101.

Karmarkar, Uday S. (1991), "Quality Management", Working Paper, Center for Manufacturing and Operations Management, William E. Simon Graduate School of Business Administration, University of Rochester, Rochester, New York 14627.

Karmarkar, Uday S. and M. Lele (1991), "Strategic Issues in Distribution Systems Design", Working Paper, William E. Simon Graduate School of Business Administration, University of Rochester, Rochester, New York 14627.

Karmarkar, Uday S. and M. Lele (1989), "The Marketing/Manufacturing Interface: Strategic Issues", Working Paper No. CMOM 89-10, Center for Manufacturing and Operations Management, William E. Simon Graduate School of Business Administration, University of Rochester, Rochester, New York 14627.

Karmarkar, Uday S. and Richard Pitbladdo (1990a), "Product Line Selection, Production Decisions and Allocation of Common Fixed Costs", Working Paper QM 90-10, William E. Simon Graduate School of Business Administration, University of Rochester, Rochester, New York 14627.

Karmarkar, Uday S. and Richard Pitbladdo (1990b), "Internal Pricing and Cost Allocation in a Model of Multiproduct Competition with Finite Capacity Increments", Working Paper QM 90-06, William E. Simon Graduate School of Business Administration, University of Rochester, Rochester, New York 14627.

Karmarkar, Uday S. and Richard Pitbladdo (1990c), "Quality, Class and Competition" Working Paper, William E. Simon Graduate School of Business Administration, University of Rochester, Rochester, New York 14627.

Karmarkar, Uday and Jeffrey L. Rummel (1990), "The Basis for Costs in Batching Decisions", *Journal of Manufacturing and Operations Management*, **3**:2, 153-176.

Karmarkar, Uday S., P. J. Lederer and J. L. Zimmerman (1990), "Choosing Manufacturing Production Control and Cost Accounting Systems", in *Measures for Manufacturing Excellence*, R. Kaplan (ed.), Harvard Business School Press, Boston, MA.

Karmarkar, Uday S., S. Kekre and S. Kekre (1990), "Multi-Item Batching Heuristics for Minimization of Queueing Delays", *European Journal of Operations Research*, forthcoming.

Kekre, S. and T. Mukhopadhyay (1990), "Impacts of Electronic Data Interchange Technology on Inventory, Quality and Performance of Outside Processors: A Field Study", Working Paper, Graduate School of Industrial Administration, Carnegie Mellon University, Pittsburgh, PA.

Lederer, Phillip J. and Lode Li (1991), "Pricing, Production, Scheduling, and Delivery-Time Competition", Working Paper, William E. Simon Graduate School of Business Administration, University of Rochester, Rochester, New York 14627.

Lederer, Phillip J. and Vinod R. Singhal (1992), "The Effect of Financing Decisions on the Economic Evaluation of Flexible Manufacturing Systems", Working Paper No. 89-07, William E. Simon Graduate School of Business Administration, University of Rochester, Rochester, New York 14627.

Lederer, Phillip J. and Vinod R. Singhal (1990), "The Effect of Financing Decisions on the Choice of Manufacturing Technologies", *International Journal of Flexible Manufacturing Systems*, forthcoming.

Lederer, Phillip J. and Vinod R. Singhal (1988), "Effect of Cost Structure and Demand Risk in the Justification of New Technologies", *Journal of Manufacturing and Operations Management* **1**:4, 339-371.

Lele, Milind M. and Uday S. Karmarkar (1983), "Good Product Support is Smart Marketing", *Harvard Business Review*, (November-December), 124-132.

Lev, B. (1974), "On the Association Between Operating Leverage and Risk", Journal of Financing and Quantitative Analysis, **9**:6, 27-42.

Li, L. (1989), "The Role of Inventory in Delivery-Time Competition", Working Paper Series B, No. 110, Yale School of Organization and Management, Yale, New Haven, CT, forthcoming in *Management Science.*

Li, L. and Y. S. Lee (1991), "Subcontracting, Entry Deterrence and Capacity Expansion", Working Paper, Yale School of Organization and Management, Yale, New Haven, CT.

Morton, Thomas E. and Medini R. Singh (1988), "Implicit Costs and Prices for Resources with Busy Periods", *Journal of Manufacturing and Operations Management,* **1**:3, 305-322.

Morton, Thomas E., S. R. Lawrence, S. Rajagopolan and S. Kekre (1988), "SCHED-STAR: A Price-Based Shop Scheduling Module", *Journal of Manufacturing and Operations Management,* **1**:2, 131-181.

Moskowitz, Herb and K. Kim (1990), "Quality Function Deployment: Robustness of Design Parameters for Time Based Manufacturing Deployment", Working Paper, Krannert Graduate School of Business, Purdue University, West Lafayette, Indiana.

Moskowitz, Herb, K. Kim, A. Dhingra and G. Evans (1991), "Fuzzy Multi-Criteria Methodology for Quality Function Deployment", Working Paper, Krannert Graduate School of Business, Purdue University, West Lafayette, Indiana.

Mussa, Michael and Sherwin Rosen (1978), "Monopoly and Product Quality", *Journal of Economic Theory*, **18**, 301-317.

Noreen, Eric (1990), "Conditions Under Which Activity-Based Cost Systems Provide Relevant Costs", *Journal of Management Accounting Research*, forthcoming.

Porteous, E. L., (1986), "A Quality Control Model with Learning Effects", *Operations Research*, **34**:1, 137-144.

Porteous, E. L., (1985), "Undiscounted Approximations of Discounted Regenerative Models", *Operations Research Letters*, **3**, 293-300.

Raturi, A. S. and V. R. Singhal (1990), "Estimating the Opportunity Cost of Capital for Inventory Investments", *OMEGA The International Journal of Management Science*, **18**:4, 407-413.

Rosen, Sherwin (1974), "Hedonic Prices and Implicit Markets: Product Differentiation in Pure Competition", *Journal of Political Economy.*

Rummel, J. L. (1989), "Cost Models for Batching Decisions", Ph.D. Dissertation, William E. Simon Graduate School of Business Administration, University of Rochester, Rochester, New York 14627.

Schmenner, Roger W. (1988), "Behind Labor Productivity Gains in the Factory", *Journal of Manufacturing and Operations Management*, **1**:4, 323-338.

Seidmann, A. and E. Wang (1992), "Electronic Data Interchange: Competitive Externalities and Implementation Policies", Working Paper, William E. Simon Graduate School of Business Administration, University of Rochester, Rochester, New York 14627.

Singhal, Vinod R. 1988, "Risk and the Value of the Firm", *Journal of Manufacturing and Operations Management*, **1**:1, 4-43.

Singhal, Vinod R. and A. S. Raturi (1990), "The Effect of Inventory Decisions and Parameters on the Opportunity Cost of Capital", *Journal of Operations Management*, **9**:3, 406-420.

Srinivasan, K., S. Kekre and T. Mukhopadhyay (1991), "Impact of Supplier Material Flow Complexity and Information Technology Support on JIT Performance: An Empirical Investigation", Working Paper, Graduate School of Industrial Administration, Carnegie Mellon University, Pittsburgh, PA.

Stalk, G. JR. (1988), "Time - the Next Source of Competitive Advantage", *Harvard Business Review*, July-August, 41-51.

Von Wassenhove, Luk (1991), personal communication.

Zimmerman, J. (1979), "The Costs and Benefits of Cost Allocation", *The Accounting Review*, **LIV**:3.

Zipkin, Paul (1991),"Does Manufacturing Need a JIT Revolution", *Harvard Business Review*, (January-February), 4-10.

A DECISION ANALYSIS APPROACH FOR COORDINATING DESIGN, MANUFACTURING AND MARKETING IN NEW PRODUCT DECISIONS

Rakesh K. Sarin
Anderson Graduate School of Management
University of California Los Angeles
Los Angeles, CA 90024

In the late 1800's, Faber pencils dominated the market. By one estimate, by the end of the 19th century Faber made 250 million pencils per year. L. & C. Hardmuth Company, facing stiff competition from Faber, designed a top-quality pencil that would sell for three times the price of any other pencil in the market. They named it Koh-I-Noor after the great diamond of history and painted the pencil golden yellow to suggest the Oriental source of the then-finest graphite. The pencil was a huge success. Was it simply designing and making a better pencil or was it also selecting a name and a color to connote quality and value that made the Koh-I-Noor pencil so successful? This example, though of a relatively simple product, demonstrates that a creative fusion of marketing insights with design and high quality production could reap a firm substantial success in the market place.

Today it is widely recognized that coordination across functional areas, from design through marketing and product service, is critical to a firm's long-term success. In response, a variety of tools such as concurrent engineering, quality function deployment (QFD), etc., has been developed to facilitate integration across traditional functional boundaries. These approaches have produced success stories in a wide variety of industries. To implement these approaches, however, it is necessary that research and development (R&D), manufacturing, and marketing groups agree on a common set of objectives and criteria. Since these groups often have different and sometimes conflicting objectives, a systematic approach is needed to facilitate consensus among them.

In this paper I discuss an approach for coordinating the efforts of R&D, manufacturing, and marketing groups that is well-known to a rather small group of management scientists called decision analysts but has not yet gained the wider recognition and acceptance that I believe it deserves. The approach is useful at the planning stage for creating a pre-decision assent. Thus, it complements concurrent engineering and QFD, which seem to have a wider scope covering the entire range from planning to implementation. The decision analysis approach is particularly useful in quantifying the concerns of various groups from R&D through marketing in new product development decisions and in creating a sensitivity among these groups towards each other's concerns. The approach has a relatively simple systematic structure and therefore it can be applied consistently to a wide range of problems.

1. A Real-Case Example

A few years ago I was part of a consulting team involved in helping a company evaluate several possible designs for a sophisticated high technology product. The product when completed was to provide geological and drilling information in a real time mode to facilitate decision making.

The marketing group preferred Design A, which could be developed quickly and which caused the least interference with the way users (drilling operators) were accustomed to conduct their business. Since such a product was unavailable at the time to the industry, a quick entry into the market to gain first mover advantage was considered very important.

The R&D group preferred Design B, on which they had already made some progress. They also preferred a more sophisticated product which included a wider range of sensors, and a more patient approach to entry into the market to ensure a high degree of product reliability.

The manufacturing group favored a design configuration that was easy to manufacture. For example, one concern of the manufacturing group was to avoid the use of non-magnetic collars, as these tend to be harder to machine.

The decision analysis approach was used to quantify the objectives relevant to each group and to facilitate consensus about the preferred design among the groups. I will now present the basics of this approach in the context of this case example.

2. The Problem

An oil service company was facing the problem of selecting the design and configuration of a new product for development and eventual introduction to the market. The product would provide information that would be useful in guiding the drilling process so as to increase the probability of finding oil.

The development of the product would require approximately $50 million of investment in R&D.

At the time that the company was involved in this new product decision, none of its competitors offered such a product. At least two or three competitors, however, were also engaged in developing similar products, each pursuing its own design and development strategy. The industry norm at that time was to take measurements when the drill bit was retrieved. The purpose of the product under development, which is called by the generic name "measurement while drilling" (MWD), was to provide information to the operators on a continuous basis through the down hole sensors and a transmitting medium (which believe it or not is mud, and hence the name mud pulse telemetry). Therefore MWD products had a potential to considerably improve both the quality of information and the efficiency of drilling. The company perceived that it was in a race to come out with the first generation of MWD products quickly.

It was unclear, however, which of the several possible designs the company should pursue. Further, there was no agreement on which basic sensors must be included in the product first offered to the market.

The broader strategic issue for the company was to develop a sequential product development plan and concomitant resource allocations.

It was evident that over time a newer generation of product would replace the older one. However, a previous decision to follow a particular design and development strategy would limit the future possibilities in a specific way. The initial decision to select a design for the first market entry was therefore an important one.

An MWD product could be made in two basic designs. In Design A, called "Insert," a sensor was located around the circumference of the collar. In Design B, called "Sonde," a sensor was located in an instrument that hung in the middle of the collar. It was possible to locate some sensors in the Insert mode and some in the Sonde mode. If, however, a sensor was developed for location in the Insert or Sonde mode, converting it to the other mode in a future generation of products would be expensive and time-consuming. Sensors of both types were in different stages of development. Some merely required prototype building and testing, whereas others required basic research to solve some difficult technical problems before the prototype construction could begin. The amount of time needed to develop some of the sensors was therefore highly uncertain. Since the sensors were required to operate in extremely harsh environments close to the drill bit, the reliability, i.e. ability to function properly without failure, was also very important. One of the design engineers remarked that the technology required for MWD products was more sophisticated than for the sensors built for space exploration. The basic research, design, engineering, prototype development, and laboratory and field testing all had to be coordinated and successfully completed to deliver a marketable product.

The marketing group preferred an all-Sonde design with only some basic sensors included. Their primary concern was to enter the market quickly, as they felt that the market window would close rapidly if the competitors introduced their products first. The R&D group had already made some progress in designing and developing some sensors in Insert mode. They preferred a more sophisticated product that could provide a variety of information to the drill operator. Though they realized the importance of an early entry into the market, they were concerned about the reliability of the product in field setting. In their view a failure of the product, if it was not properly integrated and tested, could tarnish the image of the company and make consumer acceptance of their successive products more difficult.

To choose a preferred design which is in the best interest of a company, the various groups involved in the decision making process must first agree on the objectives or common success factors relevant to the decision. R&D groups often tend to place a greater emphasis on technical capabilities of the product. In contrast, marketing groups feel a greater pressure to generate sales and tend to emphasize quick entry into the market and product proliferation through market segmentation. In this application, since there was no product to sell, several energetic young MBA's hired by the company were primarily engaged in preparatory and planning activities. It is understandable that they were eager to launch a basic version of the product into the market quickly. The R&D group had a different pressure. Having spent several million dollars each year for several years, they did not wish to be embarrassed by a rudimentary product that might fail in the field. It is difficult to reach an agreement by debating over which alternative is the best, because each alternative design is indeed the best from some vantage point. We therefore first had to convince the various parties involved in the decision to take a step back and reach some agreement on the objectives and criteria upon which alternative designs must be evaluated.

3. Critical Success Factors

The most important step of decision analysis is identification of objectives, criteria, and attributes that are relevant for evaluating decision alternatives. How do we identify and quantify the relevant heirarchy of objectives, criteria, and attributes?

One approach is to begin with the broad objectives such as profitability, market share, etc., and continue refining each till the lowest level attributes are defined. This top-down approach requires involvement of all concerned groups. In this application we first interrogated R&D, marketing, and manufacturing people separately to identify the attributes that were most relevant from the perspective of their own group. The groups were then brought together and agreement was reached on the list of attributes that ought to be used in the evaluation process.

In Figure 1, the list of attributes that was finally agreed upon by various participants in the decision making process is shown. The four mid-level criteria shown in Figure 1 measure the concerns:

- When will the product be ready to market?
  (Timing)
- What will the product do?
  (Performance)
- How well will it perform?
  (Reliability)
- How much will it cost?
  (Cost)

The lowest level attributes are precisely measurable, though some may be measured on a subjective scale. For example, time to first market entry is measured simply in months from the time of decision making. Rig interference, however, is measured on a five point scale. Each level of this subjective scale was precisely defined. An important part of this approach is to incorporate both quantitative and qualitative criteria in analysis.

While the process of identifying relevant attributes and their measurement scales is largely an art, some guidelines are helpful:

- the list of attributes should be <u>comprehensive</u> to include all concerns of the decision makers,

- the attributes should be <u>non-redundant</u> so that double counting is avoided,

- the attributes should be <u>measurable</u> either on a quantitative or a subjective scale,

- to the extent possible the attributes should be <u>independent</u>.

Figure 1

Objectives and Attributes of the Decision Problem

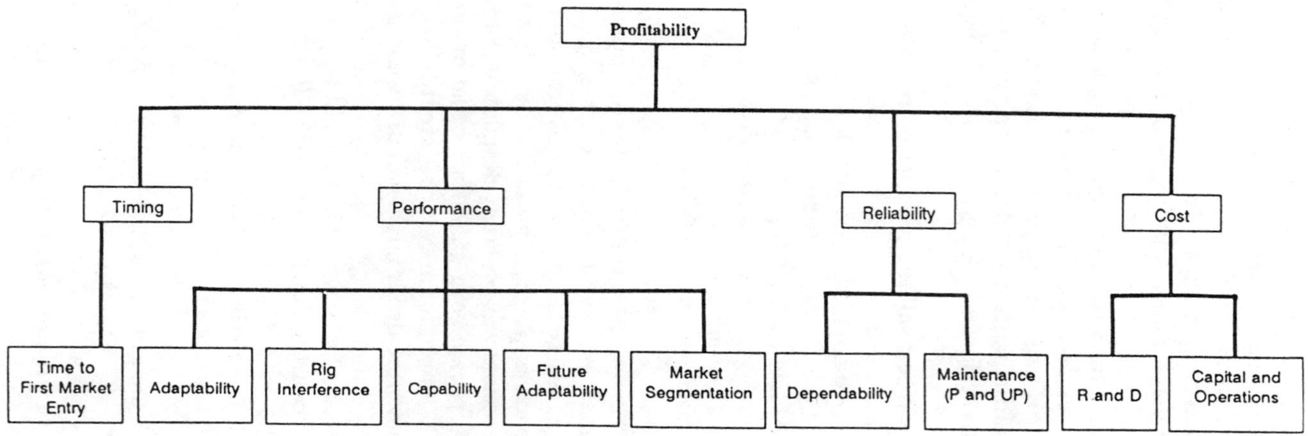

Independence among attributes simplifies analysis. In some cases independence can be achieved by redefinition or by combining attributes.

It is remarkable to note that, while all of the attributes listed in Figure 1 were identified by the participants themselves, they had not deliberated along these dimensions prior to our involvement. This process of identifying objectives and attributes was helpful in improving communication among participants. To everyone's surprise, however, it also helped in developing additional design alternatives that they had not previously thought about.

4. Decision Alternatives

The immediate decision facing the company was the choice of a Sonde or an Insert design and the sensors that must be included in the initial product. While the development work would continue on several sensors in parallel, greater resources would be devoted to those sensors that were chosen for the first market entry. Additionally, the product had to be able to be accommodated in four collar sizes (9", 8", 7", 6 1/4"), and this consideration presented some design choices as well.

The design engineers had generated over one hundred possible alternatives. A careful screening based on development cost, development time, technical feasibility, etc., led to a reduction to nine contending design alternatives.

It is to be noted that several design alternatives that were not thought out in early investigation were created after the groups identified the relevant attributes. In our experience, new decision alternatives often are suggested by the process of structuring objectives. Additionally, something pleasant happened. The discussion that was focused on Sonde (the marketing group's favored design) versus Insert (the R&D group's favored design) shifted to which sensors should be in the Insert mode and which should be in the Sonde mode. At this stage the two groups were far from reaching a consensus on the preferred design but they began to appreciate the advantages of each design--a definite bit of progress. Though manufacturing had a smaller role in this project, the participants also developed an understanding of ease of manufacturing and maintenance issues in the choice of a final design.

There is an eighty-twenty rule in decision analysis as well, except it is in the reverse order for the decision maker than for the analyst. The eighty percent of the value to the decision maker has already accrued in the two steps (identification of objectives and attributes and generation of decision alternatives) described above. But from an analyst's viewpoint all we have done is this: There are n alternatives $A_1,...,A_n$ (n = 9 in this case example) and each is to be evaluated on m attributes $X_1,...,X_m$ (m = 10 in this case example). The eighty percent of the value an analyst derives is from quantitative modeling that is yet to be described. These interests must be re-aligned to achieve a wider use of management science techniques.

5. Evaluation Model

Once we have specified decision alternatives $A_1,...,A_n$ and the attributes $X_1,...,X_m$, we need an evaluation model to rank order the alternatives. The steps of the decision process are not so sequential as they seem in our description. During the evaluation process, it is quite possible that new alternatives may be added to the list or that some attributes may require refining. But let us proceed.

The overall attractiveness of an alternative is measured by an index (called value function). Let us denote the scores or outcomes or consequences of an alternative A_i by the vector $(x_1^i,...,x_m^i)$. Then,

$$V(A_i) = f(x_1^i,...,x_m^i), \quad (1)$$

where V denotes the value function that assigns to each A_i a real number. The function f is a real-valued function that assigns to each multi-attribute consequence vector a real number. The model given by (1) is not useful unless we are able to decompose f into some simple forms. Welcome to the field of multiattribute preference theory! In this theory some simple verifiable conditions on the decision maker's preferences are specified that permit us to decompose f in some easily assessed functional forms. Two of the simplest forms are given below:

Additive Preference Function:

$$V(A_i) = w_1 f_1(x_1^i) + w_2 f_2(x_2^i) + ... + w_m f_m(x_m^i), \quad (2)$$

where w_j, j = 1 to m, measures the relative importance weight of the attribute x_j, and f_j is the single attribute preference function defined over x_j that measures the relative attractiveness of various levels of attribute x_j. In model (2), we usually set

$$\sum_{j=1}^{m} w_j = 1 \text{ and } f_j(x_j^o) = 0 \text{ and } f_j(x_j^*) = 1,\ j = 1 \text{ to } m, \text{ where } x_j^o \text{ and } x_j^*$$

are the worst and the best scores on attribute j.

Multiplicative Preference Function:

$$V(A_i) = \frac{(\prod_{j=1}^{m} (1 + k k_j f_j (x_j^i)) - 1)}{k}, \quad (3)$$

where k and k_j, j = 1 to m, are scaling constants and f_j is the single attribute preference function similar to the model (2) above. In this model,

$$k > -1 \text{ and } 1 + k = \prod_{j=1}^{m} (1 + k k_j).$$

The additional scaling constant k measures the complementarity among attributes. Both of the models (2) and (3) have been widely used in numerous applications.

In the case example discussed here, model (2) was judged to be the appropriate one and therefore we will focus on how we assessed the f_j's and w_j's required to calibrate this model.

In many decision problems, as was true in our case example, the consequences of an alternative on some attributes cannot be determined with certainty. For example, given a concept design alternative, the amount of time needed to develop the product and introduce it to the market is highly uncertain. Before we discuss calibration of model (2), we describe some of these key uncertainties.

6. Key Uncertainties

The consequences of a new product decision are seldom known with certainty. In our case example, the implications of alternative design choice for the MWD product on the attributes time to market entry and reliability in the field were judged to be highly uncertain.

To determine time to market entry we need to estimate (i) how long it will take to develop each sensor included in a given design alternative; (ii) how long it will take to manufacture and test the product. An example of the assessed probability distribution for developing a sensor is given in Figure 2. The manager in charge of the sensor was interviewed to elicit this probability distribution. Similar probability distributions were assessed for each sensor. Since the development work on these sensors could proceed in parallel, the probability distribution for completing all sensors was easily determined using a simple computer program. The time for testing and manufacturing was assumed to be deterministic.

Figure 2
Input to Timing Model
Cumulative Probability Distribution for Time to Develop a Sensor

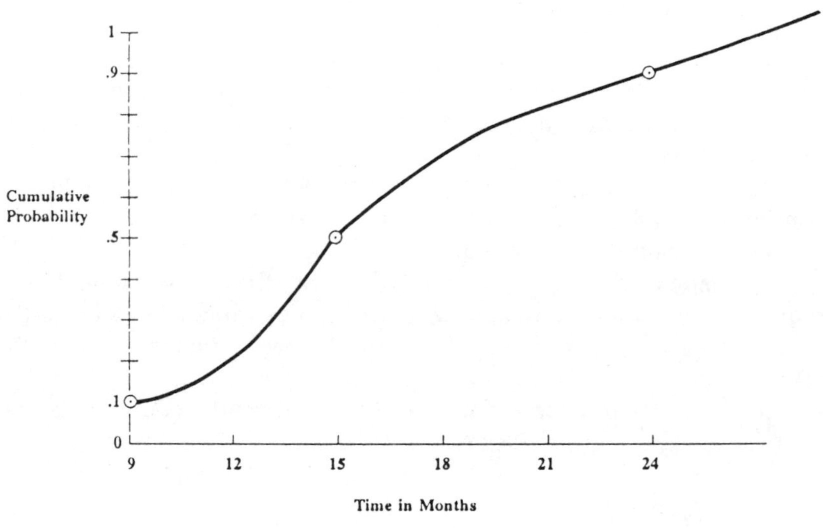

The estimation of the reliability of the system requires the probability of failure for each sensor. If these probabilities are independent, then the probability that the system will work is simply the product of each sensor's probability of working. In our case example these probabilities were not independent. In Table 1 a typical input required to assess the system success probability is shown. The arrows indicate the data elicited. A computer program was used to estimate the system success probability from the marginal and joint probabilities.

Table 1
Input to Reliability Model

Sensor S_2 \ Sensor S_1	Success	Failure	
Success	90	4	94◂
Failure	3	3◂	6
	93◂	7	100

In our case example, the consequences of a failure varied greatly depending on the sensors that failed. For example, a failure of Pulser Driver meant that no data could be transmitted on a real time basis but it could be recorded and evaluated when the drill bit was retrieved. If, however, a sensor such as Gamma-Ray failed, then only one piece of information was missed. On the other hand, if the power generator failed, then no data could be recorded or transmitted. In the most extreme case, all collected data could be lost. It was therefore important not just to produce a probability number but also to describe to the decision makers various consequences that could result under alternative scenarios.

The output of the time to market entry and reliability models depends on the input data supplied by individual managers-in-charge of each sensor program. Our approach of elicitation reduces the impact of "wishful thinking" but cannot eliminate strategic mis-representation. Managers often work with plans and budgets. It is therefore natural for them to reflect these planned consequences in their responses. Thus a manager states that there is a 90 percent chance of completing a sensor within one year because that was his target. In our approach we emphasize that we are interested in his present estimates based on all available information to date and not in the planned values specified some time ago. We also present him with some bets and ask him to choose between them. The bets, though hypothetical, aid a person in introspecting about his true beliefs. There were several instances in our case example where a manager specified say t months as a median (50 percent chance of above or below t) of completion time. However, when asked to take an even bet below or above t, he chose to bet above t. Thus, he really believed that the median was larger than t. We adjusted t till he felt equally comfortable in betting on either side of t.

But these managers are smart, and once they see the process they can manipulate their responses accordingly. We therefore collected the data in two stages. In the first stage each manager was interviewed independently. In some cases a manager called in a few engineers who were working on the sensor. In the second stage we involved the group manager who oversaw the development of several sensors as well. The group manager could ask for a rationale if he thought the response in stage 1 seemed inappropriate. There were surprisingly few changes in this case study. In other studies, however, we have found that this process has resulted in considerable revisions in the original estimates.

The only surprise we found was that even though some sensors were being developed in the Insert mode, the time to start afresh and develop them in Sonde mode was not much longer. The impact on the time to develop the entire system was even smaller if these sensors were not on a critical path. As we will see later, this observation led to a further narrowing in the difference in opinion of the R&D group that favored Insert mode and the marketing group that favored the Sonde mode at the beginning of the study.

7. Assessment of Preference Function

The assessment of preference function (2) or (3) requires single attribute functions f_j's and importance weights or scaling constant k_j's (as well as k if model (3) is used).

7.1. Assessment of Single Attribute Preference Functions

We assessed single attribute preference functions, f_j's, by interviewing individual managers. For example, to assess f_j for the attribute of reliability, the R&D manager was interviewed. Since the time to market entry was considered an extremely important attribute, we interviewed the vice president in charge of the division to elicit the utility function for this attribute. We describe here the steps used for assessing the utility function for the attribute time to market entry. The procedure used for other attributes was similar and will not be discussed here.

We first discussed the scope of the study with the vice president of the division, to whom all of the managers in this study reported. We explained to him the procedure we were using in evaluating alternative designs. We then discussed how the attribute time to market entry is defined. We wanted to make sure that he understood that the attribute definition included not only the development of the product, but also the testing and manufacturing lead times. We explained to him how the time to market entry for each of the nine proposed concept designs was estimated. These preliminaries took about 45 minutes of our two-hour session.

We then discussed in a qualitative manner the impact on the company as the time to market increases. He reiterated to us the viewpoint that the marketing manager had advanced earlier, that the market window will close rapidly if the company fails to introduce some version of the product to the market in a timely manner. He indicated, for example, that an increase in time to market entry from 1983 end to 1984 end would be much worse than from 1984 end to 1985 end. Even though we had estimated the time to market entry in months from the date of decision, we found it helpful to the vice president if we carried out our discussions and assessments using calendar months. The two framings of the question (in months or in calendar time) are equivalent, but the latter was easier to relate to by the managers in charge of developing individual sensors.

At this point we are ready to assess u(t), the utility function for attribute time to market entry. Since the earliest time to market entry for any of the concept designs was 1983 end, we set u(t = 1983 end) = 1. The vice president was insistent that some version of the product must be introduced by 1985 end. To ensure that he really meant that u(t = 1985 end) = 0, we proposed to him a hypothetical alternative that received best scores on all attributes except it was not available to market until March 1986. He felt comfortable in rejecting this alternative and in fact asserted that a delay in market entry beyond mid-1985 seriously compromises his company's competitive position. We set u(t = 1985 end) = 0.

We now know the general shape of the utility function (decreasing convex) and the two end-points, u(t = 1983 end) = 1, u(t = 1985 end) = 0. To assess the utility of intermediate consequences such as 1984 end, we asked the vice president to indicate his preference between two alternatives that had identical consequences on all attributes except the attribute time to market entry. Alternative 1 would have been available for market entry by the end of 1984. Alternative 2 had a 0.5 chance of market entry by 1983 end and a 0.5 chance of market entry by 1985 end. He preferred alternative 2, which was consistent with his qualitative reasoning that there would be a greater value loss in delaying from 1983 end to 1984 end than from 1984 end to 1985 end. We now modified the alternative 2 so that there was a p = 0.4 chance of market entry by 1983 end and (1-p) = 0.6 chance by 1985 end. He still preferred alternative 2. When we set p = 0.2 in alternative 2, he preferred alternative 1. We now know that u(t = 1984 end) was between 0.2 and 0.4. By varying p we found that at p = 0.3 he was indifferent between alternative 1 and alternative 2. Thus, u(t = 1984 end) = 0.3. We assessed the utilities of two other points (mid-1984 and mid-1985) using the method described above. Given the five points on the utility curve, we fitted an analytical form that best approximates these assessments. Define x = number of months to first market entry from 1983 end. Then,

$$u(x) = \left(\frac{24-x}{24}\right)^{1.74}.$$

In figure 3 the utility function for the attribute time to market entry is shown.

Figure 3
Utility Function for the Attribute "Time to Market Entry"

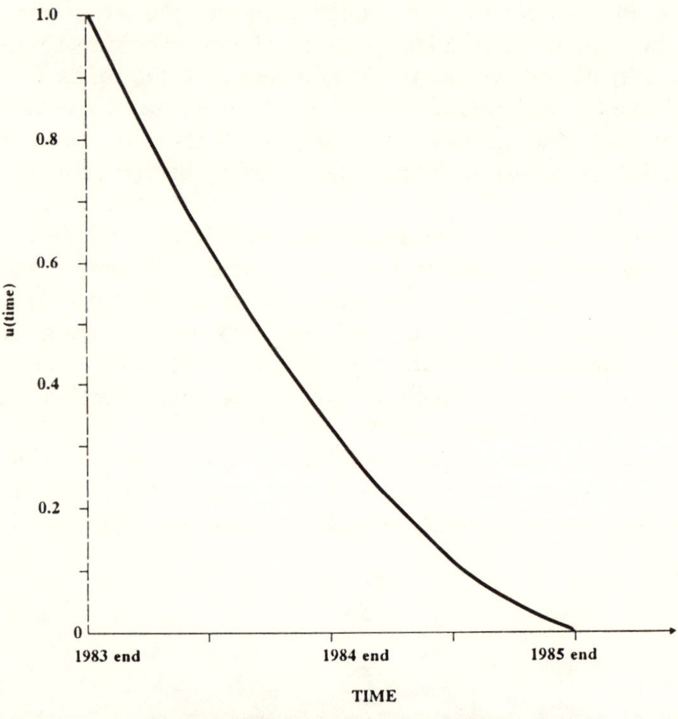

For the attributes on which outcomes were assumed to be deterministic, we used a rating procedure to obtain the utility function. The appropriate managers in these cases simply rated on a 100-point scale the desirability of various consequences. The worst consequence was assigned a utility level 0 and the best 100. The intermediate consequences were assigned levels between 0 and 100 based on the decision maker's subjective evaluation of "goodness" of the consequence. Even when the rating scale was used, we checked for consistency through appropriate lottery-type questions. For example, if a consequence was assigned a rating of 70, we verified that the decision maker was indifferent between receiving this consequence for sure versus a lottery which yielded the best consequence with a 0.7 chance and the worst consequence with a 0.3 chance.

In our study the process for assessing utility function for each attribute took about one to two hours. Three-fourths of the time in each of these sessions was devoted to qualitative discussion that included making sure that the decision maker understood the precise definition of the attribute and his subjective feeling toward the impact of achieving alternative consequences on the attribute. In this study as well as several others that we have conducted, we find that the process of assigning utility levels to consequences goes rather smoothly once the decision maker has understood the process.

7.2 Importance Weights

Assessment of importance weights for each attribute requires trade-offs between consequences on two or more attributes. In either the additive model (2) or multiplicative model (3), the importance weights, k_j's, can be assessed by considering trade-offs between two attributes at a time. We used a participatory group procedure where all the managers, including the head of R&D and the head of marketing, met as a group to provide their trade-offs.

The first step is to identify the most important attribute. In our case study this attribute was time to market entry. The importance weight for the most important attribute will be denoted k_1. Now, to assess k_j for the attribute X_j, we seek x_1 on attribute X_1 so that the decision maker expresses the indifference:

$$(x_1, x_j^0) \sim (x_1^0, x_j^*).$$

In our example, the group has to decide how many months in the time to market entry will they give up to improve the consequence on the attribute X_j from the worst to the best. By model (2), the above indifference implies

$$k_1 u_1(x_1) = k_j. \tag{4}$$

Since we have already assessed u_1, the above procedure will yield (m-1) linear equations by varying i = 2 to m. We know $k_1 + ... + k_m = 1$; thus there are m linear equations and m unknowns $k_1, ..., k_m$. These are easily solved to obtain the importance weights for each attribute. Another procedure is to first obtain k_1 by seeking p so that the decision maker is indifferent between:

$(x_1^*, x_2^0, ..., x_m^0)$ and a lottery yielding

p: $(x_1^*, x_2^*, ..., x_m^*)$

(1-p): $(x_1^0, x_2^0, ..., x_m^0)$.

Then $k_1 = p$. Now k_j's are directly assessed using (4), since we know both k_1 and $u_1(.)$.

In the multiplicative model an additional trade-off is needed since there are (m+1) scaling constants. We will not discuss the assessment of multiplicative model here.

The importance weights or scaling constants for the ten attributes in our study are shown in Table 2.

Table 2
Importance Weights for the Attributes

Time to Market Entry	0.333
Dependability	0.168
Planned Maintenance	0.102
Cost*	0.087
Capability	0.071
Adaptability to Collar Sizes	0.055
Market Segmentation	0.055
Future Adaptability to Accommodate Sensors	0.055
Rig Interference	0.055
Unplanned Maintenance	0.019
k = 0.0 (Additive Model)	

* the cost differential among the nine design alternatives was small; a composite measure was therefore used.

Since the trade-offs provided by the group may not be precise, the exact importance weights in Table 2 are used for the preliminary evaluation of alternatives. Sensitivity analysis by varying these importance weights is important to establish the robustness of the recommended alternative.

To our surprise, in our study the group had a high degree of agreement on the trade-offs from which the importance weights are derived. In some cases, however, it is conceivable that two or more distinct sets of importance weights may emerge. In such cases it is useful to first do evaluation using each set of weights independently. We have found that even in such cases a large number of alternatives are eliminated regardless of which set of weights is employed. Further, the disagreement on the weights for a large number of attributes may not matter to the final ranking of alternatives. It is only on a few attributes that differences in opinion must be reconciled. By highlighting areas of agreement, the environment is made more conducive to focus on a narrow set of issues where an open discussion must take place.

8. Evaluation of Design Alternatives

We now have all the input to evaluate the nine design alternatives. To involve all of the participants in the decision process, a meeting was called that included the vice president of the division, the head of R&D and some of his key managers, and the manufacturing manager. At this point no one knew which alternative would be selected, though the marketing group suspected that the decision process was biased towards the R&D group and a predominantly Insert design would be recommended.

To keep the focus of the group on the objectives, we sought the consensus of the group on the decision problem--to select a product for design and development for the initial market entry. We then reviewed the objectives and the list of attributes. The group was then given an opportunity to specify trade-offs among key attributes. Some small changes from the initial input did occur and were incorporated in recalculation of the importance weights.

The group was then presented with the ranking of alternative designs. This ranking is shown in Table 3. The third column in the table indicates the length of time the market entry must be accelerated for an alternative to become as good as the top ranked alternative. This information is useful since the differences in expected utilities of various alternatives shown in Column 2 have no intuitive interpretation. In applications where cost is the important criterion, the differential cost (rather than differential time as in this example) could be used to show managers how various alternatives differ from one another. In Table 3 alternatives 2A and 4 were very close. The rankings were produced for several different scenarios. For example, in one scenario it was assumed that an important sensor called Pulser Driver could be completed sooner if additional resources were expended to reduce its development time. In this scenario alternative 2A remained at the top. It was found that the ranking of alternative 2 was most robust under a variety of sensitivity analyses.

Table 3
Ranking of Alternative Concept Designs
in Order of Expected Utility

Alternative	Expected Utility	Differential Time (months) $\sim\Delta$
2A	0.5078	0.0
4	0.5076	0.0
1A	0.5037	1.3
2B	0.4828	1.6
3B	0.4453	4.1
5	0.4393	4.4
1B	0.4382	6.0
3A	0.4265	6.1
3C	0.4003	6.5

Based on this information, the group selected the alternative 2A. Essentially it is an all Sonde alternative, except for two sensors in the Insert mode that would be introduced to the market in January, 1985.

Recall that the time to market entry is uncertain. We therefore presented to the group the probability distribution for time to market entry for the selected alternative. There was a 0.5 chance that the product would be ready for introduction before January, 1985 and a 0.9 chance it would be ready before June, 1985. The group felt that only a 50% chance of market entry before January, 1985 was undesirable or too risky from the marketing strategy viewpoint.

The discussion then shifted to what had to be done to ensure January, 1985 market entry. An examination of the critical path revealed that the design and development time for a particular sensor had to be reduced. The manager in-charge of this sensor offered suggestions to reduce this time, and under this alternative scenario the probability of market entry before January, 1985 increased to 0.75 and before April, 1985 to 0.95.

The dependability of the recommended alternative was 0.77. Thus there was 0.23 chance that at least one sensor would fail. The probability that the data was recorded but might not be transmitted on a real time basis was 0.87. The group discussed measures that had to be taken to improve the dependability of the recommended alternative.

Toward the end of this evaluation session the group had achieved consensus on pursuing the recommended alternative. More importantly, however, the group had identified resource allocation priorities to ensure that the recommended design was developed and manufactured for introduction to the market before January, 1985. The recommended design turned out to be a hybrid of the initially preferred designs of the marketing and the R&D groups. The vice president of the division said at the conclusion of the meeting that this was the first time he had a good "tummy feeling" about where the company was heading on the measurement while drilling project. He then proceeded to have lunch with the manager in-charge of the sensor whose development had to be accelerated to achieve January, 1985 market introduction.

The case study described here was conducted as a consulting project and we did not pursue the implementation cycle for the decision. The annual reports of the company, however, reveal that in 1985 the first of the company's MWD product was introduced. In 1986 mud pulse telemetry to provide real time information was introduced. In 1987 some advanced sensors such as a neutron porosity tool and formation density tools were introduced. The formation density tool was the industry's first, and it has been found to be useful in the higher cost, more critical exploratory drilling. These sensors were under design and development at the time of the study. The company's MWD products have had good success in the market and the company has continued to add additional sensors over time. These enhancements have led to applications of the product in specialized market segments, for example in high angle and horizontal wells.

9. Summary

In this paper we described a decision analysis approach for a new product decision. The approach facilitated consensus among R&D, marketing, and manufacturing in the choice of a product design that was selected for development and initial introduction to the market. The approach also aided the managers in identifying key resource allocation decisions to ensure the product market entry by the target date.

The steps of decision analysis are summarized below:

Step 1. Structure the Problem
- generate decision alternatives
- define objectives
- select attributes
- determine chronology of solution process

Step 2. Assess Consequences and Associated Uncertainties

Step 3. Determine Preference Structure
- verify independence conditions to select an evaluation model
- assess single attribute preference functions
- assess importance weights

Step 4. Synthesize Information
- rank alternatives
- sensitivity analysis

In some ways the above steps represent a "common sense" approach to problem solving. The approach, however, provides a structure which facilitates a dialogue among various participants involved in the decision process. The outcome of a decision analysis approach is not merely the ranking of decision alternatives, but also insight to the managers on critical issues in the decision problem. This approach has a great potential to assist managers in coordinating efforts of various groups in new product decisions.

Acknowledgement:

The case study reported in this paper is based on a consulting project. The author is thankful to Dr. Kesh Nair and Alan Sicherman who were part of the consulting team and several engineers and managers of the company who provided inputs to the study.

MODELS FOR PRODUCTION AND OPERATIONS PLANNING

MULTI-PRODUCT CO-PRODUCTION IN MANUFACTURING AND SERVICES

Gabriel R. Bitran
Sloan School of Management
M.I.T. Cambridge, MA 02139.

Sriram Dasu
Anderson Graduate School of Management
U.C.L.A. Los Angeles, CA 90024.

and

Stephen M. Gilbert
Weatherhead School of Management
Case Western Reserve University
Cleveland, OH 44106.

Abstract: A co-production process is one that produces several products simultaneously. Examples of co-production problems abound in the manufacturing sector. Problems similar in structure also arise in the service sector and have been referred to as yield management problems. Here the supply (e.g., number of hotel rooms) is fixed, but there are several types of customers to whom this resource can be allocated. In this paper we review the research that has been done on co-production problems in both service and manufacturing sectors. We are concerned with studies that consider variabilities in the yields. In the manufacturing sector, if yields are variable then the proportions of different products obtained in a batch are random variables. In the service sector variability in yields is due to the uncertainty in the mix of customers who will actually consume the resource.

1. Introduction: Traditional production and inventory models assume that, although demand may be stochastic, production is completely deterministic. Given the order quantity, the fraction of non-defective output is known. These

models clearly have broad applicability, but are not appropriate for processes which are characterized by large amounts of yield variability.

Because the industries involved in emerging technologies are notorious for having large and variable yield losses, a great deal of attention has recently been paid to the subject of uncertain yields in manufacturing. In the "high tech" industries, rapid product and process innovation often result in large yield losses. In the manufacture of certain semi-conductors, it is not unusual for a process to average over 50% defective parts. The processes in this industry tend to be extremely sensitive to changes in temperature, cleanliness, purity of materials, etc. Slight variations can have large effects upon the performance characteristics of the finished product. Since the rate of innovation is high, processes often become obsolete before the relationships between parameter settings and defect rates are well understood. Thus, there is often a great deal of uncertainty associated with the number of units that will be defective in a given process batch.

The subject of uncertain yields in manufacturing has been approached from several perspectives, including: system design, product design, production and inventory policy, and quality control. Most of the research is focused on processes that produce units of a single product which are either good or bad. Yano and Lee (1989) provide a comprehensive summary of the work that has been done in this area.

There are many situations where the "yield" from a process is more than simply a one-dimensional measure of the fraction defective. For example, if products are defined according to some quantitative measure of performance, then, due to variability of the process, individual units in each batch can have different characteristics and be classified as different products. We refer to processes that produce multiple products simultaneously as "co-production" processes.

Examples of multi-product co-production abound. Semi-conductor chips are produced in wafers. Each wafer contains several chips. Often chips from the same wafer may exhibit different electrical properties and may be classified as different products (e.g., Leachman 1986, Bitran and Dasu 1989). In fiber-optics, fibers are sold on the basis of length. During the manufacturing of these fibers, they are drawn out until they break. A long fiber can substitute for a short one, but not vice-versa. Each of these problems involves the production of a product family where there is uncertainty regarding the amounts of the individual products that will be obtained.

Models of co-production processes are closely related to a set of problems arising in the service industry known as "yield management" problems. In many service operations (e.g., hotel, airline, rental agency), the facility co-produces the availability of a perishable but renewable resource. For instance in the hotel industry a variety of types of reservations is made by customers who wish to “consume” the resource (the room) at some pre-arranged future time. The portion of customers who actually show up to claim their reservations represents the yield. Although there is no uncertainty associated with supply (the number of hotel rooms, airplane seats, rental cars, etc. is fixed), management controls the uncertainty in demand by limiting the quantity and mix of reservations that it will issue.

The purpose of this paper is to discuss the research that has been done on co-production problems in both manufacturing and services. We are primarily interested in variable yield models and restrict the scope of our survey to studies of production planning and inventory control problems. Clearly the prime objective of management should be to minimize or eliminate yield losses from a manufacturing process; nevertheless, there is considerable value in studying production and inventory models which treat yield as an exogenous variable. Yano and Lee (1989) cite three reasons for studying production and inventory models that approach yield loss as a given rather than a controllable variable: First, the results can be used to enhance short-term operational efficiencies. Second, process improvement decisions can be made more effectively if the system-wide effects of these decisions are well understood. Finally, in industries that experience high rates of technological innovation, it may be necessary to consider yield variability as a long-term parameter for purposes of production and capacity planning.

In Section 2, we discuss the multi-product co-production problem arising in the manufacturing sector. Studies in the service sector are described in section 3. Finally, in Section 4 we discuss some directions for future research.

2. Random Yields in Manufacturing

Figure 1 contains a taxonomy of random yield problems that arise in practice. Problems are first divided between those in which yield is single or multi-dimensional. In the single dimensional case, units of output are either good or bad. In the multi-dimensional case (i.e. co-production), they can be classified as one of several products depending on their level of performance. Within co-production, problems can be differentiated according to whether the various products are substitutable and whether there are alternative processes or parameter settings available to produce the products in different proportions. Although in practice problems may involve both substitutable products and alternative processes, models developed to date tend to emphasize one or the other.

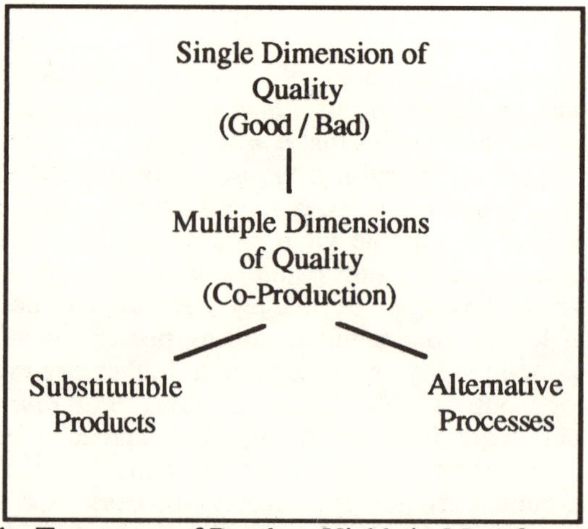

Figure 1: Taxonomy of Random Yields in Manufacturing

Given the objective of this paper we will focus primarily on co-production problems. Since yield variability is of significant concern, in the next sub-section we briefly describe different approaches that have been used for modeling variability in yields. For ease of exposition we restrict this discussion to the single dimension of quality. The models proposed for "one dimensional" processes can easily be (and have been) generalized to model co-production processes. Co-production problems will be discussed in greater detail in section 2.2.

2.1 Models for Random Yields

Yield models can be differentiated into two broad classes, depending on whether or not they explicitly refer to the state of the production process. When the state of the process is time invariant and static, the distribution of defective items in one batch is independent of the yield outcomes in previous batches. However, the quality of individual items within the same batch may be dependent.

In many models of time invariant processes, each item in a batch is assumed to be non-defective with probability p. Consequently the total number of good items has a binomial distribution. Here, both the mean and variance of the number of good items are proportional to the size of the batch. A variant of this model is one in which the expected number of non-defective items is proportional to the batch size, but the standard deviation is constant . For example, if Q units are processed, then the number of non-defectives is $pX + \varepsilon$, where p is a constant (generally between 0 and 1) and ε is a random variable with zero mean and finite variance. One of the first applications of this class of models was performed by Giffler (1960). Many others, including Silver (1976), Mazzola, McCoy, and Wagner (1987), have followed.

Under the binomial yield model there is no dependence between the quality of different items. However, in practice, there are many applications where the quality of an item is correlated with that of other units in the same batch. Albin and Friedman(1989), Singh, Abraham, and Akella (1990), Avram and Wein (1990), provide examples in the semi-conductor industry. Albin and Friedman cite empirical evidence that individual items in integrated circuit (IC) manufacturing may not be good or bad independently. They claim that a significant fraction of defects results from scratches, spattered ink, or dust particles which settle on the surface of a wafer. These "particulate defects" tend to cluster. Thus the variance in the number of defects in a given field of view (FOV), or neighborhood, tends to be high. If the quality of a circuit were independent of the qualities of neighboring circuits, we could expect the number of good circuits in any neighborhood to have a Poisson distribution with a squared coefficient of variation equal to 1. However, Albin and Friedman claim that in many cases this squared coefficient is close to 10. Instead of modeling defects with the Poisson distribution, they suggest the Neyman Distribution. This is equivalent to modeling the number of clusters as a Poisson random variable with mean = λ, and the number of defects in each cluster as a Poisson random variable with mean = Θ. Only when $\lambda >> \Theta$ is the Poisson a reasonable approximation to the Neyman.

An alternative model of yields in static processes assumes that the fraction of defectives in a batch is a random variable that is independent of the size of the batch. For example, if Q units are processed, then the number of non-defectives is pQ, where p is a random variable that typically lies in the range [0,1]. The mean and standard deviation of the number of good units is proportional to the size of the batch, and the yield is said to be "stochastically proportional". This model is most appropriate for batch processing applications where variations in the process parameters affect all items in a batch equally. For example, in semi-conductor wafer fabrication, the purity of materials and the level of cleanliness tend to affect all of the units in a batch equally. Stochastically proportional yield models can also be used as reasonable approximations to binomially distributed yields when batch sizes are large. Recall that in the binomial model, the variance of the number of good units is proportional to the size of the batch Q (versus the square of Q in the stochastically proportional model). Thus, when batch sizes are large, reasonable approximations to a binomial distribution can be obtained by specifying different distributions for the fraction defective in different ranges of Q.

An attractive feature of the "stochastically proportional" yield model is that it facilitates analysis. If the inventory holding and shortage costs are convex functions, then production planning problems can often be formulated as convex programs. For example, consider a variation on the traditional EOQ model in which the amount received may be less than Q, the amount ordered. Shih (1980) has shown that if the yield is stochastically proportional to Q, then the expected cost per unit time is a convex function of the order quantity Q. Thus, the optimal solution can be obtained through straightforward calculus. Even in discrete time models, if the yield can be modeled as stochastically proportional to the order quantity, then production planning problems can be formulated as convex programs (e.g., Bitran and Dasu 1989, Henig and Gerchak 1990). For example, if backlogging is permitted and there is no set-up cost, then the discrete time (periodic review) problem can be formulated as the following dynamic program:

$$f_t(I)=\min_{Q\in X} \{ h(I,Q) + E_p E_D f_{t-1}(I+Qp -D) \}, \qquad 2.1.1$$

where: $f_t(I)$: the minimum cost for a t period problem; $h(I,Q)$: expected cost incurred in the current period given that the starting inventory is I and Q units are processed; I : starting inventory level in period t; E_p (E_D) : expectation with respect to the yield (demand); and X : a convex set. If $f_0(.)$ and $h(.,.)$ are convex functions, then a straightforward inductive argument can be employed to show that $f_t(.)$ is convex for all t.

Several authors, including Gerchak et al. (1988), Lee and Yano (1988), Lee(1991), Henig and Gerchak (1990), and Gerchak and Wang (1991), have utilized the convexity of the optimal cost function to derive properties of the optimal policies. The optimal policies are characterized by a set of critical numbers. For instance, Henig and Gerchak(1990) show that for problem 2.1.1. there exists a number n_t, such that only if the starting inventory in period t is less than n_t will the optimal order quantity in period t exceed zero.

In the second broad category of yield models, the state of the process is not assumed to be constant. Instead, it is assumed that the process shifts between states, typically in a Markovian fashion. The time spent in each state is a random variable, and the likelihood of producing a defective is state dependent. Gregory and Dov (1967) propose such a model for a situation in which the setup of a machine is "bad" with some probability. If the setup is "bad", the machine produces defectives until the state of the machine is detected. State dependent process models are also appropriate for systems which are subject to deterioration. Lee (1991), Porteus (1986), and Vachani (1969) provide examples of such situations. A significant difference between the two categories of yield models is that in the second category product/process inspection decisions can also be taken into account. For example, Lee and Rosenblatt (1989) study a process where it is possible to inspect the products prior to the completion of the batch. Because inspection is costly and may provide only partial information about the state of the machine, the lot size and the inspection policy must be jointly determined.

2.2. Multi-Product Co-Production

The yield of a process cannot always be completely characterized by the quantity of non-defectives. Often the individual units in a production batch demonstrate different performance levels and may be viewed as different products. In some cases these products may be substitutable. As stated earlier (Figure 1), models of co-production problems in manufacturing can be separated according to whether they consider alternative process parameters and whether they consider substitutability among different products.

Consider first a process that can be operated with only one parameter setting. Each unit of operation of the process in period t results in $p_{i,t}$ units of product i, i =1,...,n. Such a co-production problem was first described by Evans (1969) who studied an n-product "by-production" process. He assumed that the demands were stochastic, the "production coefficients" $p_{i,t}$ were deterministic, a set-up cost was incurred if an order was placed, single period holding and backorder costs were convex in the inventory position, and the products were not substitutable. Under these assumptions he showed, for the backlogging case, that the optimal ordering policy is a generalization of Scarf's (s,S) policy. (As is the case for independently produced products, the problem becomes much more difficult to analyze under lost sales than under backlogging.) Let $Q_n(I)$ be the optimal order quantity in period n when the starting inventory vector is **I**. Then Evans shows that: $Q_n(\mathbf{I})\mathbf{p} = Q_n^*(\mathbf{I}) - \mathbf{I}$ if $\mathbf{I} \leq Q_n'(\mathbf{I})$ and $Q_n(\mathbf{I}) = 0$ if $\mathbf{I} \geq Q_n'(\mathbf{I})$, where $Q_n^*(\mathbf{I})$, $Q_n'(\mathbf{I})$ and **p** are n-vectors and are the parameters of the optimal policy. Further $Q_n^*(\mathbf{I})$ and $Q_n'(\mathbf{I})$ are such that $Q_n'(\mathbf{I}) = Q_n'(\mathbf{I} + B\mathbf{p})$ and $Q_n^*(\mathbf{I}) = Q_n^*(\mathbf{I} + B\mathbf{p})$ for all real B. Observe that $Q_n'(\mathbf{I})$ is equivalent to s and $Q_n^*(\mathbf{I})$ is equivalent to S.

Evans assumed that the "production coefficients" can be modeled in terms of their expected values. In cases where the physics of the production process are well understood and controlled (e.g., oil refining), the coefficients of variation of the yields of the various products are low, and this is a reasonable approach. However, in other cases, the process is not so well understood and variability cannot be ignored. Such situation are often encountered in the semi-conductor industry.

In practice, when products are differentiated on the basis of quality, there often exists a hierarchy of substitutability. That is, customers will be satisfied with a specified level of performance or better. They are perfectly happy to receive a better product than they have specified. Managers must decide when to satisfy unmet demand with higher quality products and when to simultaneously hold high quality items in inventory and low quality items in backlog. When we have co-production with substitutable demands, two sets of decisions have to be made. In addition to determining the quantity to be processed, management must also determine how to allocate the available products to different customers.

The question of determining a down-grading policy in the presence of co-production and stochastic yields was first addressed by Bitran and Dasu (1989). Their work was motivated by a process for manufacturing semi-conductor devices that has nested performance specifications. This creates a situation in which product i can be substituted to satisfy demand for product j for $j \geq i$. The inventory flow balance constraints for a 3-product version of the model are depicted in Figure 2. Note that this model is similar to the one studied by Evans except for the presence of the $W_{i,t}$ arcs, which represent the downward substitution of product i for i+1.

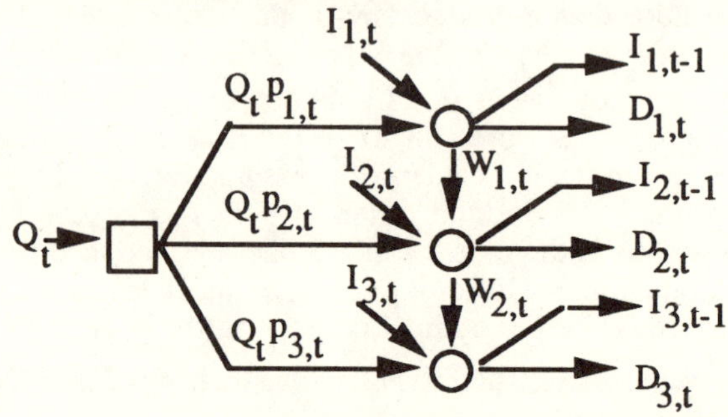

Figure 2 - Inventory Balance with Downward Substitution

Bitran and Dasu modeled the problem as a dynamic program in which there are two stages for every production period. In the first stage (morning stage) the quantity to be processed is determined and in the second stage (afternoon stage), after the yield outcomes are observed, the items are allocated to the customers. The demands were assumed to be deterministic but time varying. A formulation of this problem under the assumption that the lead times are zero is given below.

<u>Morning Problem</u>:

$$M_t(\mathbf{I}_t) = \text{Min } \{cQ_t + E_{\mathbf{p}}[A_t(\mathbf{I}_t, Q_t, \mathbf{p})] \mid Q_t \geq 0\}$$

<u>Afternoon Problem</u>:

$$A_t(\mathbf{I}_t, Q_t, \mathbf{p}) =$$

$$\text{Min} \quad \sum_{i=1}^{n} h_i I^+_{i,t} + s_i I^-_{i,t} + \beta M_{t-1}(\mathbf{I}_{t-1})$$

Subject to: $I_{i,t} - I_{i,t-1} + p_i Q_{i,t} + W_{i-1,t} - W_{i,t} = d_{i,t}$ for t = 1,.T, and i =1,n.

$$I^+_{i,t} = \text{Max}(I_{i,t}, 0), \quad I^-_{i,t} = \text{Max}(-I_{i,t}, 0), \quad W_{i,t} \geq 0 .$$

The morning problem, $M_t(\mathbf{I}_t)$ in period t, involves the selection of a lot size to minimize the cost of production which is equal to cQ_t plus the expected cost of an optimal solution to the afternoon problem $A_t(\mathbf{I}_t, Q_t, \mathbf{p})$. Note that vectors are indicated by bold face and expectations are taken with respect to the yield vector $\mathbf{p}$. After the process is run, the yield of each product is observed. The decisions (W_{it} ; i=1,...,n-1) in the afternoon problem of period t are the numbers of units that should be downgraded from product i to i+1, for i = 1,...,n-1. Here s_i and h_i denote the per unit costs of backordering and holding product i for one period. ß represents the per period discount factor.

If the yield outcomes can be modeled as discrete random variables with a finite number of outcomes, then the finite horizon problem is a linear program. However, the size of the program grows vary rapidly in the number of periods. Hence, approximation procedures become necessary. One approach is to aggregate the yield outcomes in all but the first few periods and to solve finite horizon problems on a rolling horizon basis. Further computationally efficiencies can be gained by exploiting the embedded network structure of the coefficient matrix. Nevertheless, since we are confronted by a computationally burdensome problem, there is a need to gain further insight into the structure of the problem. For this purpose Bitran and Dasu analyze a two period, two product problem and observe that the optimal down-grading policy has a simple intuitive structure. In the penultimate period it is optimal to down-grade item 1 if the net demand for that item in the final period is less than a critical fraction of the total net demand in the final period. Based on this observation they propose a class of heuristics. The heuristics consist of two modules that correspond to the morning and afternoon problems. The quantity decision is made such that the demands in the next period are satisfied with some probability, and the allocations are made such that the demands for the higher grade items in the next period do not exceed a critical portion of the total demand. An example of heuristic belonging to this class is:

Production Decision: $Q_t = \text{Max}_i \{ \sum_{j=1}^{i} d^*_{t,j} / f_i \}$

Allocation Decisions: Continue to allocate any excess of item k to customer k+i provided :

$$\sum_{j=1}^{m} d^*_{t-1,j} / d'_{t-1} \leq \theta_m, \text{ for } m = k \text{ to } k+1-i,$$

where $d^*_{t,j}$ is the net demand for item i, d'_t is the total net demand in period t, and f_i and θ_m are parameters of the heuristic.

Clearly the performance of the heuristic will depend considerably on the procedures employed to determine the parameters -- f_i and θ_m . Note that in the example given above the quantity and allocation decisions are made taking into account only the demand in the next period. Also, the production decision is made so as to satisfy a service level constraint under the assumption that all excesses are downgraded. By altering (i) the horizon over which the demands are kept in balance, (ii) the procedure employed to define the service level constraint, and (iii) the procedure employed to determine the parameters, several heuristics can be developed for this class of problems. For further details the reader is referred to Bitran and Dasu (1989), Bitran and Leong (1990), and Bitran and Gilbert (1992).

One of the problems encountered while developing and testing approximations procedures is that it is difficult to get good bounds on the optimal solution value. The primary contribution of Bitran and Gilbert is a bound on the value of an optimal solution to the dynamic program and the development of a set of heuristics. The heuristics determine production quantities and substitutions on the basis of fractiles of the yield distributions. The bound that Gilbert and Bitran derive is based on the idea that a decision maker who knows all of the random outcomes a priori can make better decisions than one who must make decisions before observing the uncertain events. The value of this bound is that it can be used as a benchmark to compare the performance of heuristics. Computational experiments indicate that the proposed heuristics perform well vis-a-vis the bound for a broad range of parameters.

Bitran and Leong (1991) approximate the yield distribution with a multinomial, continuous distribution. They construct a large, non-linear mathematical program that has constraints of the form: Prob $[I(i,t) \leq 0] \leq \alpha_{it}$, where α_{it} is a fraction that is close to unity. Because these service level constraints are non-linear, Bitran and Leong focus their efforts upon developing linear approximations to them. They also propose and test a set of heuristics which are obtained from solutions to the linear approximations. These heuristics have the same general structure described above.

While the work of Bitran and co-authors focuses on discrete time models, Gerchak and Wang (1991) analyze a two product version of the substitutable co-production problem in the context of an EOQ model. Under the assumption that surplus of the high grade product is always used to reduce shortages of the lower grade, they show that when the yields are stochastically proportional to the batch size, the expected cost function is convex. Thus, they are able to specify the optimal ordering policy. However, the strength of their results is limited by the fact that it is not necessarily optimal to automatically "downgrade" surpluses of the higher grade to reduce backorders of the low grade.

Let us now turn to the second class of co-production problems, where there are multiple process alternatives but the products are not substitutable. Often, when co-production arises in practice, the multivariate distributions for the yields are dependent upon the process parameters. Each set of parameters defines a different process and implies a different distribution for the yields. The level at which each process is to be run depends upon requirements of each product. Thus, it is necessary to decide not only how much to produce, but which of the alternative processes to run.

Such a system was first modeled by Deuermeyer and Pierskalla (1978). They study a production system which involves two alternative processes -- process A and B. Each unit of input to process A results in p_1 and p_2 units of products 1 and 2 respectively. Process B produces product 2 only. Two industrial examples of such a system are cited. The first is the manufacture of transistors, which can be classified into "bin-splits" according to quality -- p_j would represent the fraction of a batch of transistors which fall into bin j. The second example is oil refining where, for each barrel of crude, p_j units of hydrocarbon j are produced. In their analysis, Deuermeyer and Pierskalla model an n period problem with stochastic demand. As in the work of Evans, they use expected values for the production coefficients p_1 and p_2. For this version of the problem, they derive an optimal policy that is defined in terms of four regions on a plot of the inventory vector. Depending upon the inventory position in the current period, it is optimal to operate: neither process, A only, B only, or both A and B.

Motivated by problems in the semi-conductor industry, Ou and Wein (1991) study a problem where the yields are stochastic. They consider n processes, each of which is targeted at producing a given product in a hierarchically ordered set of products but may produce lower level products with some probability. Yields are modeled with a generalized binomial distribution. That is, a unit of output from process k will be a type i product with probability $p_{i,k}$, where $\sum_{i=1}^{n} p_{i,k} \leq 1$ for k = 1,...,n. Thus, the variance of the number of units of product i is proportional to the total number of units that have been processed. The approach of Ou and Wein is to approximate each inventory process ($I_{i,t}$) by Brownian motion, and the production problem is modeled as a Brownian control problem. The decision variables are the cumulative amount of time each process is operated and the cumulative idle time for the machine. Ou and Wein derive optimal control policies for this approximation, and use them as the motivation for developing heuristics. The policy dynamically tracks the weighted inventory process, which represents the machine time that is already invested in the inventories. A critical level is determined. When the weighted inventory process exceeds this level, the machine is idled. Otherwise the machine is allocated to a process that is targeted at one of the products that is in danger of being backordered. If multiple items are in danger of being backordered, then the selection of the best process is based on the ratio of the backorder costs to the expected time to service completion.

3 Co-Production in the Service Industry

Another application of multi-product co-production manifests itself in the service industry. Airlines, car rental agencies, hotels, etc. sell the availability of renewable resources to their customers, often by accepting advance reservations. A reservation provides a customer with the right to purchase the availability of an airline seat, rental car, hotel room, etc. during some pre-specified period in the future. When the service products are differentiated, the reservation process is a form of co-production. The service facility "co-produces" the availability of a family of differentiated products. The policies that guide the reservations process have significant bearing upon the profits of the service facility.

As shown in Figure 3, the simplest reservations management problems involve only a single type of reservation. When a variety of types of reservations is involved, the problems can be classified as co-production. Reservations can be differentiated in several ways: 1) Timing of reservation requests. Often prices vary according to how far in advance a reservation is booked; 2) Type of facility. Depending upon the intended use of the resource, different reservations may specify different physical characteristics, e.g., a one bed versus a two bed hotel room; 3) Duration of use. Different customers may require access to the resource for different lengths of time; 4) Arrival behavior. Customers with different types of reservations may demonstrate different behavior in terms of their arrival at the service facility. For example, customers who have non-refundable airline tickets are much more likely to show up for their flight than those who do not; Hotel customers whose reservations expire at 6 p.m. tend to arrive earlier in the day than those who have guaranteed reservations.

Depending on the specific reservations problem, the relative importance of these four dimensions of product differentiation may differ. For example, in the airline industry, the timing of reservation requests is frequently the primary basis of product differentiation; discount fares are offered to passengers who are willing to book reservations far in advance of the date of travel. The problem is to determine limits on the sale of inexpensive tickets in order to maximize the expected revenue from a given flight. The solution requires trading off higher revenue sales against the likelihood of unsold tickets. In practice these decisions have to be made dynamically taking into account the reservations that have already been made and the likelihood of some these reservations being cancelled or not being honored.

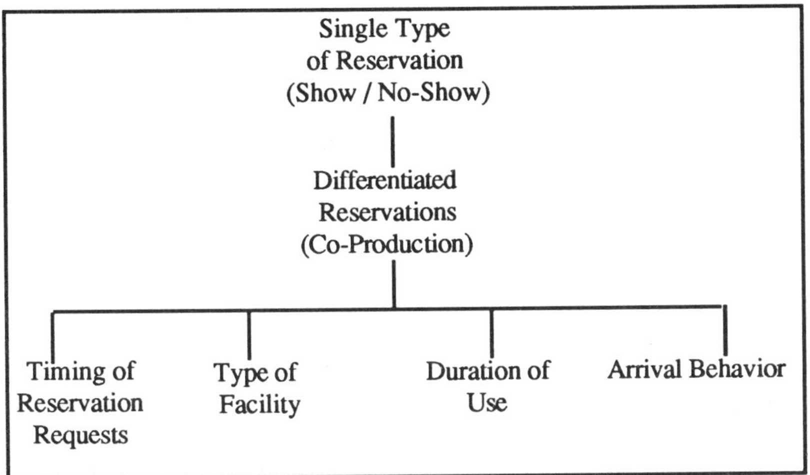

Figure 3: Taxonomy of Random Yields in Reservations Planning

Rothstein (1974) and Belobaba (1989) provide detailed reviews of mathematical approaches to this problem. Several authors (e.g., Littlewood 1972, Belobaba 1987, Brumelle et al. 1990) have taken a marginal pricing approach. In these models, it is generally assumed that low fare classes purchase before high ones. The idea is to "protect" seats from sale to a given fare class i in such a way that the marginal revenue (with respect to the number of seats withheld) from sales to higher paying classes is exactly equal to the fare for class i. However, the emphasis in these research efforts is on the problem of allocating a fixed number of tickets to random demands for different fares. The impact of random cancellations and no-shows receives little attention.

In certain situations management can either increase or decrease the availability of certain types of reservations in response to the relative rates at which they are demanded. For example, if demand is low, managemnt may increase the availability of inexpensive reservations. Liberman and Yechiali (1978) study an instance of this problem that they have observed in a hotel. At each decision point, management can either cancel previously confirmed reservations, or acquire additional bookings at costs (by providing travel agent discounts) that depend upon the time remaining prior to the date of reservation. The objective of Liberman and Yechiali's analysis is to develop a policy that maximizes the expected profit on a given reservation date. Although they consider random cancellations and no-shows, they make the simplifying assumption that the distribution of the fraction of reservations which cancels or fail to show up in any given period is independent of the total number of reservations. Thus the expected profit function is concave in the number of reservations, and the optimal policy in each period is characterized by two critical values. Depending upon the number of current reservations, management should either increase/decrease the level of bookings, or do nothing. The primary contribution of Liberman and Yechiali's work is to provide a quantitative framework that conceptualizes the management of a spectrum of types of reservations that are differentiated according to the time of booking.

In many situations, reservations can be differentiated according to the type of facility that is required. The facilities may differ in terms of levels of luxury or performance, e.g., single versus double hotel rooms. Products that are differentiated in this way are often substitutable. For example, a coach class passenger will gladly accept a larger, more comfortable first class seat; a business traveler with a reservation for a single hotel room is happy to stay in a double.

Instances of these problems have been modeled as dynamic programs in which the stages are the time periods preceding the date of reservation, and the states are the numbers of each reservation type. In each stage, a random number of reservation requests and cancellations is received. The controls are the limits on the number of each type of reservation to accept. Examples of such models are provided by Ladany (1976), who examines a hotel with single and double rooms, and Alstrup et al. (1986), who studies an airplane with luxury and economy seats. Both analyses recommend solving the model through some form of enumeration.

Another way in which reservations may be differentiated is according to the duration of use. The two most common manifestations of this type of differentiation are: multiple day stays in hotels and multiple leg flights in airlines. The problem arises when demand is higher for one day (flight segment) than for others. Given that demand for reservations exceeds capacity on a particular day (flight segment), the service facility can increase the utilization on adjacent days (flight segments) by giving preference to customers who request reservations for multiple days (flight segments).

Shlifer and Vardi (1975) study a situation in which an airline must take reservations for a two leg flight. Reservations are taken for three product packages: segment 1, segment 2, and both segments. Their results are based on normal approximations of the distributions of cancellations and no-shows. Given the revenues associated with each, as well as the penalties for denied boardings, they show that a revenue maximizing policy can be expressed in terms of fractiles of a standard normal distribution. Depending upon the current bookings, the optimal policy is to accept reservations for both segments, segment 1 only, segment 2 only, or neither. However, because the model does not consider the distribution of future demand for reservations, the policy may be myopic. For example, one would expect that a good policy would be less willing to accept type 2 reservations if the demand for type 1 reservations is expected to be high.

The final means of differentiating among service reservations that we have identified involves the behavior of the arrivals of customers on the date of reservation. One example is the probability of showing up. The probability that a customer will fail to show up can depend on the type of reservation held. An airline passenger with a non-refundable ticket is much more likely to show up than one with an unrestricted ticket; a hotel customer who has guaranteed a reservation with a credit card is more likely to show up than one who has not.

An instance of this problem in an airline booking system is identified by Shlifer and Vardi (1975). They model two types of customers who fail to show with different probabilities. In parallel to their study of multi-leg flights, their analysis is based on normal approximations of the distributions of cancellations and no-shows. Again, the optimal policy is to increase one or the other type of reservation, neither, or both.

Another way in which reservations may be differentiated according to the behavior of customer arrivals involves the timing of those arrivals. A common link between most of the research that has been done in reservations planning is the assumption that all customers arrive simultaneously on the pre-arranged reservation date. Although this may be valid in the airline industry, where passengers are required to arrive at the gate 30 minutes prior to boarding, it is not be appropriate for hotels. In practice, the arrival of customers to a hotel is dispersed over several hours of the afternoon and evening, and different types of customers tend to arrive earlier or later than others. Since different customer types pay different prices, and show up with different probabilities, the problem of allocating rooms to customers as they arrive may not be trivial. Bitran and Gilbert (1992b) specifically address this problem. They point out that not only do different types of customers tend to arrive during different periods of the day, but that the costs of failing to honor a reservation increase late in the evening. Thus, there may be situations in which it is advantageous for a hotel to "walk" customers early in the evening in order to decrease the likelihood of having to do so later on. Bitran and Gilbert consider three types of customers, walk-ins, guaranteed reservations, and 6 p.m. hold reservations. Modeling the room allocation problem as a stochastic dynamic program, they derive a policy that is optimal with respect to an expected revenue objective function. Based on the insights gained from this analysis, they develop heuristic procedures for accepting reservations.

4. Conclusions and Directions for Future Research

The class of problems that we have identified is held together by the fact that they all involve the allocation of a simultaneously produced, and possibly substitutable, family of products to a set of customer types in the presence of uncertainty. Examples of these problems can be found in both services and manufacturing.

Future research could be focused on relaxations of two common assumptions: 1) Downgrading decisions do not affect demand. 2) Yield distributions are known. In practice, substituting high grade products to satisfy demand for lower grades may tend to shift demand toward the lower grades in the long run. This is most likely to be the case when one customer purchases both grades and the client can easily determine when a high grade item has been downgraded. Understanding the long run effects of downgrading upon demand for the high grades could have significant impact upon the way in which these decisions are made.

The second assumption, that yield distributions are known, is often not valid in practice. Since random yield problems tend to arise in processes which are subject to high rates of technological innovation, there is often inadequate data to obtain a good characterization of the yield distribution. Each production run provides valuable information about the underlying distribution. Incorporating the value of this information into a lot sizing model would be useful for dealing with nascent processes.

REFERENCES

Albin, S. L. and D. J. Friedman, "The Impact of Clustered Defect Distributions in IC Fabrication," Management Science, Vol 35, No. 9, pp. 1066-1078, 1989.

Alstrup, J , S. Boas, O. B.G. Madsen, and R. V. Valqui, "Booking Policy for Flights with Two Types of Passengers," European Journal of Operations Research, Vol. 27, pp. 274-288, 1986.

Avram, F., and L. Wein, "A Product Design Problem in Semi-Conductor Manufacturing," Working Paper, MIT Sloan School of Management, 1990.

Belobaba, P.P., "Application of a Probabilistic Decision Model to Airline Seat Inventory Control," OR Practice, Vol. 37, No. 2, pp. 63-73, 1987.

Belobaba, P.P., "Airline Yield Management - An Overview of Seat Inventory Control," Transportation Science, Vol. 21, pp. 183-197, 1989.

Bitran, G. R., and S. Dasu, "Ordering Policies in an Environment of Stochastic Yields and Substitutable Demands," M.I.T. Sloan School Working Paper #3019-89-MS, 1989, to appear in Operations Research.

Bitran, G.R., and S.M. Gilbert, "Co-Production Processes with Random Yields in the Semiconductor Industry," Working Paper, Weatherhead School of Management, Case Western Reserve University, 1992.

Bitran, G.R., and S.M. Gilbert, "Managing Hotel Reservations with Uncertain Arrivals," Working Paper, Weatherhead School of Management, Case Western Reserve University, 1992b.

Bitran, G. R., and T.Y. Leong, "Deterministic Approximations to Co-Production Problems with Service Constraints," M.I.T. Sloan School Working Paper #3071-89-MS, 1990.

Brumelle, S.L., J.I. McGill, T.H. Oum, K.Sawaki, and M.W. Tretheway, "Allocation of Airline Seats Between Stochastically Dependent Demands," Transportation Science, Vol. 24, pp. 183-192, 1990.

Deuermeyer, B. L. and W. P. Pierskalla, "A By-Product Production System with an Alternative," Management Science, Vol. 24, No. 13, pp.1373-1383, 1978.

Evans, R., "Inventory Control of By-Products," Naval Research Logistics Quarterly, Vol. 16, No. 1, pp.85-92, March 1969.

Gerchak, Y., R.G. Vickson, and M. Parlar, "Periodic Review Production Models with Variable Yield and Uncertain Demand," IIE Transactions, Vol. 20, No. 2, pp. 161-177, 1988.

Gerchak, Y., and K. Wang, "Co-production Models with Random Functionality Yields," Working Paper, Department of Management Science, University of Waterloo, Waterloo, Canada, 1991.

Giffler, B., "Determining an Optimum Reject Allowance," Naval Research Logistics Quarterly, Vol. 7, pp. 201-206, 1960.

Gregory, W. R. and A. B. Dov, "On the Determination of Optimal Shrinkage Allowance in a Job Shop," The Journal of Industrial Engineering, Vol. 18, No. 4, pp. 284-288, 1967.

Henig, M. and Y. Gerchak, " The Structure of Periodic Review Policies in the Presence of Random Yield," Operations Research, Vol. 38, No. 4, pp. 634-643, August, 1990.

Ladany, S., "Dynamic Operating Rules for Motel Reservations," Decision Sciences, Vol. 7, pp. 829-840, 1976.

Leachman, R.C., "Preliminary Design and Development of a Corporate-level Production Planning System for the Semiconductor Industry," Working Paper, ORC 86-11, University of California, Berkeley, 1986.

Lee, H. L., "Lot Sizing in a Production Process with Defective Items, Process Corrections, and Rework," Stanford University Working Paper, Department of Industrial Engineering and Engineering Management, August, 1991.

Lee, H. L. and C. A. Yano, "Production Control in Multi-Stage Systems with Variable Yield Losses," Operations Research, Vol. 36, No. 2, pp. 269-278, 1988.

Lee, H. L. and M. J. Rosenblatt, "Simultaneous Determination of Production Cycle and Inspection Schedules in a Production System," Management Science, Vol. 33, No. 9, pp. 1125-1136, September, 1987.

Liberman, V. and U. Yechiali, "On the Hotel Overbooking Problem - An Inventory System with Stochastic Cancellations," Management Science, Vol. 24, No. 11, pp. 1117-1126, 1978.

Littlewood, K., "Forecasting and Control of Passenger Bookings," AGIFORS Symposium Proceedings, Vol. 12, pp. 95-117, 1972.

Mazzola, J. B., W. F. McCoy, and H. M. Wagner, "Algorithms and Heuristics for Variable-Yield Lot Sizing," Naval Research Logistics Quarterly, Vol. 34, No. 1, pp 67-86, 1987.

Ou, J., and L. M. Wein, "Dynamic Scheduling of a Production/Inventory System with Bi-Products and Random Yield," M.I.T. Operations Research Center Technical Report #250-91, May, 1991.

Porteus, E.L., "Optimal Lot Sizing Process Quality Improvement and Setup Cost Reduction," Operations Research, Vol. 34, No. 1, pp. 137-144, 1986.

Rothstein, M., "Hotel Overbooking as a Markovian Sequential Decision Process", Decision Sciences, Vol. 5, pp. 389-404, 1974.

Shih, W., "Optimal Inventory Policies When Stockouts Result from Defective Products," International Journal of Production Research, Vol. 18 No. 6 pp. 677-686, 1980.

Shlifer, E. and Y. Vardi, "An Airline Overbooking Policy," Transportation Science, Vol. 9, pp. 101-114, 1975.

Silver, E. A., "Establishing the Re-Order Quantity When the Amount Received is Uncertain,"," INFOR, Vol. 14, pp. 32-39, 1976.

Singh, M., C. Abraham, and R. Akella, "A Wafer Design Problem in Semiconductor Manufacturing for Reliable Customer Service," IEEE Transactions of CHMT, Vol. 13, No. 1, March 1990.

Vachani, M., "Determining Optimum Reject Allowance for Deteriorating Production Systems," Naval Research Logistics Quarterly, Vol. 16, No. 3, pp. 275-286, 1969.

Williams, F. E. "Decision Theory and the Innkeeper: An Approach for Setting Hotel Reservation Policy," Interfaces, Vol. 7, No. 4, pp. 18-31, 1977.

Yano, C. and H. L. Lee, "Lot Sizing with Random Yields; A Review," Working Paper, The University of Michigan, Department of Industrial and Operations Engineering, 1989.

STOCKLESS AND FAST PRODUCTION: REVIEW AND RESEARCH AGENDA

L. Joseph Thomas
Johnson School of Management
Cornell University

Abstract

Stockless Production (or *Just-in-Time* or *Zero Inventory*) and *Fast Production* (or *Time-Based Competition* or *Competing Against Time*) are important topics in the practice of operations management and in the fields of *Operations Research* and *Management Science* (OR/MS). Both topics will be reviewed here. There are three main goals for this paper:

1. Present a current (but not exhaustive) list of research references on these topics as of January 1, 1992;
2. Discuss the importance of process design as well as control in Stockless and Fast Production; and
3. Give opinions about the value of research to date and about the research that needs to be done.

I. Introduction

It is a pleasure to contribute to a conference and volume in honor of Elwood Buffa. Professor Buffa has positively affected the field of operations management and countless students, now managers. In this paper, I will discuss one important way in which we need to relearn a message that Professor Buffa has been delivering in his books for over 30 years.

I will discuss two "hot" topics in operations management: *Stockless Production* (or *Just-in-Time*) and *Fast Production* (*Time-Based Competition*). These two concepts are important (in my opinion), related (to an OR audience at least), and easy to explain. I will examine the research that has been done and needs to be done, and how we in the field of operations management can contribute to teaching and research of these two topics.

Stockless Production (or Just-in-Time) is described in hundreds of references, including Schonberger (1986), and McClain et al. (1992). Where it is used and how to implement it are discussed, for example, in Billesbach (1991), Inman and Mehra (1991), and Zipkin (1991). The basic idea is that if variation in production time can be driven out (including driving setup time to zero), no inventory is needed in the production process, and production can be "pulled" through (quickly) by demands. The benefits in practice can be dramatic, as discussed in the above references. Research questions include: (1) exactly how to manage JIT when some variation remains, (2) how much reduction in variation (including setup time reduction) is cost effective to purchase, and (3) what competing concepts for controlling production may be effective in some situations?

Fast Production (Time-Based Competition) is described by dozens of references, including Stalk and Hout (1990) and Blackburn (1990). Time-Based Competition goes by many other names, and includes distribution, design, and some other areas as well as production. Both competitive advantage and cost reduction can be obtained. Research questions include (1) how best to relate production with other areas in order to reduce time and serve the customer (this is both an O.R. and a behavioral science research area), (2) how to improve control techniques to reduce information and decision lags, and (3) the relation among production decisions such as lot size, sequencing, and inventory and the effect on response time.

Several issues should be faced head on. Is operations research (O.R.) obsolete? (My answer is no, I hasten to add.) Does O.R. research tackle issues such as Stockless and Fast Production by trying to find ways to make them sufficiently hard to understand that articles can be accepted in prestigious journals. (There may be a grain of truth in that one, but I think that important problems exist that can best be analyzed using O.R. tools.) In our teaching, are we reduced to a "set of bullets" as the most important items? (The bullets are important, but some underlying principles are even more so.)

White (1991) gives four components of necessary and sufficient conditions for the existence of OR/MS. I agree with all four, but the first two are most relevant for this discussion. The first two are: (1) solve real-world problems, and (2) employ cross-disciplinary, team-oriented approaches. I believe that Stockless and Fast Production offer an opportunity to fulfill both of these. The problems are from the real world (although the research is not likely to have the impact that the basic concepts, derived from practice, have had); the problems require cross-functional approaches in practice; and it may be possible to do good cross-disciplinary research on them.

Wagner (1988) argues that O.R. has been successful because useful concepts have been added to the business planning language. I agree, and I believe that Stockless and Fast Production (SAFP) offers an opportunity to develop new concepts that enhance both our understanding of and the usefulness of the basic ideas. In this paper I will propose a few such concepts.

How important is SAFP as viewed by managers? Miller and Kim (1990) and Miller et al. (1991) present comparative data across geographic regions, showing what the top strategic imperatives are for manufacturing, as viewed by manufacturing executives. A top priority in all regions is quality. Stockless Production is thought to enhance quality, because of the fast feedback that is available. Also, Miller and Kim (1990) say that managers must be prepared to move "Beyond the Quality Revolution," to goals such as fast production. And on-time delivery (related to but not the same as fast delivery) is the second priority in the United States, Europe, and Japan, according to Miller et al. (1991). Delivery speed makes the top five only in Europe, apparently because there is a great opportunity to improve speed after December 31, 1992, when the E.E.C. officially becomes one market.

The outline of this paper is as follows. Section II analyzes design and control issues in SAFP. Section III presents some research on Stockless Production and alternatives thereto. Section IV examines research in Fast Production. Finally, Section V gives my opinion about research and teaching agenda related to SAFP. Appendix I gives a mnemonic device to help in remembering the key components of Just-in-Time.

II. Design Versus (?) Control

Buffa (1961) emphasized the need for operations managers to focus on design issues as well as control. In the intervening 31 years, production research in O.R. journals has focused largely on control issues, occasionally solving the wrong problem.[1] Many people are again paying attention to design issues, and to coordinating design and control, relearning the lessons of Buffa (1961) and others.

Design occurs in many forms. Product design should focus not only on customer needs, but also on strategic imperatives such as speed of response and ease of manufacture. Our focus is on process design, and this also occurs in many forms. In an overall sense, process design must be tied to strategic

---

1 In addition to Type I and Type II errors, Type III and IV errors are important in manufacturing systems. The Type III error is solving the wrong problem, and the Type IV error is solving the right problem, but too late to be useful.

imperatives. In "the small," process design must focus on ways to achieve those same goals, designing each work station, the tasks, the worker's range of decision, and so on, in a manner consistent with the strategic goals. Only then can day-to-day control be managed effectively, using the same strategic goals as guidelines.

Figure 1, from McClain et al. (1992), depicts the connection among Strategy, Design, and Operations (day-to-day control). The arrows go both ways because what we are able to do day-to-day should affect the strategy, as well as the reverse. Quality, broadly defined, is shown pervading all activities.

I believe that Stockless Production is a design approach more than a control approach. A manufacturing unit can operate in Just-in-Time[2] fashion only if quality of products is very high, if setup times are low, and if production times are not highly variable. Then a very simple control procedure works. In other words, change the process design and focus of the workforce until the answer we had in mind is correct, namely small batches, low inventory, and pull production; do not solve the problem, but change the problem to fit our answer.[3]

Fast Production is also attacked best at the design phase. Eliminate activities that are unnecessary and save that time, for example. Rather than controlling things as they are, change the things. Of course, the things to change must be based on strategic choices, as suggested by Figure 1, not orthodoxy. Hausman and Montgomery (1990) discuss an automobile example where money was spent chasing a strategic goal, option flexibility, that did not matter to the consumers.

In determining when to use SAFP, that is, when to change the process so that SAFP can be achieved, some managers have used what Nadler (1991) and others call "Breakthrough Thinking." Nadler gives six contrasting characteristics of "Breakthrough" and "Conventional" thinking in design. Three of those six are:

Breakthrough	**Conventional**
Purpose oriented to find right problem	Technique oriented for problem as stated
Copes with soft data	Insists on hard data
Tolerant of ambiguity	Insists on firm statements of work and specifications

---

2 Here, I use Just-in-Time in the more restricted sense of JIT control; that is, using kanban control or some other method, not JIT in the sense that some people use, where JIT equals manufacturing excellence.

3 This could get managers in trouble if the "problem" cannot be changed sufficiently. Perhaps we should have a Type V error: using a predetermined answer, when you have not been able to make the question match that answer.

O.R. approaches seem to be strongly on the "conventional" side. SAFP, used appropriately, seems to fit "breakthrough" thinking better. "Purpose oriented" – Find the way to meet customer needs better, rather than find the "optimal" schedule for the current system. "Soft Data" – How much will we save? Who knows; let's find out.

I have argued that SAFP lies in the design realm first, control realm second. Further, while SAFP seems to be exactly the kind of issue that would have excited the cross-disciplinary groups that successfully began O.R. in the 1940s, much of current O.R. research seems ill-equipped to deal with SAFP. In spite of that negative view, I will argue in the next two sections that many good pieces of research have been done on SAFP by O.R. professionals and that much interesting and useful research remains to be done.

Figure 1. * Integrating Strategy, Design, and Operations

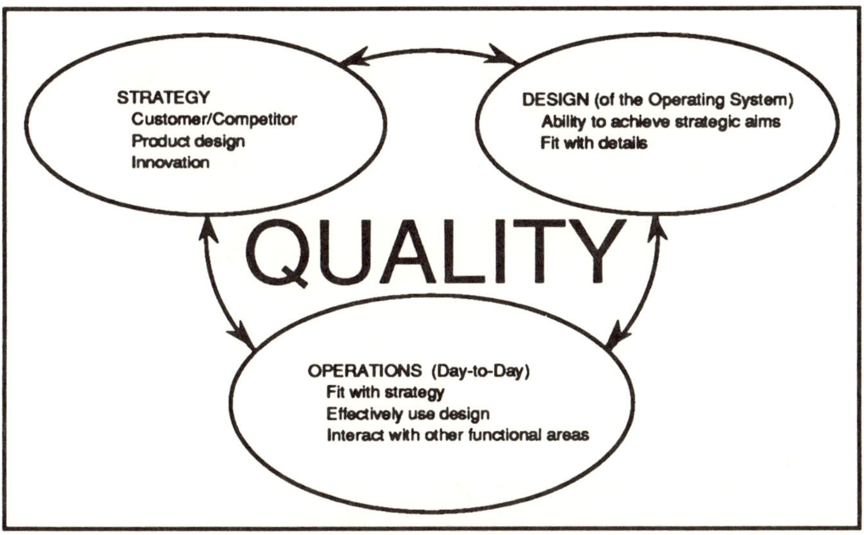

* From McClain, J., J. Thomas, and J. Mazzola, *Operations Management: Production of Goods and Services*, Englewood Cliffs, NJ, Prentice-Hall, 1992, 3rd ed.

III. Stockless Production

Stockless production (SP/JIT) means everything from having very small inventories to manufacturing excellence. Most definitions include some design aspects, in which firms continually improve the production system in order to reduce inventory while improving cost and customer service. For this discussion, I will use SP/JIT to denote the more narrow concept of reducing inventory and using some "pull" control technique. SP/JIT has been successful in practice and that success has caused research. I will examine only a few of the many papers in this area and give opinions about research that needs to be done. Groenevelt (1992) gives a detailed review of the O.R. literature related to JIT.

One major category of research is: When should SP/JIT be used? (Some practitioners I know consider that a dangerous question to ask, let alone answer.) Papers such as Krajewski et al. (1987) address the issue. The answer, of course, is: Use SP/JIT when the system can be changed enough (lower setup times and costs, reduced number of components, reduced variation of several types) to allow fast service with less inventory. (For those who think of SP/JIT as the process of making those changes, rather than some resulting method of inventory control, SP/JIT should be used everywhere.) Spence and Porteus (1989) analytically study the amount of process change (using setup cost as the item to change) to purchase. Several authors give opinions about how or when MRP and SP/JIT can coexist, including McClain et al. (1992) and Karmarkar (1989).

Deleersnyder et al. (1989) discuss three levels of problems and, thus, research areas associated with SP/JIT: (1) identification of flow lines, or designing the factory, perhaps using cellular manufacturing, (2) flowline loading, where a planning system gives a (hopefully feasible) load to each flowline, and (3) operational control, perhaps using kanbans. Deleersynder et al. (1989) analyze problem three in the presence of two sources of variation: machine reliability and demand. Hopp and Spearman (1991) also examine the case of machine failure, using a "CONWIP" approach, which is described below.

The appropriate control mechanism and parameters to use are likely to be different when production time variation is the main source of variation. (Of course, machine failure is just an extreme form of production time variability.) No matter what the source of variation, one area in which O.R. has contributed and can continue to do so is determining how small the inventories in SP/JIT should be. (Everyone knows that zero inventory is inappropriate; the question is how small is small?)

Some variation will always remain in manufacturing. The main reason for this is that once someone understands exactly how to make something (particularly a high-tech product), someone else will want to make a product that delivers more function and that is, therefore, harder to manufacture. Once we drive variation out, we will introduce more variation.

When production time variation is dominant, Conway et al. (1988) use simulation to show that small inventories suffice even for large coefficients of variation, using serial production lines. Suggested storage capacities are also given. They also discuss buffer placement (the middle of the line is often best), as do Hillier and So (1991). Baker et al. (1990) generalize the work to assembly systems. Glasserman and Yao (1991) analytically study buffer allocation in production lines, and they are able to characterize the optimal allocation under some conditions. Several other authors, including Buzacott (1989), Mitra and Mitrani (1990, 1991), Zipkin (1989), and references in all of the papers mentioned here study buffer allocation when production time is variable.

Different control mechanisms for managing the small inventories are discussed in several papers. Denardo and Tang (1988) use a periodic review model and develop target inventories for buffers. Woodruff et al. (1990) describe CONWIP, a method for maintaining constant work-in-process inventory in portions of a production line. Muckstadt and Tayur (1991) study conditions under which either kanban or CONWIP control are efficacious, and also show that CONWIP and kanban control are the extreme cases of a finite set of control strategies of that category.

SP/JIT should be implemented with employee involvement, utilizing ideas for change from all employees. Ostolaza et al. (1990) analyze ways to expand job definitions by allowing adjacent workers in a serial production line to help each other, balancing production "on the fly." This concept is studied in situations with very little buffer space and first-come, first-served buffers by McClain et al. (1991). Work is traded between adjacent workers to keep each individual buffer near a target level. While the concept is similar to CONWIP, there are some important differences. Each buffer is individually managed. Second, the buffer is managed only by the upstream worker, but with each buffer managed by one person the entire system performs very well.

Finally, there is an important, basic concept coming out of this work. In a serial system with limited buffer space, that space is most useful when it has some inventory (not zero!) to keep the downstream workstation from being starved, and some space, to keep the upstream workstation from being blocked. For a wide range of cases, "a half-full buffer is a joy forever.*" In some queuing situations, queues are not managed, they are a result. However, if the queues can be managed in a serial production line, a small, half-full buffer can increase throughput while keeping flowtime small.

There are several other research topics in SP/JIT. Karmarkar and Kekre (1990) study the effect of batching decisions in kanban systems. Determining the number of kanbans that are needed in a deterministic situation, to minimize inventory but avoid blocking, is discussed by Bitran and Chang (1987) and Bard and Golany (1991). Bitran and Chang examine assembly systems. Bard and Golany examine the multiproduct case. Stochastic and deterministic models are discussed in Golhar and Stamm (1991) and Goldberg (1991).

What important research related to Stockless Production remains to be done? What research results need to be taught to practitioners and what practical issues need to be understood by researchers and practitioners? The first question will be postponed to the last section of this paper. Here, I will comment briefly on the two questions in the second sentence.

First, pull control mechanisms, in very small batches, do not apply everywhere, and some of the above research helps us understand when to use a pull system. Second, there are pull systems other than single kanban control, and these alternatives can outperform kanban control in some situations. Third,

---

* with apologies to John Keats, who in *Endymion* wrote, "A Thing of Beauty is a Joy Forever."

small inventories are sufficient in many situations, and some of the above research helps to indicate how small and where. The alternative of empirically testing "where and how small" is often intractable. If an empirical approach led to slightly too much or too little inventory in most locations, there would be a negligible cost penalty. But if an important inventory was omitted, the cost penalty could be large.

The research to date is insufficient in that it does not include the constant improvement that is possible if workers are properly motivated. For example, "unreasonable" reductions in inventory (or variation or number of defects) are sometimes very reasonable because of the improvement in the system that may result. Analyzing these issues will take a different type of research, to be discussed in Section V.

Is the research to date valuable? Absolutely. However, it is dangerous if researchers or practitioners use it to make detailed decisions, without considering the management and motivational issues mentioned in the preceding paragraph. The research to date is a source of concepts and understanding, not a blueprint for action, in my opinion. See Silver (1992) for further discussion of this issue.

IV. Fast Production

Fast Production implies getting products to customers quickly. "Competing with Speed" or "Time-Based Competition" includes far more than just production. In this section we will focus on fast production, but also briefly discuss supply, distribution, and design. Only by integrating the entire value-chain can time-based competition be effective, as discussed by Zipkin (1990) and Lee (1992). Also see Keegan (1990) and Sharman (1991) for discussion of some of the logistics issues involved, and Clark and Fujimoto (1989) for comments and data on design speed.

At least three definitions are necessary to discuss fast production. *Flow time* (or *throughput time*) is the time that a particular raw material takes from the time it enters the production system until it is a finished product. *Cycle time* is the time between successive completions of finished products, and *customer response time* is the time between a customer order and the time when the customer receives the item. Any of these can be the focus of a time reduction effort, but flow time and customer response time are more appropriate. Cycle time is really more a capacity decision.

Customer response time is usually the focus of time reduction. For most businesses, service is not off the shelf (see Lalonde and Zinszer, 1976 and Lalonde et al. 1988), but customers want fast service. One way to do that is to hold inventory. Another way is to reduce flow time. Combinations of inventory and flow time reduction are also possible.

As O.R. researchers know, flow time is effectively reduced by reducing inventory. Thus stockless production and fast production are connected by Little's law:

$$L = \lambda W, \tag{1}$$

where

L = the amount of inventory in the system (or "length" of the queue)

λ = the rate at which units are processed by the system (equal to the arrival rate if no units are turned away), and

W = the time a unit spends in the system, or waiting time."

One of O.R.'s main obligations is to explain this relationship to future managers, in all of its generality and importance. $L = \lambda W$ does not imply zero inventory, as stated in Section III, but it does give one major reason to reduce work in process in either service or manufacturing settings.

Describing O.R. research on fast production is difficult, since many parts of the literature discuss response time. Much of the queuing literature analyzes response time. For example, Wein (1990 and 1991) describes models with a production basis that consider scheduling, sequencing, and due-date setting. Duenyas and Hopp (1991) analyze the problem of "quoting customer lead times," that is, what promise can be kept? Survey research shows that even today, keeping a response-time promise is at least as important as fast delivery. Some of the models described in more detail below utilize the type of analysis described in these three papers.

Another area of O.R. research related to response time is the area of multi-echelon inventory control. Much of that literature does deal with a response-time objective rather than off-the-shelf service. That literature is surveyed in Axsater (1992) and Federgruen (1992a). This paper will concentrate on production research.

I will comment on two kinds of O.R. research – control of production lead times and real-time scheduling. In discussing MRP systems, McClain et al. (1992) point out that MRP lead times are often grossly too large, leading to large inventories and large flow times. They also point out that lead times, capacity planning, and lot sizing are inexplicably tied together. Several areas of research spring from these connections.

Dobson et al. (1987) and Karmarkar (1987) analyze the relationship between batching and flow times, showing that batches can be used to control time. Billington et al. (1983) and McClain et al. (1989) present optimization models that compute minimum lead times subject to capacity constraints. They do deal in "time buckets," something that causes part of the lead-time problem for MRP-type systems.

One way to give fast customer response time is to have large finished goods inventories. Another way is to have a large amount of "excess" capacity so that products can be built to order quickly. Combinations of inventory and capacity can also be used. Thomas and Weiss (1986) proposed keeping inventory in items that not only had high, but also predictable demand patterns and give priority in scheduling to other items. The "Stable A" items have variable response times in terms of production but good service because of the inventory. The other items have production priority and thus fast production response. The situation is basically a priority queuing situation (as studied by Wein, 1990 and 1991, and hundreds of other authors) with actual service depending on both inventory and production.

Zheng and Zipkin (1990) analyze a multi-echelon inventory system in which production is scheduled at a central facility based on the current inventory levels at all locations. Items are processed first from the longest queue (LQ), and improved performance results. The idea in Thomas and Weiss (1986) is that many items would have no inventory at all. Priority is determined based on putting backorders (all low-demand items plus "Stable A" items when they run out) before any inventory replenishment. Sox et al. (1991) examine the model in more depth, both analytically and using simulation. Three models are analyzed: FIFO, in which all orders have equal priority; BOP or backorder priority policy, in which stockorders and backorders are each handled on a FIFO basis, but the backorder queue has priority; and ANT or anticipatory policy, in which backorders still have priority, but stockorders are prioritized based on anticipating which stock item will reach backorder status first. The goal is to give service within T periods (a common objective) rather than just to decrease service time. Of course, T can be selected to be a small value.

As expected, BOP and ANT give improved performance compared to FIFO, in general. However, the comparison depends on several things, including the service time T that is required. For very small T (zero, for example), a priority scheduling system makes no sense. Also, the advantage of ANT over BOP is very small in most cases. One graph from Sox et al. is shown below as Figure 2.

Figure 2. Comparisons of Rules at 90% Utilization and 20 Units of Inventory

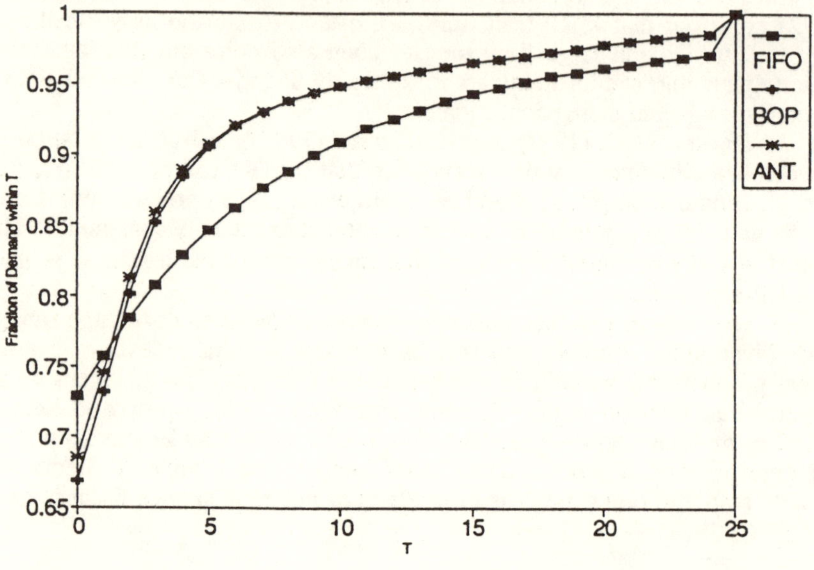

Sox et al. (1991) also find that BOP and ANT not only increases the fraction served within T (customer response time) for a given total inventory, but also the second moment of customer response time is reduced. The efficacy of these policies depends on many factors including how large and small demand rates are, how variable individual demand rates are, and how busy the facility is. If the facility is not busy, the firm has a capacity buffer and does not need an inventory buffer. The utilization of different types of buffers is discussed briefly in Section V and in Schwarz (1992).

Fast Production research examines marketing as well as operations issues. It is insufficient in that market response to different service policies (customer response times, for example) is not explicitly included. It is helpful in that cost saving can be achieved while meeting the externally selected marketing goal.

The final topic reviewed here is real-time scheduling or "leitstands." Real-time scheduling involves models and/or computer programs for production control that treat time as continuous, rather than in discrete "buckets." The potential reduction in lead time is huge, since lead times need not be integer multiples of days or weeks. The reduction is made possible by real-time scheduling software such as described in Conway and Maxwell (1991).

One problem from an O.R. viewpoint is that, in a multi-stage manufacturing process, heuristics must be used on the fly to make the production schedule rather than an optimization approach (which can be used in a discrete time model). Roundy et al. (1991) discuss a price-directed approach to making real-time decisions. The efficacy of heuristics in real-time scheduling as opposed to optimization of discrete time models is an interesting research question, discussed in Section V.

O.R. research on time-based competition has been insufficient in that it does not examine the main way that customer response times have been improved in practice, namely by carefully analyzing exactly what the current process is, then simplifying it. Connecting O.R. and process simplification is discussed in Section V, along with other items on a research agenda.

V. Suggestions for Research in Stockless and Fast Production

O.R. research to date in SAFP has contributed in several ways. The implications of Little's law were understood long before SAFP became popular; Little's law contributes to understanding SAFP. Research has added important and practical concepts, such as controlling queues individually or in an aggregate form, rather than allowing them to be a result, and managing inventory and production together in innovative ways to achieve fast service. When to use kanban control versus CONWIP or other alternatives, and how to manage "tightly coupled" production systems (see Baker, 1992 for more discussion) are both areas of useful research. The design issues raised in Section II have been dealt with only lightly, regarding investing in reduction of cost parameters. This research helps us better to understand the issues, but more should be done.

In my opinion, three major categories of research need work related to SAFP. In (my) increasing order of importance, they are (1) research to fill in gaps in the literature above, (2) some "straight O.R. topics" related to the above, and, most important, (3) cross functional research related to SAFP.

The "straight O.R. topics" are of three types. First, models for integration along the supply chain can be very important; see Cohen and Lee (1988) and Federgruen (1992b). Second, neither stockless nor fast production eliminates the need for planning. (Many Japanese companies do have 12-month, 6-month, and 1-month plans that lead to orders released to a factory on a just-in-time basis.) Hierarchical planning models (see Hax and Meal, 1974, and Bitran and Tirupati, 1992) should be developed that lead to good just-in-time schedules. This research can include both model-building and analysis as well as empirical research. The analysis can be connected to research on when kanban (or other) control is appropriate.

The third exciting O.R. topic is when and how to use real-time scheduling. Real-time scheduling can reduce lead times, but will suboptimization lead to poor solutions? In what circumstances is it appropriate to heuristically solve the right problem (real time) or "optimally" solve the wrong problem (discrete time). Also, what planning models fit well with real-time schedulers?

The most important future SAFP research, in my opinion, will be cross-functional in nature. We should practice what we preach* and work across organizational boundaries. Karmarkar (1992) addresses some cross-functional research topics. More needs to be done.

First is work with our organizational behavior colleagues. In practice, the main benefits of SAFP are obtained by enlisting everyone to find ways to eliminate wasted time and effort. "Lean manufacturing," a term coined by Womack et al. (1990) to include all of SAFP, is achieved by continuous improvement. What environments lead to successful improvement efforts? How can small, detailed successes be spread throughout the organization? How and why do workers react to different control procedures and/or improvement programs? How does design (product, process <u>and</u> organizational) help a firm to achieve lean manufacturing?

Second, SAFP should lead to marketing benefit. Fast customer response is believed to increase market share and/or allow a premium price for products and services. Some more research should be done to connect marketing analysis with operations management decisions. In addition to references noted above, Little (1991) describes interesting research possibilities between operations and marketing. Davis and Maggard (1990) analyzes some data to see how customers actually feel about waiting, in a two-stage process. Empirical

---

* In an interesting article, Runkle (1991) challenges academics to practice what they preach, concentrating on research and teaching with impact, rather than just complaining about managerial performance.

work is needed to validate production-marketing models and to assess the value of fast response.

Cross functional work involving accounting with SAFP will also be important. What managerial performance measures fit with JIT and cause better performance? See Young and Selto, 1991, for some preliminary work in this area. For many years it has been known that cost accounting has caused trouble for operations managers. See Gantt (1915)! There are good ideas around (see Cooper and Kaplan, 1988, for example) regarding how to collect useful managerial accounting measures, but empirical work tied with operations management variables is in its infancy.

Some of the above research will be very different in style from most O.R. research, and some will be similar. I believe we should be open to new types of research. In our teaching, we need to convey the simplicity of the concepts in SAFP (the "set of bullets" mentioned in the introduction) and convey the challenge and fascination of making those things really work, utilizing ideas from several sources (not just standard operations management) and utilizing people in many different positions in a firm. Both the research and practice look very interesting.

The ideas we need to remember from Elwood Buffa's writings are: first, to be driven by problems and methods from practice and second, to concentrate on design issues as well as control issues. SAFP implies a strategic look at what is needed, and companies achieve the strategic goals largely by designing the product, process, and organization so that control becomes easier.

As a way to remember some of the basic ideas from SAFP, I include an appendix on "Just-in-Time," a mnemonic device for making the ideas stick.

Appendix - A Mnemonic Device for Stockless Production

JUST IN TIME*

JUST IN TIME, I NEED PARTS JUST IN TIME,

JUST WHEN MY STOCK ON HAND IS RUNNING LOW.

FASTER PACE, NO STOCK HELD "JUST IN CASE,"

HELPS ME TO WIN THE RACE, AND MAKE SOME DOUGH.

(repeat first part of tune)

E.O.Q., THAT'S NOT THE THING TO DO,

AND SAFETY STOCK JUST HIDES PROBLEMS FROM YOU.

LONG LEAD TIMES, WE'LL CUT THEM DOWN JUST FINE,

AND WE'LL MAKE OUR VENDORS MIND THE SYSTEM TOO.

(second part of tune)

QUALITY MUST BE AT NEARLY PERFECT LEVELS,

AND INVENTORY MUST BE ZERO, AND THEY SAY

SET-UP TIME NEAR NOTHING IS SUBLIME,

AND THAT MAKES JUST-IN-TIME THE WAY TO GO.

© Joe Thomas, April 1985

*Sung to the tune "Just in Time," by Jule Styne, © Betty Comden, Adolph Green, and Jule Styne, 1956.

References

Axsater, S. (1992), "Continuous Review Policies for Multi-Level Inventory Systems with Stochastic Demand," in Graves et al. (eds.).

Baker, K. (1992), "Models of Tightly Coupled Production Systems," in Sarin, R. (ed.).

Baker, K., A. Powell, and D. Pyke (1990), "Buffered and Unbuffered Assembly Systems with Variable Processing Times," *Journal of Manufacturing and Operations Management*, Vol. 3, No. 3, pp. 200-223.

Bard, J., and B. Golany (1991), "Determining the Number of Kanbans in a Multiproduct, Multistage Production System," *International Journal of Production Research*, Vol. 29, pp. 881-895.

Berkley, B.J., and A.S. Kiran (1991), "A Simulation Study of Sequencing Rules in a Kanban-Controlled Flow Shop," *Decision Sciences*, Vol. 22, No.3, pp. 559-582.

Billesbach, T. (1991), "A Study of the Implementation of Just-in-Time in the United States," *Production and Inventory Management Journal*, Vol. 32, No. 3, pp. 1-4.

Billington, P., J. McClain, and L. J. Thomas (1983), "Mathematical Approaches to Capacity-Constrained MRP Systems: Review, Formulations, and Problem Reduction," *Management Science*, Vol. 29, No. 10, pp. 1126-1141.

Bitran, G., and L. Chang (1987), "A Mathematical Programming Approach to A Deterministic Kanban System," *Management Science*, Vol. 33, No. 4, pp. 427-441.

Bitran, G., and D. Tirupati (1992), "Hierarchical Production Planning," in Graves et al. (eds.).

Blackburn, J. (1990), *Time-Based Competition*, R.D. Irwin, Homewood, Illinois.

Buffa, E.S. (1961), *Modern Production Management*, John Wiley, New York.

Buzacott, J. (1989), "Queuing Models of Kanban and MRP Controlled Production Systems," *Engineering Costs and Production Economics*, Vol. 17, pp. 3-20.

Clark, K., and T. Fujimoto (1989), "Overlapping Problem-Solving in Product Development," in Ferdows, K. (ed.), *Managing International Manufacturing*, Amsterdam, North-Holland.

Cohen, M. and H. Lee (1988), "Strategic Analysis of Integrated Production-Distribution Systems: Models and Methods," *Operations Research*, Vol. 36, No. 2, pp. 216-228.

Conway, R., and W. Maxwell (1991), *XACT Production Planning System: User's Guide*, C-Way Associates, Lansing, New York.

Conway, R., W. Maxwell, J. McClain, and L. J. Thomas (1988), "The Role of Work-in-Process Inventory in Serial Production Lines," *Operations Research*, Vol. 36, pp. 229-241.

Cooper, R. and R.S. Kaplan (1988), "Measure Costs Right, Make the Right Decisions," *Harvard Business Review*, Vol. 66, No. 5, pp. 96-103.

Davis, M.M., and M.J. Maggard (1990), "An Analysis of Customer Satisfaction with Waiting Times in a Two-Stage Service Process," *Journal of Operations Management*, Vol. 9, No. 3.

Deleersnyder, J., T. Hodgson, H. Muller, and P. O'Grady (1989), "Kanban Controlled Pull Systems: An Analytic Approach," *Management Science*, Vol. 35. pp. 1079-1091.

Denardo, E., and C. Tang (1988), "Pulling a Markov Production Line, Parts I, II, and III," Technical Reports, Yale School of Management.

Dobson, G., U. Karmarkar, and J. Rummel (1987), "Batching to Minimize Flow Time on One Machine," *Management Science*, Vol. 33, pp. 784-799.

Duenyas, I. and W.J. Hopp (1991), "Quoting Customer Lead Times," Tech. Rep. 91-14, Department of IE and MS, Northwestern University.

Federgruen, A. (1992a), "Centralized Planning Models for Multi-echelon Inventory Systems Under Uncertainty," in Graves et al. (eds.).

Federgruen, A. (1992b), "Recent Advances in Production and Distribution Management," in Sarin, R. (ed.).

Gantt, H.L. (1990), "The Relation Between Production and Costs," *American Society of Mechanical Engineers*, Spring Meeting, 1915. Reprinted in *Journal of Manufacturing and Operations Management*, Vol. 3, No. 3. (Also see William Kent's comment.)

Glasserman, P., and D. Yao (1991), "Structured Buffer-Allocation Problems in Production Lines," Working Paper, IE/OR, Columbia University.

Goldberg, H. (1991), "Japanese Just-in-Time Production/Inventory Management from an Operations Research Perspective," School of Industrial Management, New Jersey Institute of Technology," Newark, NJ.

Goldratt, E. (1988), "Computerized Shop Floor Scheduling," *International Journal of Production Research*, Vol. 26, No. 3.

Golhar, D., and C. Stamm (1991), "The Just-in-Time Philosophy: A Literature Review," *International Journal of Production Research*, Vol. 29, pp. 657-676.

Graves, S., A. Rinnooy Kan, and P. Zipkin (1992), eds., *Handbook in Operations Research and Management Science, Volume 4: Logistics of Production and Inventory*, North-Holland, Amsterdam.

Groenevelt, H. (1992), "The Just-in-Time System," in Graves et al. (eds.).

Hausman, W.H., and D.B. Montgomery (1990), "Making Manufacturing Market Driven," Research Paper 1003, Dept. of IE and EM, Stanford University.

Hax, A., and H. Meal (1975), "Hierarchical Integration of Production Planning and Scheduling," in *TIMS Studies in the Management Sciences, Vol. I, Logistics*, M. Geisler (ed.), New York, North Holland-American Elsevier.

Hillier, F., and K. So (1991), "The Effect of Coefficient of Variation of Operation Times on the Allocation of Storage Space in Production Line Systems," *IIE Transactions*, Vol. 23, No. 2.

Hopp, W. and M. Spearman (1991), "Throughput of a Constant WIP Manufacturing Line Subject to Failure," *International Journal of Production Research*, Vol. 29, pp. 635-655.

Inman, R.A., and S. Mehra (1991), "JIT Applications for Service Environments," *Production and Inventory Management Journal*, Vol. 32, No. 3.

Karmarkar, U. (1987), "Lot Sizes, Lead Times and In-Process Inventories," *Management Science*, Vol. 33, No. 3.

Karmarkar, U. (1989), "Getting Control of Just-in-Time," *Harvard Business Review*, Vol. 67, No. 5.

Karmarkar, U. (1992), "Research in Manufacturing Strategy: A Cross-Functional Perspective," in Sarin, R. (ed.).

Karmarkar, U. and S. Kekre (1990), "Batching Policy in Kanban Systems," *Journal of Manufacturing Systems*, Vol. 8, No. 4.

Keegan, K. (1990), "Xerox Supply-Chain Integration," *PRTM Insight*, Vol. 2, No. 3, Fall.

Krajewski, L., B. King, and L. Ritzman (1987), "Kanban, MRP, and Shaping the Production Environment," *Management Science*, Vol. 33, No. 1.

Lalonde, B., and P. Zinszer (1976), *Customer Service: Meaning and Measurement*, NCPDM (now Council of Logistics Management), Oak Brook, Illinois.

Lalonde, B., M. Cooper, and T. Noordewier (1988), *Customer Service: A Management Perspective*, Council of Logistics Management, Oak Brook, Illinois.

Lee, H. (1992), "Design for Supply Chain Management: Methods and Examples," in Sarin, R. (ed.).

Little, J.D.C. (1991), "Operations Research in Industry: New Opportunities in a Changing World," *Operations Research*, Vol. 39, No. 4.

McClain, J., L.J. Thomas, and J.B. Mazzola (1992), *Operations Management: Production of Goods and Services*, Englewood Cliffs, NJ, Prentice-Hall.

McClain, J., L.J. Thomas, and C. Sox (1991), "'On-the-Fly' Line Balancing with Very Little WIP," Johnson Graduate School of Management, Cornell University, WP91-01.

McClain, J., L.J. Thomas, and E. Weiss (1989), "Efficient Solutions to a Linear Programming Model for Production Scheduling with Capacity Constraints and No Initial Stock," *IIE Transactions*, Vol. 21, No. 2.

Miller, J.G., and J.S. Kim (1990), *Beyond the Quality Revolution: U.S. Manufacturing Strategy in the 1990's*, B.U. Manufacturing Roundtable, Boston University.

Miller, J., J. Kim, A. DeMeyer, K. Ferdows, J. Nakane, and S. Kurosa (1991), *Factories of the Future: Executive Summary of the 1990 International Manufacturing Futures Survey*, B.U. Manufacturing Roundtable, Boston University.

Mitra, D. and I. Mitrani (1990), "Analysis of a Kanban Discipline for Cell Coordination in Production Lines I." *Management Science*, Vol. 36, pp. 1548-1566.

Mitra, D. and I. Mitrani (1991), "Analysis of a Kanban Discipline for Cell Coordination in Production Lines, II: Stochastic Demands," *Operations Research*, Vol. 39, No. 5.

Muckstadt, J.A., and S.R. Tayur (1991), "A Comparison of Alternative Kanban Control Mechanisms," Tech. Report, S.O.R.I.E., Cornell University.

Nadler, G. (1991), "Design Teams: Breakthroughs for Effectiveness," *Design Management Journal*, Vol. 2, No. 2.

Ostolaza, J., J. McClain, and L.J. Thomas (1990), "The Use of Dynamic (State-Dependent) Assembly-Line Balancing to Improve Throughput," *Journal of Manufacturing and Operations Management*, Vol. 3, No. 2.

Roundy, R., W. Maxwell, Y. Herer, S. Tayur, and A Getzler (1991), "A Price-Directed Approach to Real-Time Scheduling of Production Operations," *IIE Transactions*, Vol. 23, No. 2.

Runkle, D.L. (1991), "Taught in America," *MIT Management*, Fall.

Sarin, R. (1992), ed., *Perspectives in Operations Management: Essays in Honor of Elwood S. Buffa*, Kluwer Academic Publishers, Hingham, Massachussetts.

Schonberger, R. (1986), *World Class Manufacturing: The Lessons of Simplicity Applied*, The Free Press, New York.

Schwarz, L. (1992), "The Information/Control/Buffer Portfolio: An Operations Management Paradigm," in Sarin, R. (ed.).

Sharman, G. (1991), "Global Integrated Logistics," in *Integrated Logistics: Perspectives for the 1990s*, McKinsey and Company, New York.

Silver, E. (1992), "Modeling in Support of Continuous Improvement Toward Achieving World-Class Operations," in Sarin, R. (ed.).

Sox, C., J. McClain, and L.J. Thomas (1991), "Jellybeans: Storing Capacity to Satisfy Customer Demands Within a Specified Time Limit," Working Paper, JGSM-Cornell University.

Spence, A., and E. Porteus, "Setup Reduction and Increased Effective Capacity," *Management Science*, Vol. 33, No. 10.

Stalk, G., Jr., and T. Hout (1990), *Competing Against Time*, The Free Press, New York.

Thomas, L.J., and E. Weiss (1986), "An Integrative Approach for Production Scheduling and Capacity Determination," in *Production Management: Methods and Applications*, B. Lev (ed.), North-Holland (TIMS Series), Amsterdam.

Wagner, H.M. (1988), "Operations Research: A Global Language for Business Strategy," *Operations Research*, Vol. 36, No. 5.

Wein, L. (1990), "Dynamic Scheduling of a Multi-Class Make-to-Stock Queue," MIT Sloan School.

Wein, L. (1991), "Due Date Setting and Priority Sequencing in a Multiclass M/G/1 Queue," *Management Science*, Vol. 37, No. 7.

White, J.A. (1991), "An Existence Theorem for OR/MS," *Operations Research*, Vol. 39, No. 2.

Womack, J.P., D.T. Jones, and D. Roos (1990), *The Machine that Changed the World*, New York, Rawson.

Woodruff, D.L., M.L. Spearman, and W.J. Hopp (1990), "CONWIP: A Pull Alternative to Kanban," *International Journal of Production Research*, Vol. 28, No. 5.

Young, S.M., and F.H. Selto (1991), "The Roles of Accounting, Production and Behavioral Variables in Determining Manufacturing Performance in a Just-in-Time Environment," College of Business, University of Colorado.

Zheng, Y.S. and P. Zipkin (1990), "A Queuing Model to Analyze the Value of Centralized Information," *Operations Research*, Vol. 38, No. 2.

Zipkin, P. (1989), "A Kanban-Like Production Controlled System: Analysis of Simple Models," Working Paper, Columbia University, Graduate School of Business.

Zipkin, P. (1990), "Optimizing the Supply Chain: A Model for Operations-Marketing Planning," Working Paper RWP-90-9, GSB, Columbia University.

Zipkin P. (1991), "Does Manufacturing Need a JIT Revolution?" *Harvard Business Review*, Vol. 69, No. 3.

MATHEMATICAL MODELS OF RETAILER INVENTORY SYSTEMS: A REVIEW

Steven Nahmias
Stephen A. Smith
Santa Clara University

Introduction

Retail inventory management is an important and challenging research area. Based on performance measures such as turns per year, the efficiency with which retailers manage their inventory lags well behind other sectors, such as manufacturing. However, retail inventory management must deal with unique problems and complexities. By the nature of their business, major retailers hold stock at many geographically dispersed locations at anywhere from several hundred to thousands of stores for the typical retailing chain. Retailers are also faced with the problem of managing an extremely large number of line items; several hundred thousand SKUs (stock keeping units) is not uncommon. Until recently, demand rates at the store level have been reported only in aggregate form, and often with considerable delay and inaccuracy. This made it nearly impossible for retailers to accurately determine the inventory position of an SKU at a store, let alone estimate the probability distribution of weekly demand or the service level experienced by customers on individual items.

More powerful data processing technology, the decline in the cost of on-line electronic storage, and the widespread adoption of point-of-sale scanners at the cash register now provide the information processing tools to manage retailers' inventories much more efficiently. Many major manufacturers and suppliers who sell to retailers now use electronic data interchange for purchase orders and shipment coordination, and offer more flexibility in providing quick response and frequent deliveries. However, few retailers use inventory management methods tailored to the modern retailing environment. We believe that with the improved information transfer technology, retailers are now in a position to implement more sophisticated inventory control systems and realize significant savings through effective inventory management.

Inventory management models have been an important application area in operations research for many years and the number of published papers on inventory management is well into the thousands. Only a very small percentage of these papers deal specifically with the two echelon retail system we treat, but many others are potentially applicable to retailing. For the type of retail inventory system that we consider, we summarize both those papers that deal specifically with retailing and other papers that appear to be closely related to this application. We also evaluate the strengths and weaknesses of the existing approaches in this context.

We consider a two echelon inventory structure consisting of a distribution center supporting N retail outlets or stores, as illustrated in Figure 1. This type of structure is common to most major retailers today, although there are many variations in the procedures for operating it. The distribution center (DC) receives shipments from suppliers and distributes them to the stores, and may hold inventory as well in some cases. Indeed, the appropriate role for DC stock is currently viewed as a key strategic issue by some retailers. At a roundtable discussion held at Santa Clara University, major retailers reported that their DC's served as few as fifty and as many as eight hundred stores. Retailers generally locate the DC so that the stores it supports are within one day's travel distance by truck.

Figure 1 shows the special type of multiechelon inventory system we consider. Demand is assumed to originate at the first echelon (stores) and is transmitted to the second echelon (DC). Units flow in the opposite direction, from the DC to the stores. In some cases the DC holds stock and in some it serves as a central location from which stock is distributed to the stores.

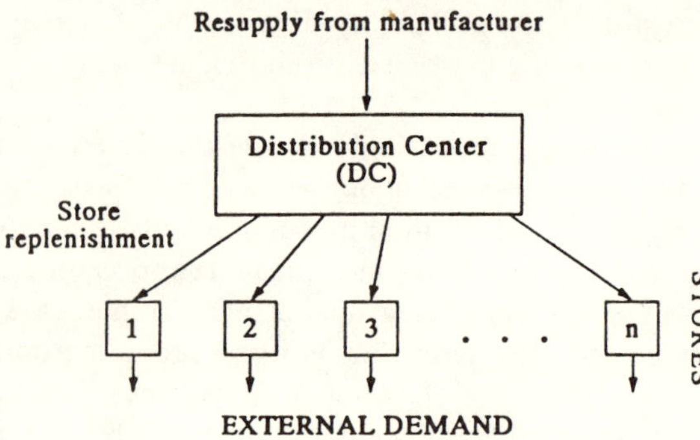

Figure 1. Two Echelon DC/Store System

There are several possible motives for introducing one or more DC's between the supplier and the store. One is that it could be less costly to hold inventory at the DC than at the retail store; stores may be located in malls or shopping centers in densely populated neighborhoods where lease space is expensive. Furthermore, intermediate locations serve as a buffer against uncertainty. Since demands are typically uncertain at the stores, it is desirable to effect a periodic redistribution of stock. In retailing, direct inter-store transfers are rare. The redistribution problem is solved by retaining a proportion of stock at the central warehouse to be shipped when needed. Finally, transportation costs could be reduced when there is a DC. Transportation economies of scale result from shipping from the suppliers to the DC in bulk.

The multiechelon inventory problem was first recognized from studies of military logistics and has played a major role in the management of materiel in the armed forces. (See Demmy and Presutti(1981), for example). Some inventory positioning problems faced by retailers today are similar to those faced by military logisticians, so the early models formed the basis for those used today. On the other hand, a substantial part of the modeling effort in military systems was directed at so called reparable items, like aircraft engines, which have very different characteristics from retail items (Nahmias(1981)).

Virtually all retailers face either full or partial lost sales. When a demand cannot be filled from available stock there are two possibilities: either the demand is backordered to be filled at a later time, or the customer looks elsewhere and the sale is lost. The vast majority of papers on inventory modeling assume full backordering of demand. However, full backordering is not an appropriate assumption for retailers in most situations. Some portion of the customers whose demand cannot be met from available stock will choose to go elsewhere to make their purchase, resulting in partial lost sales. The appropriate backorder assumption depends on the type of item being considered. Basic goods, such as shirts, are typically lost sales items, while heavy items, such as large appliances, are typically backorder items. Other items may be a mixture of backorders and lost sales.

While lost sales, which we believe is common in retailing, has been treated in the literature, another characteristic of retail systems that has received little attention in the inventory literature is substitution. It is not uncommon for consumers to purchase other brands or colors of products when their first choice is out of stock. We are not aware of any work on the substitution problem in the context of retail inventory control.

In this review we restrict attention to stochastic demand only. We note that there is a substantial literature on the deterministic problem. Roundy(1985), most notably, has obtained some remarkable results. However, our interest is in policies that protect retailers against uncertainty. While deterministic demand may be a reasonable approximation for some multiechelon production systems, it is not likely to be an accurate assumption in retailing.

This review will be organized as follows.

1. First we briefly consider "classical" multi-echelon inventory models even though these models may not be directly applicable to retailing. The early researchers in this area developed the concepts and techniques which formed the basis of the work that followed.

2. The second section looks at the early work on the DC/store system. Most of the early work considered only single period models. Considerable work has been done on the problem of finding optimal distribution and allocation policies in a two level system when only a single planning period was considered.

3. Several researchers have treated the retailer problem under the assumption that the DC holds no stock. That is, the DC acts only to consolidate the store's orders. Even when the DC holds no stock, there can be substantial advantages in coordinating system wide replenishments from a central location.

4. Next we review models where the DC can hold stock. This allows added "risk pooling" among the stores in that the DC stock can be used to balance the store inventories between DC replenishments.

5. The fifth section treats lot size reorder point (Q,R) models in the context of the DC/store system. These models form the basis for several commercial inventory systems and have played an important role in the development of the theory.

6. When inventory levels are reviewed periodically rather than continuously, (s,S) rather than (Q,R) policies are appropriate. In this section we explore (s,S) policies in retailer inventory systems.

7. Full or partial lost sales is an appropriate assumption in most cases. However, the vast majority of the models we review here assume full backordering of excess demand. The reason is that when there is a positive order lead time and sales are lost, optimal policies are complex. This section discusses the complexities that arise from assuming lost sales.

8. Finally, we conclude with a discussion of the applicability of the models reviewed and directions for future research.

1 Classical Models of Multiechelon Inventory Systems

One of the first multiechelon models was due to Clark and Scarf(1960). They assumed that the echelon structure was a pure series system consisting of N echelons with stock being transmitted from echelon n to echelon n-1 for $2 \leq n \leq N$. Demand for the product originates at echelon one only. They also assumed that the cost of transshipping from echelon n to echelon n-1 is a linear function of the amount shipped. Finally, positive lead times for shipping from one echelon to the next lower one were assumed.

The Clark and Scarf study was important for several reasons. It was the first to characterize the form of an optimal policy for a reasonably general stochastic demand, multiperiod, multiechelon model. It also was important for introducing the concepts of echelon stock and implied shortage cost, which formed the basis for the analysis of more complex systems. By incorporating an explicit penalty at echelon n for not being able to satisfy the order placed by echelon n-1, the authors were able to decompose the N echelon problem into N single echelon problems each having a simple optimal policy. This approach was used by others in subsequent work.

Although a seminal work in the development of the theory of multiechelon inventory systems, the assumptions required make it unlikely that anyone would use the Clark and Scarf(1960) model to manage a real system. One limitation is the assumption that all replenishment costs are proportional to the size of the replenishment order. When economies of scale (i.e., fixed ordering costs) were included at all locations Clark and Scarf(1962) were able to compute only approximately optimal policies. Furthermore, the simple serial system considered has limited applicability in practice.

Bessler and Veinott(1965) extended Clark and Scarf's serial system to a general arborescence (unidirected trees). They assumed only proportional costs of ordering and that each facility could order from an exogenous supplier. They also assume full backordering of demand at each facility and a policy that allows redistribution of stock among the outlets at the end of each period. Although the authors obtain interesting results for their model, it is also a poor model for a retail system. In retailing, external demand occurs only at the retail level, and resupply from outside only at the DC level. Furthermore, assuming full backordering of excess demand at the retail level is unreasonable. Most systems experience either full or partial lost sales. Rosenfield and Pendrock(1980) discuss the limitations of these approaches for managing real distribution systems.

Computational issues surrounding the Clark and Scarf model were explored by Federgruen and Zipkin(1984b). They showed that the form of the optimal policy established by Clark and Scarf for the finite horizon problem also holds in the infinite horizon versions of the problem for both discounted and average cost cases. They further establish simpler computational formulas in the infinite horizon case. Similar simplifications were known for many years for the single product single facility problem (for example, Karlin(1958).

A class of multiechelon inventory models was developed to deal with the problem of determining suitable stocking levels for so called "reparable" items. Reparable items are ones that are (almost all) recycled back into the system after some random period of time required for repair. The prototype model of this type is METRIC, reported in the literature by Sherbrooke(1968). The METRIC model is a culmination of a multi-year effort by several researchers, many at the RAND Corporation, who studied the problem of determining suitable stocking levels for expensive reparable items. These inventories are valued in the multibillion dollar range and their availability is crucial for successful military operations. METRIC and several of its successors form the backbone for many military inventory control systems. However, these models bear little resemblance to those appropriate for retailers. Clark(1972) provides an expository survey of these models as well as a discussion of work on related deterministic problems. The study of the particular multiechelon system that arises in reparable item systems has been the subject of much research. Since these models, by and large, are not applicable to retailers, we will not discuss them further in this review. For a comprehensive review of the work done up to 1981 on this problem we refer the interested reader to the survey paper by Nahmias(1981). More recent results on the reparable inventory problem can be found in Albright(1989).

2 Single Period Models

Many of the early efforts at modeling the DC/store system pictured in Figure 1 focused on a single period only. These studies can be considered as extensions of the newsboy model (Nahmias(1989), for example), in which both ordering and redistribution policies are considered. While the assumption of a single planning period makes these approaches only marginally applicable, they were important in building an understanding of the nature of the issues faced when managing such a system.

Allen(1958) was the first to consider a one period version of the system pictured in Figure 1. Allen's is actually a model of a single echelon problem consisting of K locations. The objective of the analysis is to determine an optimal redistribution of stock among the locations; no replenishment decision is included. Allen assumes that the demands at the locations are random variables with continuous probability density functions. Let S_k^0 be the starting stocks at the locations and S_k the stock levels after redistribution, for $1 \le k \le K$. Then

$$S_k = S_k^0 - \sum_{j=1}^{K} x_{kj} + \sum_{j=1}^{K} x_{jk}.$$

Interpret x_{kj} as the amount shipped out of location k into location j and x_{jk} as the amount shipped into location k from location j. The objective of the analysis is to find the optimal values of the variables x_{kj} and x_{jk} to minimize the total cost of transshipment and shortage at the K locations. If we assume that f_k is the probability density of demand at location k, and p the unit cost of a shortage, it follows that the objective function to be minimized is

$$p\sum_{k=1}^{K} \int_{S_k}^{\infty} (t-S_k)\, f_k(t)\, dt + \sum_{k=1}^{K} \sum_{j=1}^{K} c_{kj} x_{kj}.$$

Allen supplies an iterative algorithm for finding the optimal redistribution of stock among the locations. He makes the observation that a shipment from location k to location j will occur only when the shortage probability is higher at j than at k.

Simpson(1959) considers a model similar to Allen's. He assumes that there are Q units to apportion among N outlets. He shows that the optimal allocation is the one that equalizes the stockout probabilities at each of the locations.

Although both Allen's and Simpson's models are too simple to be useful in most situations, they were the first to recognize the importance of the concept of equal runout probabilities. The concept of equal runout times and equal runout probabilities also arises in later work.

Gross(1963) considers a similar one period model but allows for both replenishment and redistribution at the retail outlets. Gross considers a single supply point he called M, (similar to the DC) and assumes that no stock is retained at M. Demand at the retail stores is random with specified continuous probability distributions. At the end of the period each outlet could place an order to the supply depot and/or effect redistribution of stocks among the retail outlets. Both replenishment costs and transshipment costs are proportional to the size of the reorder or shipment. Depending on the relationship of the cost parameters and the probability of stocking out at each location, one takes one of the 12 possible combinations of actions: order for location 1 only, order for location 2 only, order for both locations, order for neither location, permuted with the three choices of transshipping from location 1 to location 2, from location 2 to location 1, or not transshipping at all. For the case of two locations the optimal policy can be described by six regions in the plane whose axes correspond to the starting inventories at each location. Within some regions, one of several actions may be optimal depending on certain relationships between costs and demand probabilities.

Similar models were considered by Krishnan and Rao(1965) and Das(1975). Krishnan and Rao's work is very similar to Gross(1963). Das's work is less general than that of Gross in that he considered only two locations, but more general in that he allowed for two subperiods and assumed that there would be a redistribution of stock between the subperiods. The results are similar in providing regions of two-space associated with the optimal redistribution policy.

Hoadley and Hyman(1977) consider a similar model to Gross(1963). Their model is more general in that they allow for returns to the higher echelon and disposal of stock at that echelon as well as ordering from the stores to the DC as well as transshipment of inventories among the retail outlets. They derive a form for the objective function of the problem and indicate that the problems they formulate could be solved by known search techniques for nonlinear optimization. No structural results about the form of an optimal policy are given, however.

Eppen(1979) considers a somewhat different issue in a single period model. He assumes N identical retail outlets that order independently according to a simple order up-to point model obtained by minimizing one period holding and penalty costs. Assuming normal demand at each location, he derives an expression for the expected cost at each facility assuming each follows an optimal policy. Now suppose that the demands among the facilities are correlated. The question is whether there are expected costs savings realized by replenishing the inventory from a central location for all the facilities. Because of variance effects, he shows that it is more economical to order for all the facilities from a central location than it is to order separately as long as the demands at the individual facilities are not perfectly correlated. When the facilities' demands are perfectly correlated, the expected costs are the same for both configurations. Although we classify Eppen's model as a single period model, we should point out that it, and many of its successors, can be effective "myopic" approximations for multi-period dynamic versions of the problem.

Suppose that the standard deviation of demand at location i is σ_i and the correlation coefficient of demands at i and j is ρ_{ij}. For a decentralized system in which each location orders independently according to an optimal policy, he shows that the optimal cost can be expressed in the form

$$K\sum_{i=1}^{N} \sigma_i,$$

while for the case of completely centralized system in which the warehouse orders for all outlets, the optimal cost is of the form

$$K\sqrt{\sum_{i=1}^{N} \sigma_i^2+2\sum_{i=1}^{N-1} \sum_{j=i+1}^{N} \sigma_i\sigma_j\rho_{ij}}.$$

where K is a constant depending on the holding and penalty costs and the standard normal loss function. It is easy to verify that these two expressions will be equal if and only if $\rho_{ij} = 1$ for all i and j and that the second expression will be smaller if the correlation of the demand at the sites is not perfect, that is, $\rho_{ij} < 1$ for at least one (i,j).

These studies have several limitations. First, they assume only a single planning period. While single period models can be used in a dynamic environment, they are often inadequate for handling system dynamics. Second, they ignore replenishment lead times. In many systems, lead times are the reason for retaining safety stock. The lead time is the response time of the system. Finally, the problem of redistribution is less important than the problem of reordering; our experience with actual systems of this type is that there is rarely lateral resupply among retail outlets. Retailers effect redistribution by retaining inventory at the DC level.

3 Models Where the DC Holds No Stock

Several researchers have considered dynamic models of the DC/store system which assume that the DC holds no inventory. This assumption is accurate for many real systems. In some cases, retailers use the DC merely as a distribution point. (Items which are not stored at the DC are referred to as "flow through" items. The most common situation is that some items are flow through items and some are not.) The first study of such a system is due to Eppen and Schrage(1981). Their model is essentially equivalent to that of Figure 1 although they use different terminology. What we call the DC they refer to as a depot, and what we call the retail outlets or stores they refer to as warehouses. However, since they assume that external demand arises only at the warehouse level, the difference is one of terminology only. We use our terminology in reporting their results.

The assumption that the DC holds no inventory does not mean that the DC serves no purpose. First, by ordering centrally, the firm can negotiate for deeper quantity discounts. Second, there are "statistical economies of scale" such as those observed by Eppen(1979) by aggregating orders. That is, unless the demand at the stores is perfectly correlated, the coefficient of variation of demand for the aggregate system will be smaller than the coefficient of variation for the demand at the individual stores. (The coefficient of variation is the ratio σ/μ.)

Eppen and Schrage assume positive lead times from both the supplier to the DC and from the DC to the retail outlets. Demand for the product in each planning period originates only at the retail level and is assumed to follow a normal distribution, although the demand parameters may differ from outlet to outlet.

The important contribution of this paper is the discovery of conditions under which a simple ordering policy is optimal. The DC must have enough stock to resupply the retail outlets so that the probability of a stockout situation arising at each retail outlet is the same. Another way of stating this condition is that there is sufficient stock at the DC to return the outlets to the same given fractile point of their demand distributions. They establish conditions under which this assumption is likely to hold: when the fixed cost of ordering from the DC is high and/or the coefficient of variation of demand at the retail outlets is moderate.

Let(μ_i, σ_i) be the mean and standard deviation of demand per period at outlet i, $1 \leq i \leq N$, L the order lead time from the supplier to the DC and M the (common) lead time from the DC to any retail outlet. Let p be the cost of a stockout and h the cost of holding, both assumed to be proportional to the ending inventory each period. Under their assumptions, the authors show that the optimal policy each period is for the DC to place an order on the supplier to bring the system wide stock up to a level y given by

$$y=(L+M+1)\sum_{k=1}^{N}\mu_k+z\left[L\sum_{k=1}^{N}\sigma_k^2+(M+1)\left(\sum_{k=1}^{N}\sigma_k\right)^2\right]^{\frac{1}{2}}$$

where z is the p/p+h fractile of the standard normal distribution. The authors contrast their results with completely decentralized systems (where the retail outlets order independently on their own) and completely centralized systems where the individual outlets are treated as one single facility. The differences arise only in the standard deviation term above (that is, the term multiplying z in the expression above). For a completely centralized system, the optimal system wide stock is:

$$y=(L+M+1)\sum_{k=1}^{N}\mu_k+z\left[(L+M+1)\sum_{k=1}^{N}\sigma_k^2\right]^{\frac{1}{2}}$$

while for the completely decentralized system the optimal system wide stock is:

$$y=(L+M+1)\sum_{k=1}^{N}\mu_k+z\left[(L+M+1)^{\frac{1}{2}}\sum_{k=1}^{N}\sigma_k\right]$$

The optimal system wide stock is greatest for the decentralized and least for the centralized system. The system stock for the DC/store system lies between these extremes. Notice that the only difference between these expressions is the term multiplying z, which can be thought of as the effective system wide standard deviation.

The authors obtain an expression for the expected total cost per period assuming one follows an optimal policy. They also explore the case where the DC orders only once every m periods. This is a reasonable policy to consider when there is a substantial fixed order cost for DC replenishment. However, it is unlikely to be an optimal policy in this case. The optimal policy is a complex function of the vectors of stocks at the DC and at each retail outlet. They obtain results similar to those above in the order every m period case.

Federgruen and Zipkin(1984a) consider the same problem but approach the solution differently. First, they formulate the problem as a dynamic program with a vector valued state variable. The dimension of this program is at least N + L, which makes it impractical to solve except for small values of N, the number of outlets, and L, the supplier/DC lead time. To circumvent the "curse of dimensionality" they recommend a myopic allocation scheme. The myopic allocation scheme requires minimizing the one period expected holding and penalty costs subject to the constraint that the total inventory allocated to the stores in any period t is equal to the order placed by the DC in period t-L. A further approximation is to relax the nonnegativity constraints.

They then show that the myopic allocation scheme is optimal for a modified version of the original dynamic programming problem. It is this modified model they recommend one use to solve the original problem. In the linear order cost case, computations suggest that their results are indeed very good. Observed errors for the cases tested were under 1%. The results were not as good when a fixed order cost at the DC was allowed. In particular, when the set-up cost for ordering at the DC was large and the coefficients of variation of demand at the stores were large and unequal, errors as high as 66% for the approximation were observed. The approximation is apparently least effective when there are many periods between ordering at the DC and there is a large difference between the demand variation at the stores. The authors report several extensions of their work as well.

Federgruen and Zipkin(1984c) consider other approaches to the problem of approximating optimal policies. In particular, they assume that the penalty and holding costs are proportional. That is, there exist positive constants p, h, and c_j, $1 \leq j \leq N$ so that $p_j = p\, c_j$ and $h_j = h\, _{cj}$ for $1 \leq j \leq N$ (this implies that the ratio of holding and penalty costs is the same at each location). The results of the later paper are useful for dealing with the problem of unequal coefficients of variation of demand at the stores. The approximation techniques and results are similar to those of Federgruen and Zipkin(1984a).

An extension of the Eppen and Schrage(1981) model to the case of correlated demands over time was considered by Erkip, et al.(1990). Although recognized as an important problem, with the exception of the work of Johnson and Thompson(1975), virtually no one has considered the impact of correlated demands over time on inventory control strategies. Let d_{jt} be the demand at retail outlet j in period t. They assume that

$$d_{jt} = R_j \hat{D}_t L_t + \epsilon_{jt}$$

where R_j is the fraction of mean total demand at store j, $\hat{D}_t$ is the forecasted (mean) total system demand, and ϵ_{jt} is the forecast error. The index variable, L_t is assumed to follow an autoregressive process of order 1 having the following representation:

$$L_t = \sigma_L Z_t + 1 \text{ where}$$
$$Z_t = \rho Z_{t-1} + \epsilon_Z.$$

Assuming that ϵ_Z has the normal distribution with mean 0 and variance $(1-\rho^2)$, it is easy to show that Z_t is standard normal and that L_t is normal with mean 1 and variance σ_L^2. The period to period demand correlation is ρ. Under the Eppen and Schrage(1981) assumptions, the authors obtain an expression for the system wide order up to level and include computations to show the effect of the demand correlation on this level. They show that the correlation can have dramatic effects on optimal stock levels. When the correlation is positive and large (near 1), the effect is to multiply the effective system wide standard deviation by as much as a factor of five.

4 Risk Pooling At the DC

The work discussed in the previous section assumed that the DC holds no stock. That is, as soon as a shipment arrives from the supplier, the entire shipment is distributed among the stores. There are clearly shortcomings in such a policy. Because of demand uncertainty at the store level, it is likely that the store's inventories could become widely out of balance prior to the time that DC receives another shipment. If goods cannot be transshipped among stores, it is possible that during the replenishment cycle some stores will run out of stock while others will be seriously overstocked. This is common in practice.

Assuming the DC holds no stock eliminates a key advantage of having a DC. When safety stock is retained at the DC, balancing of store stocks can be effected between DC replenishments. This has been labelled "risk pooling" by several researchers.

Jonsson and Silver(1987a) consider the problem of correcting the imbalance of stocks at the store level in the context of the Eppen and Schrage model discussed above. The authors assume that the DC orders once every H periods in such a way as to bring the system stock to a predetermined level, and the entire shipment is then allocated among the stores. They further assume that there is a redistribution of inventory among the stores exactly one period before the start of the next order cycle. The purpose of the redistribution is to provide a better balance of inventories among the stores to improve certain measures of service. They show that substantial improvements in system fill rate can be achieved by the redistribution, but note that the benefit of these improvements must be weighed against the costs of transshipments.

Let $B_j(t)$ be the expected number of units backordered by store j at the end of an period t, for $1 \leq t \leq H$ and $1 \leq j \leq N$ where H is the number of periods in an order cycle, and N the total number of stores. Let μ be the mean total demand of all stores in a period. Periodic demand is assumed to follow a normal distribution and is independent across time and across locations. The fraction of the total demand per order cycle that is backordered (among all the stores) is

$$\sum_{j=1}^{N} \sum_{t=1}^{H} B_j(t) / H\mu$$

One minus this quantity is used as the measure of system-wide service. (This is commonly known as the fill rate.) Expressions are derived for the expected backorders with and without redistribution and computations reported that show the advantage of redistribution. The advantages can be quite dramatic. The authors considered coefficients of variation of demand of 0.2 and 0.4 and a system wide fill rate of 0.99. For a fixed fill rate, redistribution reduced safety stock from 25% to almost 100% over the cases tested. The percentage increased with H, the number of periods in an order cycle, decreased with the reorder lead times, decreased with the coefficient of variation of the demand, and increased with the number of stores. Jonsson and Silver(1987b) considered the effect of retaining only a portion of the safety stock at the DC to be distributed to the stores to balance their inventories at the end of period H-1 (i.e., one period before the order placed by the DC is scheduled to arrive). They found that a substantial portion of the benefit of the complete redistribution assumed in Jonsson and Silver(1987a) could be achieved in this manner.

Jackson(1988) considered a problem similar to that studied by Jonsson and Silver(1987a), but assumed that the DC can effect a redistribution of stock <u>each</u>

period between DC replenishments. He assumed that this redistribution is accomplished by a ship-up- to-S policy each period. This means that as long as the DC has adequate stock, at the start of each period it will ship to each store the demand realized in the previous period. This will continue until either the DC receives a replenishment or it runs out of stock. If the DC runs out of stock before the next replenishment, then there will be one period in which there is insufficient stock at the DC to raise each store to its desired S level. A mathematical program is formulated to determine the optimal allocation in this runout period and is shown to simplify in the identical store case.

Jackson formulates an exact cost function assuming general demand and obtains a simple approximation for this function when the distribution of demand at the stores is Poisson or normal. The approximations give good results, but become less accurate as the number of stores increases. He tested from 2 to 16 stores for a four period cost function. The effect of lengthening the number of periods was not explored. The results were also compared to those of Eppen and Schrage(1981) in which the depot holds no stock. Substantial improvements (as much as 70%) in the number of backorder periods at the stores were observed when using the allocation policy described.

An extension of Jackson's work was considered by Jackson and Muckstadt(1989). They expand upon the model of Eppen and Schrage(1981) but assume zero lead times and two allocation periods within each order cycle. That is, there are two opportunities to ship stock from the DC to the stores within each order cycle. The analysis proceeds as follows: first they formulate the problem of the optimal second allocation in a period, given Q units available at the depot and (S_1, S_2, ..., S_N) at the N stores respectively. Using this solution, a cost function is formulated for the full two period problem which is shown to be convex. The general problem is formulated, but is apparently very difficult to solve. The only easily solved version of the problem assumes independent demands and two stores. The primary conclusion is that risk pooling by retaining stock at the DC appears to be attractive even when demands are independent and there are only two stores in the system. It remains to be seen whether this approach is viable for a system with a large number of stores, however.

Schwarz(1989) considers essentially the same model as Jonsson and Silver(1987a), but focuses his analysis on lead times rather than on safety stock. He treats two systems: one in which the external supplier ships directly to the stores, and one in which the DC receives the orders and then redistributes the inventory to the stores. The second case provides "risk pooling" since the presence of the DC allows limited redistribution of store stocks which have become out-of-balance over the outside supplier lead time (by postponing allocation of the order until it is received at the DC). However, the system with the DC is assumed to have an additional lead time between the DC and the store. The issue addressed in this study is: for the same system-wide fill rate and safety stock, what would be the breakeven value of the additional lead time when the DC is included? The primary result is that the value of adding the DC depends critically on the pipeline inventory costs, since the additional lead time means larger pipeline inventories. In general, retailers locate DC's near enough to stores so that the DC/store lead time is small, sometimes one day or less. Schwarz's analysis would indicate that there is a substantial advantage to including the DC between the supplier and the stores in this case.

The advantages of risk pooling were also explored in McGavin et. al.(1990). Between DC replenishments an allocation policy is specified by 1) the number of withdrawals from stock, 2) the intervals between successive withdrawals, 3) the quantity of stock withdrawn at each withdrawal, and 4) the partitioning of the withdrawn stock among the stores. This study has several advantages over previous work. It assumed lost sales at the store level, which is far more common for most low cost retail items. Second, it assumed that demand for each store was generated by a stochastic process with stationary independent increments. While this includes the Poisson as a special case, it also includes processes with variance to mean ratios larger than one such as Gaussian and gamma processes. One of the interesting results of this paper is that the optimal partitioning scheme is what they call a "balancing" scheme. This means that stock is partitioned to maximize the minimum store inventory.

The authors proceed to assume that the number of stores is infinite to circumvent the problems that arise in trying to analyze more than one withdrawal when there are a finite number of stores. Two heuristic allocation schemes are suggested from the results of the infinite store model. Their so-called "infinite retailer heuristic" seemed to perform well except when the number of stores was small, the service levels were high, and there was high demand uncertainty. (Unfortunately, many real systems have all these characteristics.) The authors conclude that policies with two well chosen withdrawals provide service levels comparable to four equal interval base stock oriented allocation policies (such as those examined by Jackson (1988)).

Kumar et. al.(1991) examine two alternative policies for replenishing and allocating inventories among N stores located along a fixed delivery route. The DC is restocked every m periods after a fixed lead time of T periods. The DC holds no inventory. Under a "static" policy, each replenishment order is allocated simultaneously at the instant of receipt at the DC. Under the "dynamic" policy, allocations are determined sequentially upon arrival of the delivery vehicle at each store. Simulation results show that the dynamic policies give far better results.

A unique model of the two level system we discuss in this review was developed by Graves(1989). He considers a system in which final demand at each retail site is generated by independent Poisson processes (other demand cases are considered as well), there are known transshipment lead times from each supply point, and the system is a general distribution topology. This last assumption means that the system structure is more general than that pictured in Figure 1. Each site has a single supplier, but may have several successors. This structure, known as an arborescence, was also considered by Bessler and Veinott(1965). He further assumes that replenishments are made only at fixed times at each location, and the size of the replenishment is equal to the number of units demanded at that location since the last replenishment. What makes the model interesting is the concept of virtual allocation. Each time a demand occurs, that demand triggers a replenishment request at the facility's supplier which then triggers another request to the next supplier and so on, all the way up the supply chain. This assumes that an information system is in place which allows each transshipment site to observe in real time the demand processes at all downstream retail sites.

This model is interesting in that it presents a unique way to deal with the problem of allocation when there is insufficient stock at the DC to meet orders from the stores. Virtual allocation requires that the DC track demands on a real time basis at all of the retail outlets. We are not aware of any real system possessing such sophisticated information flow technology, although it does not appear to be beyond current technology using point-of-sale data.

Graves' approach was also explored by Axsater(1990) who assumed order-up-to-S type policies. He also assumed Poisson demands and full backordering of demand. Axsater's contribution is a recursive approach for computing the holding and shortage costs at each level of the system.

Recently Nahmias and Smith(1991) developed a two level model with partial lost sales and arbitrary mean to variance ratio of the demand at the stores. They assume that the demand process is a Poisson process with a random rate parameter. Assuming that the rate parameter follows a gamma distribution, this construction gives rise to a negative binomial distribution for periodic demand at the stores. One advantage of the negative binomial distribution is that its variance to mean ratio can be any number exceeding one with proper choice of the parameters. (Our experience with actual sales data for a major store in the Bay Area showed variance to mean ratios as large as 100 or more for some product lines.)

They assume that each store is restocked according to a base stock or order-up-to S type policy. A newsboy-type model is developed to determine the optimal S level at the stores. Because of partial lost sales, the probability distribution of the total demand on the DC is quite complex. The authors derive the first and second moments of the distribution of the demand process on the DC. DC demand is approximated by a normal distribution. For the process they consider, the authors find that the benefit of retaining stock at the DC is most prominent for low demand high value items.

This model is simpler than many others in assuming that the DC/store replenishment lead time is zero. For systems with "quick response" and regularly scheduled shipments to the stores, this assumption is not unreasonable. However, this study is more realistic than others reviewed here in two respects. One is that it allows for partial or full lost sales, which characterizes the vast majority of retail systems. The other is that it allows the variance to mean ratio of demand to be larger than one.

5 Lot Size Reorder Point Systems

Probably the most widely used policy structure for managing inventories when demands are random is the lot size/reorder point, or (Q,R), policy. The policy is implemented for a single product, single echelon system in the following way: when the inventory position (on hand plus on order minus backorders) reaches a level R an order for Q units is placed. The order is assumed to arrive after a lead time τ which can be either known or random. (See Nahmias(1989) p. 222 for a discussion of the history of the development of this model.)

Since the (Q,R) model is widely used, it is only natural to consider adapting it for the DC/store system pictured in Figure 1. Deuermeyer and Schwarz(1981) were the first to do so. They assume that demands at the stores form independent Poisson processes with respective rates λ_i for $1 \leq i \leq N$. (Their results can be used for other forms of the demand process, but will only be approximate.) Each store follows an independent (Q,R) policy to place orders on the warehouse. Furthermore, the DC is also replenished according to a (Q,R) policy (with the DC lot size and reorder points being multiples of the store lot size). There is a fixed lead time L for resupply of stock from the DC to the stores. The goal of the analysis is to determine (approximate) expressions for several measures of system-wide performance. These are fill rate and expected backorders. The fill rate is the proportion of demands that can be met from available stock. They also provide expressions for the time weighted numbers of backorders at facility j.

It is difficult to analyze the process when the DC is out of stock. The natural assumption is that orders from the stores queue up at the DC and are filled on a first-come-first-served basis. This assumption leads to difficulties: one must keep track of the sequence of order placement to be sure demands are filled on a FCFS basis. The difficulty is handled in the following way. Assume that the effective lead time for store j is $M_j = L + D$, where D is the average time that an order chosen at random will wait at the DC before being shipped. An expression for D is derived based on the backorder rate at the DC. Using this expression for M_j, expressions are derived for the various measures of effectiveness discussed above.

Simulation tests showed a close match between observed simulated measures and those computed with the model. This study was important as it was the first that examined (Q,R) policies in the context of the DC/store system. However, from a practical perspective, it has limitations. One is that Poisson demand was assumed. Our experience with actual data, discussed in more detail in the last section, is that this is a poor model of actual demand patterns. Furthermore, the assumption of full backordering of demand is questionable at best. Retailers generally face partial or full lost sales. Finally, optimizing inventory levels was not considered in this study.

Even with its limitations, this study was an important first step in modeling the general DC/store problem with lot size reorder point policies and laid the groundwork for later research.

An extension to incorporate fill rate optimization was considered in Schwarz, et. al.(1985). Assume identical stores and let F_R be their common fill rate. Let S_W be the dollar investment in safety stock in the DC and S_R the dollar investment in safety stock at any store. Then the authors consider the problem of choosing the pair (S_W, S_R) to

$$\begin{aligned}&\text{maximize } F_R\\&\text{Subject to } \quad S_W + NS_R \leq S,\end{aligned}$$

where S is the total budget available for investment in safety stock. By varying values of S, the authors generate a locus of solutions in the (S_W, NS_R) plane which they refer to as a policy line. Analysis of the properties of the policy line leads to a simple piecewise linear approximation. Two heuristics are suggested: the "midband" heuristic and the "vertical" heuristic. The midband heuristic is derived from averaging upper and lower bounds on the policy line and the vertical heuristic from a deterministic version of the problem. The heuristics gave good results: the midband heuristic gave fill rates within 2% of the optimum and the vertical heuristic within 3% of the optimum for the ten cases tested.

Minimizing backorders for a similar model was considered by Schwarz and Badinelli(1988). Assuming identical stores whose expected backorders are B_R the authors formulated three optimization problems in which the goal is to choose safety stock at the stores and at the DC to 1) minimize system-wide backorders subject to a budget constraint on average system inventory, 2) minimize average system-wide inventory subject to a constraint on expected system backorders or 3) minimize system-wide backorders subject to a constraint on system-wide safety stock.

A primary contribution of this study is to demonstrate that under the stated assumptions these three optimization problems are equivalent in the sense of yielding the same optimal safety stock. The authors note the interesting result that the Deuermeyer and Schwarz(1981) model and its successors indicate that the optimal policy holds very little safety stock at the DC. Virtually all of the safety stock is held at the store level. In fact, in all the problems they consider the optimal DC safety stock is negative. This finding seems contrary to the "portfolio effect" observed by others for slightly different types of systems (for example, Eppen(1979)). McGavin et. al.(1990) speculate that this occurs because the requirement that store requests be filled on an FCFS basis at the DC precludes the ability of the DC to rebalance store stocks and thus eliminates the primary advantage of having the DC.

Svoronos and Zipkin(1988) consider several refinements of the model treated by Deuermeyer and Schwarz(1981). They obtain an expression for the variance of lead time demand at the warehouse that does not rely on asymptotic analysis and is more accurate. They consider a DC to store lead time that is composed of two components, one deterministic and one random, and derive the variance of the random component. (D&S assumed that this lead time is deterministic.) Finally, they use a mixture of Poisson distributions to estimate the distribution of DC lead time demand rather than a normal distribution. Using simulation, their results are compared to those of Deuermeyer and Schwarz(1981). The computations show a significant improvement in all cases tested.

A similar model was explored more recently by Axsater(1991). He develops a different computational method from Svoronos and Zipkin(1988) which the author claims is more efficient for large problems. He suggests three approximations which are compared numerically with inconclusive results.

6 Systems Managed With (s,S) Policies

When inventory levels are reviewed periodically rather than continuously, it is well known that for single product systems (s,S) policies are optimal (Scarf(1960)). An (s,S) policy is implemented as follows: when the starting inventory level in any period, x, is less than or equal to s, an order is placed for S-x. When inventory levels are reviewed continuously and demand is smooth, the starting inventory level when an order is placed is exactly s and the order size is S-s. By setting R=s, and Q=S-s, one sees that the continuous review (s,S) policy reduces to the (Q,R) policy. Hence, one may think of (s,S) policies as a larger class of policies that include (Q,R) policies as a special case.

The more interesting case is when inventory levels are reviewed periodically rather than continuously. In this case, the order size will vary depending on the magnitude of the overshoot of s. The first study of (s,S) policies in the context of a DC/store inventory system is due to Ehrhardt et. al.(1981). The authors assumed that each store follows an (s,S) policy placed on the DC for its replenishments and demands are reviewed periodically. Demands at the stores are assumed to be independent identically distributed random variables with arbitrary probability distributions. Full backordering of excess demand is assumed at all levels of the system, and orders are filled on a first come first served basis.

The total demand on the DC in any period is the sum of the demands at the stores. A primary contribution of this study is the observation that if the stores follow identical (s,S) replenishment disciplines, the demand pattern at the DC will be correlated in time. (Formulas for the size of the correlations are derived in Schultz(1983).) Because of these correlations, the authors note that the optimal form of the replenishment policy at the DC level is not an (s,S) policy, but still adopt that form because of its widespread popularity in practice. They suggest that the DC replenishment policy be computed by a modification of the so-called power approximation (Ehrhardt(1979)) which incorporates the period to period correlation.

Simulation was used to search for the (almost) optimal DC (s,S) replenishment policy to test the effectiveness of the approximation. For the 72 cases considered, the correlated adjusted power approximation resulted in policies that were on average 2.6% above optimal for the few stores environment and 3.4% above optimal for the many stores environment. (The numbers of stores is not specified, but the ratio of the many stores environment to the few stores environment is exactly four.) When the demand correlation is ignored, the observed cost errors are respectively 8.8% and 7.1% higher than optimal. It is interesting to note that when the correlation was included, the DC ordered more frequently and held less stock. This study is significant in being the first that recognizes that the (s,S) replenishment policy at the stores induces a correlated demand pattern at the DC and that significant errors arise if this correlation is ignored.

Recently Schneider et. al.(1990) consider the same model and focus on the issue of service levels at the DC and store levels. They treat three types of service levels: stockout occasions (α), fill rate (β), and time weighted backorders (τ). The difficult problem of determining the distribution of the lead time to the stores as a result of the DC being out of stock (which was analyzed exactly by Svoronos and Zipkin(1988) for the (Q,R) model) is dealt with in an approximate fashion via the α level of service provided by the DC. The lead time distribution from DC to store is approximated by a negative binomial. Power approximations (Ehrhardt(1979)) are used to provide estimates for the (s,S) values at the stores and at the DC. Simulations suggest that their results are good when the service levels at the DC are relatively high. For service levels of 50% or less, their results deteriorate significantly.

Svoronos and Zipkin(1991) consider continuous review (S-1,S) policies in the context of a multi-level inventory system. An (S-1,S) policy is implemented as follows: at each occurrence of a demand a unit is reordered. This policy is optimal for very high value items or for systems in which there is a small fixed cost for reordering. It is recommended for expensive reparable items, such as aircraft engines, rather than for low cost retail items (see Sherbrooke(1968), for example). The multiechelon system they consider may have an arbitrary number of layers. Only full backordering of excess demand is considered and demand at the lowest echelon is assumed to follow the Poisson distribution. However, they do allow for random transit times between levels. The transit times are assumed to follow a phase type distribution which includes many well known probability distributions. The authors explore numerically the effect of varying several system parameters on the optimal base stock levels. We believe that the three assumptions of 1) full backordering of excess demand 2) Poisson demand distributions and 3) one-for-one replenishment makes these results of only limited value to retailers.

Their results were extended by Zipkin(1991) to the case where demand at the lowest echelon follows a compound Poisson rather than simple Poisson distribution. With proper choice of the compounding distribution, one can model a demand process with an arbitrarily large variance to mean ratio. As we indicated earlier, our experience is that the variance to mean ratio of demand for some fast moving items can be as high as 100, thus making this a potentially valuable extension.

7 Modeling Lost Sales

The vast majority of the models we reviewed here assumed that excess demand at the store level was backlogged. This assumes that the customer would be willing to wait as long as it takes to have his or her order filled. This is reasonable for certain items. For example, furniture is often made to order: the store serves only as a showroom. However, for the vast majority of retail items, especially low cost items, customers will simply go elsewhere when the store is out of stock. Probably the most accurate characterization is to assume that some portion of the customers will be willing to wait and some will not. The fraction that would be willing to wait depends upon the characteristics of the item and the customer base.

When there is a positive order lead time, the lost sales problem is difficult to solve. Karlin and Scarf(1958) were the first to consider a dynamic inventory model with lost sales and established the form of an optimal order policy when the order lead time was exactly one period. They showed that the optimal policy in general was not a function of the sum of on hand and on order stocks, as it is in the full backordering case. Morton(1971) developed simple approximations for the lead time lost sales problem when there was no set-up cost for ordering. Suppose that the order lead time is λ periods. Then the approximation requires solving $\lambda+1$ backorder problems and ordering the minimum among the $\lambda+1$ solutions generated.

Nahmias(1979) developed several extensions of Morton's approximations. His approximations include adding a set-up cost for ordering, random lead times, and partial backordering. The set-up cost approximation requires solving a single dimensional dynamic program. Calculations are included which demonstrate the effectiveness of the approximations for a variety of system parameters.

In the context of continuous review, Hadley and Whitin(1963) present a modification of their heuristic lot size reorder point model which assumes lost sales. Unfortunately, the method is only approximate and is likely to be accurate only when service levels are very high.

Because of the difficulty of dealing with lost sales, few have considered it in the context of the DC/store inventory problem. Exceptions are McGavin et. al.(1990) and Nahmias and Smith(1991). In both cases, the authors assumed that the lead time for shipments from the DC to the stores was small enough to be ignored. One important direction for future research is the general two echelon DC/store inventory system with lost sales and positive order lead times from the supplier to the DC and from the DC to the stores.

8 Moving Towards Applications

Many researchers have studied the two level DC/store inventory system used by large retailers. While the theory underlying this problem has come a long way, we believe that in most cases, the assumptions underlying these models make them of only limited applicability in practice.

One common assumption is that the demand follows a Poisson distribution. Our experience is that the Poisson distribution accurately describes demand patterns for slow moving items only. The Poisson distribution has variance to mean ratio of one. When average demand exceeds about 10 units per period, variance to mean ratios increase. We performed an analysis on actual data on sales of a line of shirts sold at a store of a large retailer, and found variance to mean ratios of demand as high as 100. Also, demand is likely to be larger than sales for reasons we discuss in the next paragraph. Hence, we feel that the assumption of Poisson demand is too restrictive for most retailing applications.

Another common assumption is that excess demand is backordered. While this is a reasonable assumption for certain lines of products, such as furniture and appliances, either full or partial lost sales are common for most high demand low cost retail items such as apparel and consumer electronics. Retailers with whom we've discussed inventory management have difficulty identifying exactly what proportion of excess demands are lost: some customers may choose to special order items while others may choose to substitute similar items. However, even though the lost sales rate may be difficult to estimate, we feel that a retail inventory control model should allow some form of lost sales, either full or partial.

Several system features change from one application to another. Much of the effort on inventory modeling during the last thirty years is concerned with lead times. Safety stocks are retained to protect against demand uncertainty during the lead time for continuous review systems (Nahmias(1989)). However, for many retail systems, lead times may not be a major consideration. Deliveries from distribution centers are often made to stores on a regular schedule. Some suppliers also follow fixed delivery schedules when restocking DCs. "Quick response" systems often allow store managers to readjust order sizes right up until final delivery, making the effective reorder lead time negligible. However, this is not always the case. Fashion and/or private label items with higher than predicted demand can have long reorder lead times when supplies are scarce and demand exceeds manufacturing capacity. Special order items, such as home furnishings, are often made to order, requiring long resupply lead times. Hence, no single model can capture all the variations in lead time structure experienced by retailers.

While important strides have been made in the theory of inventory management for retailers, we believe that future efforts should be directed towards solving problems that arise in practice. Assumptions such as Poisson demand and full backordering of excess demand may be convenient analytical assumptions, but are inadequate for retailing. It is time for the research community to grapple with real issues if we expect retailers to have an interest in our work.

ACKNOWLEDGEMENT

The authors are grateful for both the financial support and the information on the current retailing environment which they have received through the Retail Management Institute, based at Santa Clara University. Dale Achabal, the RMI Director, and the major retailers who are members of the Retail Workbench have been particularly helpful in educating us about inventory problems in retailing.

References

Albright, S. C., "An Approximation to the Stationary Distribution of a Multiechelon Repairable-Item Inventory System with Finite Sources and Repair Channels," Naval Research Logistics, Vol. 36, pp. 179-195 (1989).

Allen, S. G., "Redistribution of Total Stock Over Several User Locations," Naval Research Logistics Quarterly, Vol. 5, pp. 51-59 (1958).

Axsater, S., "Optimization of Order-Up-To-S Policies in Two-Echelon Inventory Systems with Periodic Review," Unpublished manuscript, Lulea University of Technology, Lulea, Sweden, September, 1990.

Axsater, S., "Exact and Approximate Evaluation of Batch-Ordering Policies for Two-Level Inventory Systems," Unpublished manuscript, Lulea University of Technology, Lulea, Sweden, June, 1991.

Badinelli, R. D., and Schwarz, L. B., "Backorders Optimization in a One-Warehouse N-Identical Retailer Distribution System," Naval Research Logistics, Vol. 35, pp. 427-440 (1988).

Bessler S. A. and Veinott, A. F., "Optimal Policy for a Dynamic Multiechelon Inventory Model," Naval Research Logistics Quarterly, Vol. 13, pp. 355-389 (1965).

Clark, A. J., and Scarf, H. E., "Optimal Policies for a Multi-echelon Inventory Problem," Management Science, Vol. 6, pp. 475-490 (1960).

Clark, A. J., and Scarf, H. E., "Approximate Solutions to a Simple Multi-echelon Inventory Problem," Chapter 5 in Arrow, Karlin, and Scarf (eds.) Studies in Applied Probability and Management Science, Stanford University Press, Stanford, Ca., 1962.

Das, C., "Supply and Redistribution Rules for Two-Location Inventory Systems: One Period Analysis," Management Science, Vol. 21, pp. 765-776, (1975).

Demmy, W.S. and Presutti, V. J., "Multi-Echelon Inventory Theory in the Air Force Logistics Command," in Schwarz (1981).

Deuermeyer, B. L., and Schwarz, L. B., "A Model for the Analysis of System Service Level in Warehouse-Retailer Distribution Systems: The Identical Retailer Case," in Schwarz (1981).

Ehrhardt, R. A., "The Power Approximation for Computing (s,S) Inventory Policies," Management Science, Vol. 25, pp. 777-786 (1979).

Ehrhardt, R. A., Schultz, C. R., and Wagner, H. M., "(s,S) Policies for a Wholesale Inventory System," in Schwarz (1981).

Eppen, G. D., "Effects of Centralization on Expected Costs in a Multi-location Newsboy Problem," Management Science, Vol. 25, pp. 498-501 (1979).

Eppen, G. D., and Schrage, L. E., "Centralized Ordering Policies in a Multi-Warehouse System with Lead Times and Random Demand," in Schwarz (1981).

Erkip, N., Hausman, W. and Nahmias, S., "Optimal Centralized Ordering Policies in Multi-Echelon Inventory Systems with Correlated Demands," Management Science, Vol. 36, pp. 381-392 (1990).

Federgruen, A. and Zipkin, P., "Approximations of Dynamic Multilocation Production and Inventory Problems," Management Science, Vol. 30, pp. 69-84 (1984a).

Federgruen, A. and Zipkin, P., "Computational Issues in an Infinite Horizon Multiechelon Inventory Model," Operations Research, Vol. 32, pp. 818-836 (1984b).

Federgruen, A. and Zipkin, P., "Allocation Policies and Cost Approximations for Multilocation Inventory Systems," Naval Research Logistics Quarterly, Vol. 31, pp. 97-129 (1984c).

Graves, S. C., "A Multi-Echelon Inventory Model With Fixed Reorder Intervals," Working Paper #3045-89-MS, Sloan School of Management, MIT, (1989).

Gross, D., "Centralized Inventory Control in Multilocation Supply Systems," pp. 47-84 in Scarf, Gilford, Shelly (eds.), Multistage Inventory Models and Techniques, Stanford University Press, Stanford, Ca., 1963.

Hadley, G. J., and Whitin, T. M., Analysis of Inventory Systems, Prentice-Hall, Englewood Cliffs, New Jersey, 1963.

Hoadley, B., and Hyman, D. P., "A Two Echelon Inventory Model with Purchases, Dispositions, Shipments, Returns, and Transshipments," Naval Research Logistics Quarterly, Vol. 24, pp. 1-20 (1977).

Jackson, P. L., "Stock Allocation in a Two Echelon Distribution System or 'What to Do Until Your Ship Comes In'," Management Science, Vol. 34, pp. 880-895 (1988).

Jackson, P. L., and Muckstadt, J. A., "Risk Pooling in a Two Period, Two Echelon Inventory Stocking and Allocation Problem," Naval Research Logistics, Vol. 36, pp. 1-26 (1989).

Johnson, G. D. and Thompson, H. E., "Optimality of Myopic Inventory Policies for Certain Dependent Demand Processes," Management Science, Vol. 21, pp. 1303-1307 (1975).

Jonsson, H. and Silver, E. A., "Analysis of a Two-Echelon Inventory Control System with Complete Redistribution," Management Science, pp. 215-227 (1987a).

Jonsson, H. and Silver, E. A., "Stock Allocation Among A Central Warehouse and Identical Regional Warehouses in a Particular Push Inventory Control System," International Journal of Production Research, Vol. 25, pp. 191-205 (1987b).

Karlin, S., "Optimal Inventory Policy for the Arrow-Harris-Marschak Dynamic Model," Chapter 9 in Arrow, Karlin, Scarf, (eds.), Studies in the Mathematical Theory of Inventory and Production, Stanford University Press, Stanford, Ca., 1958.

Karlin, S. and Scarf, H. E., "Optimal Inventory Policy for the Arrow-Harris-Marschak Dynamic Model With a Time Lag," Chapter 10 in Arrow, Karlin, Scarf, (eds.), Studies in the Mathematical Theory of Inventory and Production, Stanford University Press, Stanford, Ca., 1958.

Krishnan, K. S., and Rao, V. R. K., "Inventory Control in N Warehouses," The Journal of Industrial Engineering, Vol. 16, pp. 212-215 (1965).

Kumar, A., Schwarz, L. B., and Ward, J. E., "Risk Pooling Along a Fixed Delivery Route Using a Dynamic Inventory-Allocation Policy," unpublished manuscript, Krannert School of Management, Purdue University, (1991).

McGavin, E. J., Schwarz L. B., and Ward, J. E., "Two-Interval Inventory Allocation Policies In a One-Warehouse N-Identical-Retailer Distribution System," Unpublished manuscript, Krannert School of Management, Purdue University, (1990)

Morton, T. E., "The Near Myopic Nature of the Lagged Proportional Cost Inventory Problem with Lost Sales," Operations Research, Vol. 19, pp. 1708-1716 (1971).

Nahmias, S., "Simple Approximations for a Variety of Dynamic Leadtime Lost Sales Inventory Models," Operations Research, Vol. 27, pp. 904-924 (1979).

Nahmias, S., "Managing Reparable Item Inventory Systems: A Review," in Schwarz (1981).

Nahmias, S. Production and Operations Analysis, Richard D. Irwin, Homewood, Illinois (1989).

Nahmias, S. and Smith, S. A., "Optimizing Inventory Levels in a Two Echelon System with Partial Lost Sales," unpublished manuscript, Santa Clara University, (1991).

Rosenfield, D. B., and Pendorck, M., "The Effects of Warehouse Configuration Design on Inventory Levels and Holding Costs," Sloan Management Review, Volume 21, Number 4, pp. 21-33 (1980).

Roundy, R., "98%-Effective Integer-Ratio Lot Sizing for One Warehouse Multi-Retailer Systems," Management Science, Vol. 31, pp. 1416-1430 (1985).

Scarf, H. E., "The Optimality of (s,S) Policies in the Dynamic Inventory Problem," in Mathematical Methods in the Social Sciences (K. J. Arrow, S. Karlin, and P. Suppes, eds.), Stanford University Press, Stanford, California, 1960.

Schneider, H., Rinks, D. B., and Kelle, P., "Allocation of Safety Stocks in a Wholesale Inventory System Using A Service Level," Unpublished manuscript, College of Business Administration, Louisiana State University (1990).

√ Schultz, C. R., "Computing Demand Properties at the Wholesale Warehouse Level," Naval Research Logistics Quarterly, Vol. 30, pp. 37-48 (1983).

√ Schwarz, L. B. (ed.), Multi-Level Production/Inventory Control Systems: Theory and Practice, TIMS Studies in the Management Sciences, Volume 16, North Holland, Amsterdam, 1981.

Schwarz, L. B., "A Model for Assessing the Value of Warehouse Risk-Pooling: Risk-Pooling Over Outside Supplier Leadtimes," Management Science, Vol. 35, pp. 828-842 (1989).

√ Schwarz, L. B., Deuermeyer, B. L., and Badinelli, R. D., "Fill-Rate Optimization in a One-Warehouse N-Identical Retailer Distribution System", Management Science, Vol 31, pp. 488-498 (1985).

Sherbrooke, C. C., "METRIC: A Multi-echelon Technique for Recoverable Item Control," Operations Research, Vol. 16, pp. 122-141 (1968).

Simpson, K. E., Jr., "A Theory of Allocation of Stocks to Warehouses," Operations Research, Vol. 7, pp. 797-805 (1959)

√ Svoronos, A., and Zipkin, P., "Estimating the Performance of Multi-Level Inventory Systems," Operations Research, Vol. 36, pp. 57-72 (1988).

Svornos, A., and Zipkin, P., "Evaluation of One-For-One Replenishment Policies for Multiechelon Inventory Systems," Management Science, Vol. 37, pp.68-83 (1991).

Zipkin, P., "Evaluation of Base Stock Policies in Multiechelon Inventory Systems with Compound-Poisson Demands," Naval Research Logistics, Vol.38, pp. 397-412 (1991).

RECENT ADVANCES IN PRODUCTION AND DISTRIBUTION MANAGEMENT

A. Federgruen, Graduate School of Business
Columbia University, New York, N.Y.

1. Introduction

Forty years ago, we experienced the birth of a new scientific discipline aimed at supporting the planning and design of production and distribution systems. This pioneering effort was led by a number of seminal papers, in particular Arrow et al. (1951), Scarf (1960), Clark and Scarf (1960), and Wagner and Whitin (1958).

Arrow et al. introduced the most basic single item inventory model with uncertain demands: inventories are inspected periodically, demands in these periods are independent and identically distributed and all cost components are linear. The authors showed that a simple base-stock policy is optimal for this model. Scarf (1960) added the complication of fixed order costs and proved that an (s,S) - policy is optimal for finite horizon models: the item's inventory position is increased to S when, at the beginning of a period, it is found to be below s. Clark and Scarf extended these results to a facility in-series system with demands at the lowest echelon only, linear costs throughout except for fixed-plus linear order costs at the top echelon facility. Wagner and Whitin addressed the single-item model with non-stationary, i.e., time-varying demands and costs, all assumed to be deterministic. These authors developed an elegant solution method whose complexity grows quadratically with the length of the planning horizon.

It is remarkable that these seminal papers appeared in the course of a few years only, and many decades after the initial insights in this area were developed: Harris (1913) introduced the classic economic order quantity (EOQ) model providing qualitative insights into the impact of economies of scale in production and order costs. See Erlenkotter (1990) for an interesting historical perspective on this paper. The basic model with uncertain demands (but no economies of scale in the cost structure) goes in actuality back to the famous nineteenth century economist Edgeworth (1888) who analyzed it in the context of cash management for banks. It is remarkable that almost all introductory Operations Management courses taught in MBA programs continue to initiate their discussion on inventory planning with a treatment of these two basic models. Likewise, every first Masters of Science or Ph.D. level course on inventory theory is likely to start with a treatment of the four above seminal papers of the fifties and early sixties.

A parallel development in the very same early sixties is the foundation of the field of Operations Management, spearheaded by El Buffa's pioneering books: Buffa (1961, 1963). These books reviewed many elementary inventory and production planning models, but distinguished themselves especially by their description of actual production/distribution systems, emphasizing the need for integration with several other planning areas. These include in particular the areas of demand forecasting, data processing and information management, human resource management and quality control. The importance of the interdependencies between these planning areas has only recently been rediscovered. Our profession is infinitely indebted to Elwood's pioneering contributions and continued leadership.

In the following 20-25 years, the fields of inventory theory and operations management diverged to a large extent. Inventory theory focused by and large on single item, single location systems; where more complex multi-level and multi-item systems were addressed, it was largely done via mathematical programming models. Such models treat all relevant information as deterministically given from the outset, thus ignoring all sources of uncertainty. Moreover, exact or even approximate solution of these models is often difficult and time consuming, in particular when a large number of integer variables are involved to reflect a variety of discrete choices. Most importantly, the solutions of these models are often of a complex structure making them difficult to implement even if they can be computed in a moderate amount of time.

Practitioners of operations management, on the other hand, focused their attention, in the same 20-25 year time interval, on the development of general production and distribution support systems, such as Material Requirement Planning (MRP) and Distribution Requirement Planning (DRP) systems, see e.g. Orlicky (1975) and Smith (1978). These describe a production/distribution systems as an acyclic network. Each node corresponds with a specific item at a specific location or at a specific stage of production. An arc between a pair of nodes (i,j) indicates that a given number of units of the item corresponding with node i is used as an input in the procurement process of one unit of the item corresponding with node j. Consider for example the network in Figure 1: 2 units of item 1, 3 units of item 2 and 4 units of item 3 are combined to produce 1 unit of item 6. One unit of item 1 is also used, along with 2 units of item 4, when producing a unit of item 5.

(MRP) and (DRP) tend to specify replenishment strategies on the basis of a simple single-pass procedure which decomposes the planning problem into as many separate single-item lot sizing problems as there are nodes in the network. (In addition, most implementations employ heuristic rather than exact solution methods for the single item lot sizing problem as well.)

More specifically, starting with the end-items for which only external demands prevail, a sequence of replenishment lots is determined on the basis of a single item lot sizing procedure. With the help of the bill of materials those get translated into demands for intermediate products and/or facilities. The remainder of the planning procedure consists of a recursive solution of single-item lot sizing models for all remaining nodes in the network.

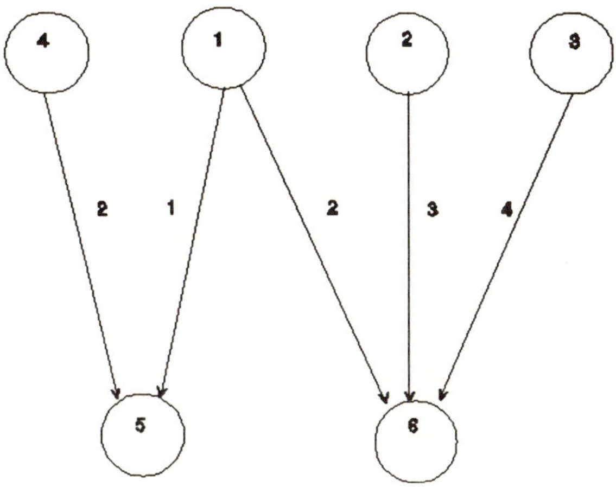

Figure 1: Production distribution network

Since the above described planning approach proceeds on a node by node basis, it is likely to result in highly suboptimal system-wide strategies. Moreover, the above planning procedure completely ignores economies of scale due to joint cost structures, or capacity limitations. As far as the latter is concerned, the more advanced MRPII systems attempt restoration of feasibility in a final stage (but do not always succeed). In other words, while the commercially available planning systems attempt to develop a feasible system-wide procurement plan, no consideration is given to balance the tradeoffs between the costs incurred at different levels of the system, to compare different feasible accommodations to capacity restrictions, or to exploit economies of scale that arise from joint cost structures.

Another example of a commercially available production support system is (OPT) which focuses on the identification of bottleneck production stages and employs heuristics to determine production schedules at those stages. The exact mechanics of these heuristic are not publicly available and their performance cannot be assessed or characterized.

Fortunately, the trend towards divergence between inventory theory as a scientific discipline and practical operations management has been dramatically reversed: in the last seven years, we have experienced rapid progress towards the development of good planning and design models, and the identification of relatively simple solutions for these models, which are efficiently computable and whose cost can be guaranteed or expected to come within a small percentage of optimality. These models and solution methods address complex production/distribution networks. Most importantly, they are used not only to identify operational replenishment strategies, but to address major tactical and strategic questions as well.

In Section 2, we give some examples of such strategic questions. The remainder of this paper is devoted to an overview of some of the recent advances in planning methods for production/distribution networks. In addition to the basic tradeoffs between setup, holding and backlogging costs for each individual item and location, those planning methods need to grapple with at least four supplementary categories of challenges:

(I) Time variations - the planning process needs to account for demands and cost parameters varying over time due to the impact of seasonal fluctuations, trends, advertising and price promotions, and the general variability of customer orders, among others.

(II) Capacity limitations - peak demands need, in the presence of capacity limitations, be anticipated by increased production and distribution activities in prior periods. As for the demand and cost parameters, capacities often vary over time due to e.g. use of temporary workers, hiring or layoff decisions, acquisition of new equipment and planned maintenance.

(III) Joint cost structures - Another category of complications arises when one or more cost components may be jointly incurred for a group of items or facilities. Joint cost structures arise in particular with respect to fixed setup costs. These joint setup costs usually represent economies of scale which an efficient replenishment strategy should try to take advantage of.

(IV) Physical dependencies - The need for careful coordination of production and distribution activities stems in addition from physical or technical interdependencies between different items or different facilities. For example, one item may be used as an input in the production process of another item, by itself or in combination with other inputs. Likewise, a specific facility (e.g. regional warehouse, retailer outlet) may be replenished by a different facility (central warehouse, regional warehouse, respectively).

The above challenges are often compounded by the prevalence of a major degree of uncertainty regarding demands, leadtimes, capacities or yields.

It has now become clear that no single class of models is capable of adequately addressing all settings one encounters in practice. On the other hand, our profession is rapidly building up an arsenal of models which cover an ever larger variety of settings. To keep the exposition reasonably succinct, we focus on a specific, relatively simple, network structure i.e., a two-echelon distribution network. This network structure is discussed throughout the remainder of this chapter under different assumptions regarding the cost structure and the demand and leadtime patterns (and the type of information available for them). Where applicable we briefly discuss how these methodologies have been extended to general production/distribution networks.

In Section 3, we discuss deterministic models and in Section 4 planning models under uncertainty.

2. Some Strategic Questions

Many companies have become aware of the need to organize or redesign their production and distribution processes so as to substantially reduce the total production and distribution leadtime, i.e., the time between the initial procurement of raw materials and externally purchased components and the ultimate delivery of the finished product to the customer. To reduce average leadtimes as well as their variability, companies select their suppliers in close geographic proximity (e.g., Toyota), and engage in tighter relationships with their suppliers emphasizing short and reliable leadtimes rather than price cuts per se; they substitute technologies by faster ones, improve the training of their work force, use faster and more reliable modes of distribution etc. These reductions in average leadtimes and their variability clearly result in reduced work-in-process and finished goods inventories for a given service level, or alternatively in improved service levels for a given investment in inventories. The *magnitude* of these efficiency and service improvements can, however, only be assessed with an adequate inventory and production planning model.

Of equal strategic importance, is a company's choice to *standardize* or *diversify* its product line. Companies move in both directions depending on the specific competitive niche which they seek to occupy in their market. Standardization clearly results in more efficient production processes and smaller aggregate inventories for a given desired service level. Once again, the magnitude of these savings needs to be assessed by an adequate production/inventory planning model.

In several industries, one observes that many companies engage in a complete redesign of their production processes to exploit commonalities between their products. An example in point is the automobile industry; many Japanese manufacturers have reduced the total throughput time (i.e., the time between initiation of a production order and final release to the customer) of their cars sold in the U.S. to 50-60 days, even for those manufactured in the Far East. While the product line consists of a large variety of car styles and options, very few distinct body and engine types are employed. Many Japanese manufacturers thus produce each body type and assemble each engine type in large batches. The subsequent diversification into the final products is obtained via a last stage installation of specific options (which sometimes occurs at the dealers). This production process exploits statistical economics of scale a term explained in Section 4, allowing for drastic reductions in safety stocks. This is in sharp contrast with the traditional production process still employed by the U.S. manufacturers in which each car type is produced by itself. U.S. manufacturers experience a total throughput time of 100 days; a U.S. manufactured car spends an average of 60 days at the dealer versus 30 days for the Japanese manufacturers. These numbers are obtained from Gonsalvez (1991). The redesigned production processes allow the Japanese manufacturers to offer a diversified line of final products while exploiting most of the benefits associated with standardization.

The pressure to compete via an efficient logistical process has given rise to new forms of cooperation between wholesale suppliers and retailers: high-level cooperation is instituted between teams of the supplier and retailer organization, involving data processing experts and sales and purchasing managers. The objective is to smooth the retailer's order sequences and to reduce uncertainty about the retailer demands to a bare minimum, by reducing information delay and improving its accuracy. This new form of cooperation has e.g., been instituted by Proctor and Gamble and many of the major retailer organizations it supplies (e.g., K-Mart, Wal-Mart). Proctor and Gamble has estimated a potential for 1 billion $ in annual savings resulting from these new forms of cooperation (see page B1 of the New York Times, July 14, 1991).

In other industries we have seen that suppliers adopt the complete inventory management function of their retailer customers: by billing the retailer at the same time it makes a sale to a consumer, the capital costs associated with the retailer's inventories are borne by the supplier. The supplier is given the responsibility to replenish the retailer's inventory at its discretion while guaranteeing a given fill rate or being charged for the lost sales or backlogs: this arrangement has become common in the industrial gas and soft drinks industry as well as among suppliers of supermarkets. It alleviates the individual retailer of its costly inventory investments; the supplier acquires the advantage of being able to determine when and in what quantities to deliver to its retailer customers. The distribution system as a whole may benefit because the supplier needs to maintain a smaller aggregate safety stock than the sum of the safety stocks required when each retailer is responsible for its own inventory. The extent of there benefits can however only be assessed by an adequate inventory/distribution planning model.

Logistical considerations have driven several industries to redesign even the structure of their distribution networks, for example by eliminating intermediate levels to reduce total leadtimes and by eliminating inventories at intermediate stocking points. An example is the Nissan automobile manufacturer which recently announced that it will start making direct sales (through mail order arrangements) to the consumer, bypassing the dealers. Finally, many industries promote substitution between similar product types, allowing them to reduce overall inventories.

3. Deterministic Models

In this section, we describe a number of recent planning models with deterministic assumptions.

Consider therefore a system with a central depot which supplies J outlets where exogenous demands for a single commodity must be filled, see Figure 2. We refer to such systems as One-Warehouse-Multi-Retailer (OWMR) systems. Inventories are reviewed continuously. The depot may, at any point in time, place an order with an outsider supplier; likewise, each retailer may, at any point in time, place an order with the warehouse.

Depot orders arrive after a lag of L time units; a shipment to retailer j ($j = 1,\ldots,J$) reaches it ℓ_j periods later. In the basic version of the model all order sizes are assumed to be unrestricted. All demands must be satisfied without backlogging. (Many of the planning models discussed below, have been extended to allow for backlogging.)

The model can also be viewed as representing a two-stage production process, involving several products instead of several locations. A depot order then corresponds to a decision to make blanks or some other intermediate product. This stage requires L time units. The intermediate product is later fabricated (in ℓ time units) into the several finished products. In the description below, we adhere to the distribution version of the model.

3.1 Models with Separable Cost Structures and Stationary Parameters

We first consider a model whose cost structure is separable across the (J + 1) - facilities. More specifically, at each retailer, as well as the warehouse, inventory carrying costs are incurred at constant (but facility specific) rates per unit and per unit of time. Variable order costs are proportional with the order sizes; in addition a (facility specific) setup cost is incurred for each individual order. The objective is to minimize long run average costs. We assume, without loss of generality that the system starts

Two echelon

distribution systems

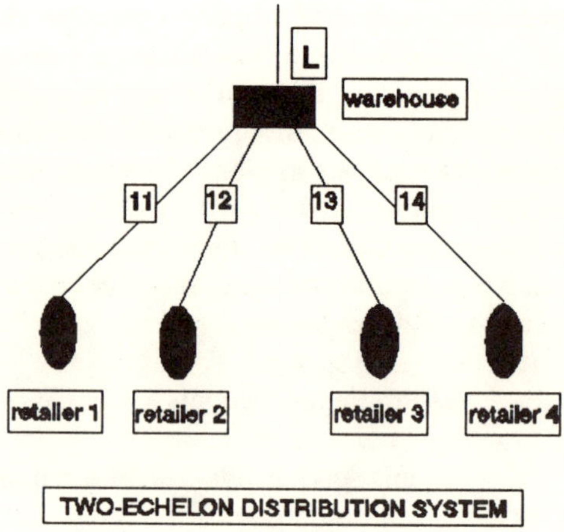

- L warehouse order leadtime
- ℓ retailer order leadtime

Two - stage

production system:

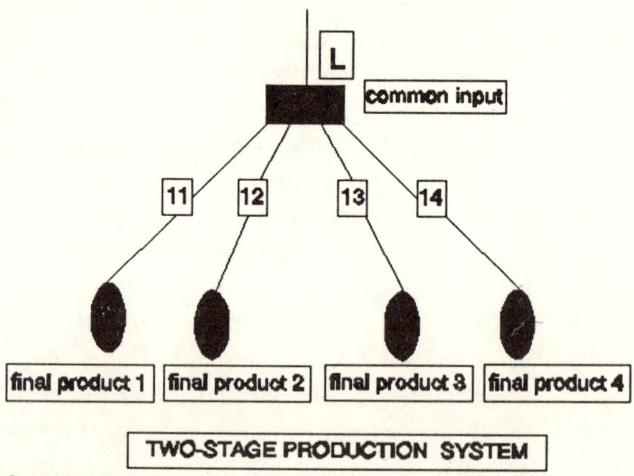

- L leadtime first production stage
- ℓ leadtime seems production stage

Figure 2

empty at time zero.

A replenishment strategy specifies all replenishment epochs and quantities for all facilities over an infinite planning horizon. The structure of an exact optimal strategy has not been identified, even for this simplest of all one warehouse multi-retailer planning models; more importantly it is clear by now that this structure is exceedingly difficult. A fully optimal strategy is therefore unattractive even if it could be computed efficiently.

Roundy (1985) established, however, that a simple so-called power-of-two-policy can be found whose cost is guaranteed to come within 2% of a lower bound for the minimum system - wide costs. Moreover, this policy can be found by a simple algorithm of complexity 0(J log J) or even 0(J) in a later implementation by Queyranne (1987). A power-of-two policy applies zero-inventory ordering, i.e., every facility places orders (if and) only if its inventory (a lead time later) is down to zero; it prescribes for each facility a replenishment interval T such that an order is placed for this facility at times T, 2T, 3T,... Moreover all replenishment intervals are chosen as power-of-two multiples of a base planning period (e.g., 1 day, 1 hour etc.).

Power-of-two policies are attractive because of the following reasons:

(1) A power-of-two policy is characterized by a number of parameters which is equal to the number of facilities (i.e., the number of replenishment intervals). In fact, the number of distinct intervals is certainly no larger than nine (say, independent of J) in any practical setting, since a larger number would imply that some facilities are replenished at a frequency more than $2^{10} = 1024$ times that of some other facility.

(2) As mentioned above, a power-of-two policy can be found whose cost is guaranteed to come within 2% of optimality. If the base planning period is fixed, the worst case optimality gap is still limited to 6% only.

(3) Power-of-two policies have often been implemented in practice because their simplicity facilitates planning, scheduling, and coordination.

Interestingly enough, all of the above results are easily extended to production/distribution networks of a general (acyclic) structure as described in Section 1, with external demands occurring at any subset of the network nodes, see Figure 1. It can still be shown, see Roundy (1986) that a power-of-two policy (defined as above, merely replacing "facility" by "node") exists whose cost comes within 2% of a lower bound for the minimum cost value.

Federgruen and Zheng (1990) building on earlier work by Maxwell and Muckstadt (1985) show that an optimal power-of-two policy may in fact be determined by a limited number of maximum flow computations in networks closely related to the original production - distribution network. Federgruen and Zheng (1990) also show that the same type of algorithm can be employed to determine an optimal power-of-two policy, even when individual upper and lower bounds prevail for the nodes' replenishment frequencies and order sizes (i.e., so-called capacitated models). Such capacity constraints arise e.g., in production settings where each product is manufactured in its own facility or work center, each with a given capacity per time unit. Capacity constraints also arise (indirectly) when the setup cost structure is described by a step function, i.e., at node i in the network a setup cost K_i is incurred whenever an order is placed for B_i units or less and a multiple of the setup cost is incurred when an order of size larger than B_i is placed. Such cost structures arise e.g., with limited container or vessel sizes in production operations or limited truck capacities in distribution settings. A third example of settings in which upper or lower bounds for the replenishment internals arise are systems in which each node's inventory needs to be stored in a facility of limited capacity, see Federgruen and Zheng (1990) for details.

The complexity of the fastest of the above mentioned algorithms is only $0(RN^2)$ with N the number of nodes in the production/distribution network and R the number of distinct routes in this network. (A route is a path in the acyclic network which starts with an arbitrary node and terminates with a node with external demands. Even though the number of routes can, in principle, be exponential in the number of network nodes, it is in most practical network structures of modest size. The fact that the complexity is linear in R is quite remarkable since the computational effort associated with the evaluation of a single power-of-two policy is 0(R) as well.) We also refer to Muckstadt and Roundy (1992) for a recent survey of this category of planning models.

3.2 Models with Submodular Joint Cost Structures and Stationary Parameters

In the above described models, the interdependency between the nodes of the production/distribution network is restricted to physical availability constraints: in the (OWMR) model, it is necessary to coordinate the warehouse replenishment strategy with that of each of the retailers, only to insure that sufficient warehouse inventories are available to cover all retailer orders. A second type of interdependency arises when economies of scale can be exploited by combining several replenishment activities (or orders) into one. The simplest such joint structure arises when a joint setup cost is incurred whenever a joint order is placed by a group of retailers, regardless of the specific set of retailers involved. The joint setup cost is incurred in addition to any retailer specific setup costs charged for each specific retailer included in the order.

The models in Section 3.1 have been extended to allow for general joint setup cost structures, merely assumed to reflect economies of scale in the sense of submodularity, i.e., the increase in the setup cost due to the addition of a new network node to a given collection of nodes is no larger than if the same node were added to a subset of this collection. Many joint cost structures satisfy this submodularity property, see Federgruen and Zheng (1988,1992) for a detailed discussion. (See however, subsection 3.3 for settings where submodularity fails to apply.)

Federgruen et al. (1992) have shown that an optimal power-of-two policy continues to be close-to-optimal in the sense that its cost is guaranteed to come within 2% of a lower bound for the minimum system-wide costs (or 6% if an arbitrary, fixed, base planning period is used). Federgruen and Zheng (1991) show that such a policy as well as the lower bound continues to be computable via algorithms which consist of a limited number of maximum flow computations in networks, closely related to the production/distribution network. The only difference with the algorithms for models with separable cost structures consists in the fact that the maximum flow computations are somewhat more involved: in addition to ensuring that the flow on each arc stays below an arc - specific upper bound (as is the case in ordinary maximum flow computations), an upper bound needs to be imposed on every subset of the collection of arcs which terminate in the network's sink. Such maximum flow problems are referred to as polymatroidal network flow problems for which efficient solution methods have been obtained, see Lawler and Martel (1982), Hassin (1982) and Tardos et al. (1985).

An interesting qualitative result is that the joint cost structure may be allocated i.e., substituted by a separable structure which is equivalent in the following sense: the original model (with joint costs) and the transformed model (with separable costs) share the same set of optimal power-of-two policies and their long-run average cost is identical under the nonseparable and separable cost structure alike. The cost allocation is obtained in the course of the above mentioned algorithms. In the context of the planning problems addressed here, this result provides a positive answer to the question whether an appropriate allocation scheme can be found for joint production/distribution cost structures which does not result in distorted incentives. See Cooper and Kaplan (1988) and Kaplan and Atkinson (1989).

3.3 One Warehouse Multi-Retailer Systems with Vehicle Routing Costs and Stationary Parameters.

In many (OWMR) - systems, goods are distributed from the warehouse to the retailers by a fleet of vehicles, combining deliveries into efficient routes. Such systems give rise to joint cost structures which in general may fail to be submodular. The methods of Section 3.2 therefore do not apply; in particular, power-of-two policies may no longer be desirable.

Assume therefore that the transportation costs between the warehouse and retailers include a fixed (leasing or renting) cost per route driven by one of the vehicles and variable costs proportional to the total (Euclidean) distance on all routes. Except for the retailers' order costs, we maintain all assumptions of the basic model of Section 3. In the production variant of the model, routing costs correspond with item dependent switchover costs in the second production stage. (The assumption that these be proportional to Euclidean distances is considerably more stringent in this setting.)

A new planning methodology for systems with vehicle routing costs has recently been proposed by Anily and Federgruen (1990,1991) based on a restriction to a class of replenishment strategies Φ with the following properties; a replenishment strategy in Φ specifies a collection of regions (subsets of retailers) covering all outlets; if an outlet belongs to several regions a specific fraction of its sales/operations is assigned to each of there regions. Each time one of the outlets in a given region gets a delivery, this delivery is made by a vehicle which visits all other outlets in the region as well (in an efficient route).

We note that a large amount of flexibility is preserved within the class Φ by allowing retailers to be assigned to several regions, i.e., by allowing regions to overlap. The restriction to the class Φ involves, admittedly, some loss of optimality, the exact magnitude of which is not yet known. On the other hand, the restriction is often externally imposed by the sales/distribution system under consideration. In many systems, the sales and delivery functions are integrated: a salesman is assigned to a given region and each salesman is required to visit the outlets in his region periodically in a given route determining replenishment quantities (in the form of definite sales or unbinding consignments, see Section 2) and delivering them as well.

Anily and Federgruen (1990,1991) develop heuristic solution methods as well as lower bounds for the minimum cost (among all strategies in Φ), computable in $0(J \log J)$ time (i.e., the same complexity order as the basic model in Section 3.1.). The first paper deals with systems in which no inventories may be kept in the warehouse; the second paper addresses systems with centralized stock. No simple characterization for the worst case optimality gap is available in this case; but it has been shown, under very mild probabilistic conditions regarding the outlets' locations, that the bounds are asymptotically accurate (as $J \rightarrow \infty$) and the heuristic asymptotically optimal (within Φ) for systems without centralized stock, and the (asymptotic) optimality and accuracy gaps are bounded by 6% for systems with centralized stock. The model and its analysis allows for a variety of constraints, e.g., upper bounds on the frequency with which the corresponding routes may be driven, possibly in combination with load capacities for the vehicles and bounds on the sales volume per region.

The asymptotic analysis has been complemented with extensive numerical studies exhibiting excellent performance, even for systems with a moderate number of outlets only.

3.4 Models with Non-Stationary Parameters

All of the models in subsections 3.1 - 3.3 assume that all cost, demand and lead time parameters are constant over time. But, as pointed out in the introduction, the planning process often needs to account for demands and cost parameters varying over time due to the impact of seasonal fluctuations, trends, advertising and price promotions and the general variability of customer orders, among others.

The following model represents the natural extension of the basic stationary model in subsection 3.1: consider a finite planning horizon of N periods, maintain all assumptions of the basic model except that all demands, holding cost rates, variable and fixed order costs are allowed to be time-dependent. The planning problem is clearly strongly NP complete as follows from Arkin et al. (1989). Moreover, until very recently no heuristic with provable optimality gaps had been identified for this problem.

Tzur (1992) recently developed an exact branch-and-bound method for the problem. The method can efficiently be used, by itself, for problems of moderate size (i.e., with 25-30 periods and several tens of retailers). The efficiency of the branch-and-bound method is due to the use of one of three new lower bounds and associated upper bounds to evaluate the nodes of the tree. In addition, recent forecast horizon results for single item dynamic lot sizing models, obtained in Federgruen and Tzur (1992) suggest that optimal or close-to-optimal initial decisions can be made by truncating the horizon after a relatively small number of periods. These are based on a new simple procedure to detect minimal forecast horizons for the general single-item model. (A forecast horizon F and associated planning horizon f implies that the decisions in the first f periods are independent of any parameters beyond the forecast horizon F.) In all of the investigated problem instances, a forecast horizon, is found in which at most three and usually only two orders are placed. It is reasonable to expect that similarly short forecast horizons continue to apply in multi-echelon systems. This together with the attractiveness of the above branch-and-bound method for relatively short problem horizons suggest that a close-to-optimal solution may be generated by partitioning the horizon into smaller intervals.

Tzur therefore develops a partitioning heuristic which can be implemented with complexity $0(JN^2 \log\log N)$. The partitioning heuristic can be designed to guarantee an ε-optimal solution for any $\varepsilon > 0$, provided that some of the model parameters are uniformly bounded from above or from below. In particular, the heuristic is asymptotically optimal as $N \to \infty$, for any fixed number of retailers J, and it remains asymptotically optimal when both J and N are simultaneously increased to infinity.

The partitioning heuristic partitions the complete horizon of N periods into several intervals, and specifies an associated (OWMR) problem for each of these. The intervals are of a size which permits the use of the exact branch-and-bound method mentioned above. The (OWMR) problem associated with a given interval is obtained from the restriction of the complete problem to the periods of that interval, with an option to choose starting conditions that appropriately complement the solutions obtained in prior intervals.

No numerical studies on the performance of the partitioning heuristic for the general (OWMR) model are available, as of yet. They are however available for the model without central stock (usually referred to as the Joint Replenishment Problem); see the numerical study in Chapter 7 of Tzur (1992). Here the partitioning heuristic has been shown to perform exceptionally well. Even with very small intervals, the average optimality gap (where measurable) is only 0.38% and in none of the problem categories considered is it larger than 0.78%. Problems with up to 30 periods and 15 retailers are solved by the partitioning heuristic on an IBM 4381 in at most a couple of CPU seconds.

The partitioning heuristic can (at least in principle) be applied to more general production/distribution networks with general time varying parameters. The only problem is that the size of the tree of an exact branch-and-bound method required to solve each interval's problem, quickly increases as more complex network structures are considered. To allow for a tractable exact solution at the leaves of the branch-and-bound tree, the order periods need to be specified in a large set of network nodes N^0. (In the (OWMR) model, only the warehouse needs to be included in N^0.) Chapter 9 in Tzur (1992) describes two approaches to circumvent this difficulty: a network decomposition approach and a network restriction approach. In the latter, one restricts the possible combinations of sets of order periods for the different nodes in the network by partitioning the nodes in N^0 into a relatively small number (H) of node sets $\{N^1,\ldots,N^H\}$ such that all nodes that belong to the same set place their orders in the same periods.

One possible way to generate this node partition is to replace all demand and cost parameters by stationary values (e.g., their average over the interval) and to identify (with the methods of subsection 3.1) a power-of-two policy which optimizes the long-run average network-wide cost for this stationary version of the problem. Recall that under a power-of-two-policy all nodes with the same replenishment interval share the same order periods. It is therefore reasonable to choose N^j as the set of nodes in N^0 with the jth largest replenishment interval under the power-of-two policy.

4. Planning Models under Uncertainty

In many settings, a major degree of uncertainty prevails about cost, demand or leadtime parameters. In this Section we discuss models capable of addressing this complication. We confine ourselves to models with fully centralized control (i.e., PUSH-systems). See Axsäter (1992) for a recent survey on planning models in semi-decentralized settings.

To incorporate demand and leadtime uncertainty, the basic model of subsection 3.1 needs to be adjusted as follows: assume inventories are reviewed and decisions taken periodically. Demands at the retailers are now described as random variables with known distributions. Demands are assumed to be independent in different periods, but arbitrary correlation patterns may prevail between demands at different retailers in the same period. Unfilled demand at each location is back ordered. (Under demand uncertainty, stockouts can not be avoided with 100% probability, unless the demand distributions are of bounded support and even then only with, in general, excessively high safety stocks.) Leadtimes may be random with arbitrary distributions, but they are assumed to be independent of the demand process. We consider the same basic cost structure as in subsection 3.1. Planning methods exist both for finite planning horizons (allowing for time-dependent cost parameters and demand distributions) and infinite planning horizons with stationary parameters and distributions.

As before, it is useful to distinguish between systems with and those without centralized stock. In the former each incoming warehouse order may be stored at the warehouse, in part or as a whole; in the latter, it needs immediately to be allocated among the retailers . For such systems, centralized ordering offers two distinct advantages: first, as in deterministic models, it allows for the exploitation of economies of scale in ordering. Note also that one could choose to decide the future allocations at the same time as the order. Postponing the allocations permits one to observe the demands in the intervening L periods, and thus to make better informed allocations. The phrase "statistical economies of scale" has been coined by Eppen and Schrage (1981) to describe this effect.

In systems with centralized stock, a third advantage is added to centralized ordering, i.e., the ability to maintain a common reserve stock, further exploiting statistical economies of scale. In systems with centralized stock, three types of decisions need potentially to be made every period; (i) whether to place a warehouse order and of what size; (ii) how much inventory to withdraw from the warehouse; (iii) the allocation of this withdrawal quantity among the retailers.

No satisfactory planning methods appear to be available to date for systems with fixed shipment costs (between the warehouse and the retailers). On the other hand, very good, efficiently computable and insightful approximations are available for systems without such fixed shipment costs. These were initiated by Eppen and Schrage (1981)'s seminal paper which deals with systems without central stock, and the special case where holding and penalty costs are identical across locations and all demands are Normally distributed. This paper requires that a warehouse ordering policy of a specific type (i.e., a base-stock policy) be followed and that incoming orders be allocated myopically, i.e., so as to minimize expected costs in the first period in which these allocations have an impact. Eppen and Schrage show how the cost of any such strategy may be approximated by a closed form cost expression and how this can be embedded in a search for the best warehouse base stock level. (When warehouse orders incur fixed costs, Eppen and Schrage permit ordering at constant intervals (rather than each period) but the analysis requires still further approximations.) Federgruen and Zipkin (1984a, 1984c) address the general model without central stock. They derive a lower bound for the minimum cost by showing that the exact multi-dimensional dynamic program collapses (exactly, or by close approximation) to a standard (stochastic) inventory problem for a single location.

It is well known from Scarf (1960) that an (s,S) - policy (with respect to the system - wide inventory position) is optimal for this approximate lower bound problem, see also Section 1. A feasible strategy for the entire (OWMR) system may be obtained by complementing this (s,S) policy with an appropriate allocation policy, e.g., the above myopic allocation policy which indeed is optimal in the approximate problem. Federgruen and Zipkin (1984c) have, however, shown that superior allocation policies can be constructed with relatively simple procedures. Extensive numerical studies have exhibited that the cost of the combined order/allocation policy comes invariably within a few percentage points of the lower bound. The conclusion is that the lower bound can be used as a simple but accurate approximation, while the combined order/allocation policy is close-to-optimal and simple to implement. (Federgruen and Zipkin have also investigated approaches with an up front restriction to a given class of order policies, as in Eppen and Schrage. Lower bounds derived in this approach only apply to the restricted class of strategies.)

The lower bound approximation is ideally suited to analyze many of the strategic questions mentioned in Section 2, in particular the impact of reductions of lead time averages or their variability, the impact of redesigned production processes to exploit commonalities between products, etcetera. This lower bound reveals that the cost performance of the (OWMR) is closely approximated by that of a single location model: one which experiences the same mean system - wide demands, but different (indeed, larger) standard deviations. For example, in the special case where all $\ell_j = \ell$, all parameters are stationary, all retailers' one-period demand standard deviations equal σ, and demands are independent across retailers, the single location model has a leadtime demand variance given by $[L\, J + (\ell + 1)\, J^2]\, \sigma^2 > [L + \ell + 1]\, J\sigma^2$, the leadtime demand variance in a system where all retailers' inventories can be pooled.

Worst case or probabilistic bounds for the optimality gap are unknown as of yet, both for J finite and asymptotically as $J \to \infty$. The first such characterizations have however, recently been obtained by Atkins and De (1991) for a related model, i.e. a two-stage assembly systems with zero leadtimes. Federgruen and Zipkin (1984b, 1991) have extended these analyses to systems with central stock. As mentioned above, no complete satisfactory approach is available for systems with fixed-plus variable shipment costs, but Chen and Zheng (1991) recently obtained a clever and tractable lower bound for such systems and a complete methodology is in progress.

4.1 Planning Models with Transshipments or Product Substitution

The above described models assume that no lateral transshipments between retailers are allowed. In many distribution systems, stockouts at one of more retailers are cleared by transshipments from retailers with positive inventories. In the multi-item version of the model such "transshipments" correspond with product substitutions where customers are offered an alternative item, in case their first preference is out of stock. Such transshipments (product substitutions) are referred to as non-anticipatory; they are only used to (partially) clear existing shortages rather than to rebalance inventory levels. See Robinson (1990) and the references therein for the systems with anticipatory transshipments.

Klein (1990) recently extended the above planning methods to allow for such non-anticipatory transshipments, assumed to incur variable costs, proportional to the amounts transshipped. Klein focuses on systems without central stock but similar methods should work for settings with central stock. Due to the transshipments, the system faces, in each period, potentially three types of decisions: (i) how much (if anything) to order by the warehouse; (ii) how to allocate any incoming order; (iii) how to transship between outlets with positive and those with negative inventories. Klein has shown that highly precise cost approximations can be obtained once again by reducing the large dimensional dynamic program to a single dimension one. (The resulting cost approximation is no longer a strict lower bound; see however below.) In extensive numerical studies, the cost approximation always comes within 1.6% of the (simulated) cost of the best of four proposed combined order/allocation/transshipment policies, and the average discrepancy is only .18%. The cost approximation and policies can be computed very fast: for systems with 5 retailers, the average computational time is only .05 seconds on an Amdahl 3090. (The computational effort is roughly proportional with the number of retailers.)

Lower bounds can be computed on the optimal cost among all strategies with fixed - interval warehouse orders: the discrepancy between the policies' cost and the lower bound is 3% in average. Allowing for non-anticipatory transshipments can reduce system-wide costs by 16% (with an average improvement in the study of 4.7%). The recommended allocation policy is markedly different from that in systems without transshipments. Once again, it is possible to approximate the cost of the entire system by that of a single location system, with mean one-period demands equal to the mean system-wide demand. The variance of one-period demand is however larger than that of system-wide aggregate demands, but smaller than that employed in models without transshipments. For the special case mentioned above, the variance of lead-time demand in the approximate single-location model is given by $[(L + \ell_r)] J + (\ell + 1 - \ell_r) J^2] \sigma^2$ for an appropriately computed number $0 \leq \ell_r < \ell + 1$.

References

Anily, S. and A. Federgruen, (1990) One warehouse multiple retailer systems with vehicle routing costs, Management Science 36, 92-115.

Anily, S. and A. Federgruen, (1991) Two-echelon distribution systems with vehicle routing costs and central inventories (to appear in Operations Research).

Arkin, E., D. Joneja and R. Roundy, (1989) "Computational complexity of uncapacitated multi-echelon production planning problems", Operatio ns Research Letters 8, 61-66.

Arrow, K., T. Harris and J. Marschak, (1951) "Optimal inventory policy", Econometrica 19, 250-272.

Atkins, D. and S. De (1991) A 94% effective heuristic for lot sizing in an assembly system facing stochastic demand, Working Paper, Faculty of Commence, University of British Columbia, Vancouver, Canada.

Axsäter, S., (1992) Continuous review policies for multi-level inventory systems with stochastic demand, Chapter 5 in Handbook in Operations Research and Management Science, Volume 4: Logistics of Production and Inventory, (eds. S. Graves, A. Rinnooy Kan and P. Zipkin), North Holland, Amsterdam, Netherlands.

Buffa, E., (1961) Modern Production Management, John Wiley, New York, NY.

Buffa, E., (1963) Models for production and operations management, John Wiley, New York, NY.

Chen, F. and Y. S. Zheng (1991) A lower bounding framework with multi-echelon stochastic inventory systems, Working Paper, Decision Sciences Department, Wharton School, Philadelphia, PA.

Clark, A. and H. Scarf, (1960) "Optimal policies for a multi-echelon inventory problem" Management Science 6, 475-490.

Cooper, R. and R. Kaplan, (1988) "Measure Costs Right. Make the Right Decisions", Harvard Business Review (September-October issue).

Edgeworth, F., (1888) "The mathematical theory of banking", J. Roy. Statist. Soc. 51, 113-127.

Eppen, G. and L. Schrage, (1981) Centralized ordering policies in a multi-warehouse system with leadtimes and random demand, in Multi-level Production/Inventory Control Systems: Theory and Practice (L. Schwarz ed.), North Holland, Amsterdam, Netherlands.

Erlenkotter, D. "Ford Whitman Harris and the Economic Order Quantity Model", Operations Research 38, 934-937.

Federgruen, A., (1992) Centralized planning models for multi-echelon inventory systems under uncertainty, Chapter 4 in Handbook in Operations Research and Management Science, Volume 4: Logistics of Production and Inventory, (eds. S. Graves, A. Rinnooy Kan and P. Zipkin), North Holland, Amsterdam, Netherlands.

Federgruen, A., M. Queyranne and Y.S. Zheng, (1992) "Simple power of two policies are close to optimal in a general class of production/distribution networks with general joint setup costs", (to appear in Math. Op. Res.).

Federgruen, A. and M. Tzur, (1992) Minimal forecast horizons and a new planning procedure for the general dynamic lot sizing model: nervousness revisited (to appear in Operations Research).

Federgruen, A. and Y.S. Zheng, (1988) "Minimizing Submodular Set Functions: Efficient Algorithms for Special Structures with Applications to Joint Replenishment Problems" Columbia University, Graduate School of Business, Working Paper, New York, NY.

Federgruen, A. and Y.S. Zheng, (1990) "Optimal power of two replenishment strategies in capacitated general production/distribution networks", Working paper, Graduate School of Business, Columbia University, New York, NY.

Federgruen, A. and Y.S. Zheng, (1991) "Efficient algorithms for finding optimal power of two policies for production/distribution systems with general joint setup costs", Working paper, Graduate School of Business, Columbia University, New York, NY.

Federgruen, A. and Y.S. Zheng, (1992) "The joint replenishment problem with general joint cost structures: general solution methods and performance bounds, (to appear in Operations Research).

Federgruen, A. and P. Zipkin (1984a) "Approximation of dynamic multi-location production and inventory problems", Management Science 30, 69-84.

Federgruen, A. and P. Zipkin (1984b) "Computational issues in an infinite-horizon, multi-echelon inventory model", Operations Research 32, 818-836.

Federgruen, A. and P. Zipkin (1984c) "Allocation policies and cost approximation for multi-location inventory systems", Naval Research Logistics Quarterly 31, 97-131.

Federgruen, A. and P. Zipkin (1992) "Approximations for dynamic two-echelon inventory systems with centralized stocks" (in preparation).

Goldratt, E. and J. Cox, (1986) The Goal, North River Press, Croton-on-Hudson, NY.

Gonsalvez, D., (1991). Integrating Sales and Manufacturing Order Management Processes, Applied Scheduling: integration of disciplines, Columbia University, October 24-25, New York, N.Y.

Harris, F., (1913) "How many parts to make at once", Factory, The Magazine of Management 10, 135-136, 152.

Hassin, R., (1982) "Minimum Cost Flow in the Set-Constraints", Networks 12, 1-21.

Kaplan, R. and A. Atkinson, (1989) Advanced Management Accounting, 2nd edition, Prentice Hall, Englewood Cliffs, NJ.

Klein, R., (1990) Dynamic multi-location inventory problems with transshipments, Ph.D. dissertation, Columbia University, New York, NY, 10027.

Lawler, E. L. and C.U. Martel, (1982) "Computing Maximal 'Polymatroidal' Network Flows", Math. Op. Res. 7, 334-346.

Maxwell, W. L. and J. A. Muckstadt, (1985) "Establishing Consistent and Realistic Reorder Intervals in Production-Distribution Systems", Operations Research 33, 1316-1341.

Muckstadt, J. and R. Roundy, (1992) Analysis of Multistage Productions Systems, Chapter 3 in Handbook in Operations Research and Management Science, Volume 4: Logistics in Production and Inventory (eds. S. Graves, A. Rinnooy Kan and P. Zipkin), North Holland, Amsterdam, Netherlands.

Orlicky, J., (1975) Material Requirement Planning, McGraw-Hill, New York, NY.

Queyranne, M., (1985) "A Polynomial-Time Submodular Extension to Roundy's 98% Effective Heuristic for Production/Inventory Systems", Working paper no. 1136, University of British Columbia, Vancouver, B.C., Canada, V6T 1W5.

Queyranne, M., (1987) "Finding 94% effective policies in linear time for some production/inventory systems", Working paper, University of British Columbia, Vancouver, Canada.

Robinson, L., (1990) Optimal and approximate policies in multiperiod, multilocation inventory models with transshipments, Operations Research 38, 278-295.

Roundy, R., (1985) "98% effective integer ratio lot sizing for one warehouse multi-retailer systems", Management Science 31, 1416-1430.

Roundy, R., (1986) "A 98% effective lot-sizing rule for a multi-product, multi-stage production/inventory system", Mathematics of Operations Research 11, 699-727.

Scarf, H., (1960) The optimality of (s,S) policies in dynamic inventory problem. In: K. Arrow, S. Karlin, and P. Suppes (Eds.), Mathematical Methods in the Social Sciences, Stanford University Press, Stanford, CA.

Smith, D., (1978) Material Requirements Planning, in Studies in Operations Management (ed. A. Hax), North-Holland, Amsterdam.

Tardos, E., C. Tovey and M. Trick, (1986) "Layered Augmented Path Algorithms", Math. Op. Res. 11, 362-370.

Tzur, M., (1992) A new planning approach for single and multi-item dynamic lot sizing models, Ph.D.-dissertation, Columbia University, New York, NY.

Wagner, H.M. and T.M. Whitin, (1958) "Dynamic version of the economic lot size model", Management Science 5, 89-96.

PERFORMANCE MANAGEMENT ISSUES IN FLEXIBLE MANUFACTURING SYSTEMS: AN ANALYTIC PERSPECTIVE

Abraham Seidmann
W.E. Simon Graduate School of
Business Administration
University of Rochester
Rochester, NY 14627

ABSTRACT

Increased competition has led to pressures for the development of manufacturing systems that recognize the need for flexibility, efficiency and economies of scope and scale. In recent years managers have started to realize that investing money in new automated facilities does not by itself guarantee performance improvement. Managed poorly, these investments may even degrade the plant's performance. This paper focuses on the major issues relevant to performance management of Flexible Manufacturing Systems (FMS). The paper highlights the distinctive nature of these modern manufacturing systems and describes their emerging role in support of the overall corporate strategy. This new role results in a more volatile workload and in an increased versatility of the required production mix. It also challenges production managers who need to quantify and monitor the effective productive capacity at their FMSs. Moreover, it is shown here why some operational decisions, which are relatively simple to model and to comprehend in a conventional manufacturing context, become more involved - and less intuitive - when carried out in the typical FMS context. All this leads to the development of a new performance management paradigm for FMSs. That new paradigm consists of the following modules: *Physical FMS Attributes, Operating Policies,* and *Order Profiles.* A critical research overview of several analytical decision models formulated recently is developed here. This overview examines the characteristic decisions within each one of the modules of the proposed performance management paradigm. The paper further evaluates the changing role of FMSs in the finished products' life cycle and identifies major research opportunities.

1. INTRODUCTION

The pace of introduction of new technologies in manufacturing has been increasing. Increased competition has led to pressures for the development of manufacturing systems that recognize the need for flexible, efficient and economies of scope and scale. Shorter product life cycles combined with rapidly changing production economies tend to make the design and redesign of manufacturing systems more frequent. What is more, production systems should follow the products as they go through their life cycle (Buffa (1984)). The appropriate interface between the production system and the products is a matter for careful strategic choice. This interface affects the choice among competing strategies that vary in terms of costs, availability, innovation, quality, or flexibility. As a result of all these problems there is an increasing need for performance evaluation and management techniques, and modeling or analytical tools, that assist in the design, evaluation, and effective management of such industrial systems. In this study, we focus on the special issues relevant to performance management of Flexible Manufacturing Systems (FMS) and on some of the analytical tools developed to help deal with them. In general, analytic models of manufacturing systems are developed to provide users with rapid and accurate answers to problems involving cost and performance tradeoffs in four major areas: *manufacturing system design, purchasing turn-key systems, continued improvements*, and *capacity management.*

The task of defining, measuring, and monitoring production capacity is typically straightforward in conventional manufacturing facilities. That task gets more involved in the flexible environment as a result of two primary sets of issues. The first set of issues is the impact of volatile, and transient production mix. Other factors confounding the capacity control tasks include pallet mix, jobs priorities, loading policies, part routing, choices of tools and processing rates, or machine grouping patterns and tool availability considerations. All these factors have direct bearing on the issue of performance monitoring. The other set of issues has to do with determining the optimal way to stretch the existing productive capacity. Given the nature of the dynamic market, and the complex interplay among the various resources, the optimal capacity expansion decision must be based upon careful quantitative analysis which integrates as many of these concerns as possible (Brown(1988)).

The complexities of these FMS environments support the conclusion that FMS capacity management should be based upon a vector of performance measures. Based upon a survey of the research literature in operations management, and based upon numerous case studies conducted by the author, a set of common measures can be identified:

* *Number of parts produced.* This includes both good and scrapped parts.
* *Machine utilization.* This includes measures by machine type and the contribution from each product type. Such metrics also indicate the time proportions dedicated for set up, processing and maintenance activities.
* *Flow times per item.* These are segmented by product type as well as by operation. The relative time proportions of net processing, queueing delays, transportation, load/unload and set up time are also important in this context.
* *Work in process (WIP) and queues.* This also includes the distribution of queues ahead of particular machines, equipment groups, or the composition of WIP per product type.
* *Due date performance.* The typical measures used in this context include the proportion of late orders and lateness contribution (or progress monitoring) at each work center.

The performance management decision options, which are relatively simple to model and to follow intuitively in a conventional manufacturing context, become more involved - and less intuitive - when carried out in the typical FMS context. Two examples illustrating the counter-intuitive nature of these decisions are given next.

We begin our examination by looking at the impact of changing the number of pallets on the FMS throughput. Doing so we denote by $TH(\mathbf{K}, \mathbf{C})$ the throughput of a single-class FMS with K pallets and a vector $\mathbf{C}$ describing the allocation of machines to work stations. It can be shown that if the FMS is modeled as a closed network of queues (Seidmann, Schweitzer, and Shalev-Oren (1987), Schweitzer and Seidmann (1992)), then the value of $TH(\mathbf{K}, \mathbf{C})$ is increasing and concave in $\mathbf{K}$. If the FMS has R types of jobs, with $\mathbf{K}_r$ dedicated pallets for class r (r=1,2,...R) then the throughput of class r, $TH_r(\mathbf{K}, \mathbf{C})$ is increasing with $\mathbf{K}_r$. On the other hand, the throughput of class m, $TH_m(\mathbf{K}, \mathbf{C})$, $m \neq r$, need not increase with $\mathbf{K}_r$. Furthermore, the aggregate system throughput $\sum_{r=1}^{R} TH_r(\mathbf{K}, \mathbf{C})$ need not increase in $\mathbf{K}_r$, for all r values. The implication is that total FMS throughput may not increase when one increases the total number of pallets. This conclusion contradicts our prior expectations from the single part type case.

Buzacott and Shanthikumar (1992)) show that $TH(\mathbf{K}, \mathbf{C})$ is increasing and concave in the number of machines in work station i, $\mathbf{C}_i$ (i=1,2,...M). Hence the FMS throughput increases if machines are pooled together to form larger work centers. If the FMS has R types of jobs, with Kr dedicated pallets for class r (r=1,2,...R) then the throughput of class r, $TH_r(\mathbf{K}, \mathbf{C})$ need not increase with C_i (i=1,2, ...M). Moreover, the aggregate system throughput $\sum_{r=1}^{R} TH_r(\mathbf{K}, \mathbf{C})$ need not increase when more machines are pooled together. In multiple product FMS the pooling of resources may not result in improved aggregated throughput. These two examples show the subtleties in the performance management of FMS, where certain decisions that are intuitively beneficial may actually result in overall inferior performance. This is due to the unexpected interaction effects among the various part types at certain bottleneck resources. All this motivates the need for developing a more comprehensive set of performance management models to improve our intuition about system response and to provide some guidance for key resource allocation decisions.

2. RESEARCH ON MAJOR FMS PERFORMANCE ISSUES

2.1. THE FMS PERFORMANCE PARADIGM

Corporate strategy determines the competitive posture of a company with respect to propositions such as development of core competency in certain manufacturing technologies, order lead times, production costs, quality assurance, degree of innovation and the rate of new product introduction, support for existing products, and the scope of products. The FMS Performance Paradigm presented in Figure 1 depicts the overall perspective for dealing with that subject. As shown in Figure 1, the major factors affecting a system's performance are: the *physical FMS attributes* (machines, layout, pallets, tools, etc.), the *operating policy* (part routing and dispatching, batching, tool allocation to machines, etc.), and the *profile of the orders* (due date assignment policies, master production schedule, release control priorities, etc.). The performance management function uses these three sets of inputs along with the corporate strategy guidelines to provide two types of outputs. The first output is a set of *feasibility indicators*. These simply signal whether certain objectives, such as on-time deliveries of particular customer orders, can be met at all. The second is an overall assessment of the *performance measures* and their trends. The next section provides an overview of analytical research efforts dealing with each one of the three major factors affecting system performance. Due to space limitations this overview is not an exhaustive one. It is presented primarily to provide an overall structure within a single integrated framework.

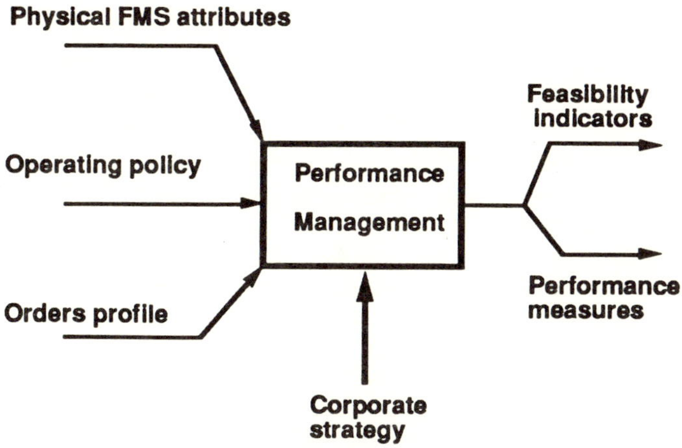

Figure 1. The FMS performance paradigm

2.2. PHYSICAL ATTRIBUTES

Capacity has become a general term with multiple interpretations in FMS research. A variety of issues are associated with planning the physical capacity attributes in FMS. These include choices in determining *processing rates*, the *number of machines*, *material handling robots*, or the *number of pallets* to be used. Representative studies of these major choices are discussed below. We begin with two recent studies treating the problem of allocating the total production capacity among the various machines in a FMS. Wein (1989) considers the capacity allocation problem for a generalized Jackson network. That network has general interarrival times and a general service time distribution, the system throughput is TH parts/unit - time, and the visit ratio for machine i (i=1,2, ...M) is V_i. The problem he considers is the determination of the variable processing rate at each machine ($1/S_i$, i=1,2, ...M), subject to an overall budget constraint of $D. A linear unit capacity cost of C_i/S_i is assumed. Heavy traffic Brownian motion approximation is used for deriving the optimal result:

$$\frac{1}{S_i} = TH\ V_i + \frac{\sqrt{c_i \sigma_i^2}}{\sum_{j=1}^{M} \sqrt{c_j \sigma_j^2}} \left(\frac{D - TH \sum_{j=1}^{M} c_j V_j}{c_i} \right)$$

The σ_i^2 values indirectly measure the variability of the queue length and of the workload at machine i. This result suggests that each machine is allocated the minimal production capacity required to satisfy its effective arrival rate, plus an added amount which is proportional to the square root of the variability parameter σ_i^2. This assures that machines associated with greater variability will be allocated a greater share of the available capacity.

The study by Schweitzer and Seidmann (1991) presents an optimization methodology for minimizing the variable capacity investment in FMS given the throughput target TH, the WIP level K, the part routes, transporter delays, and the variable production cost function. That study focused on the tooling costs and assumed a general cost function typical for metal cutting applications. The FMS dynamics are modeled by extending the Mean Value Analysis queueing network approximation of Schweitzer (1979). In the case of linear costs, it is shown by Schweitzer and Seidmann (1991) that the optimal capacity allocation is given by:

$$\frac{1}{S_i} = TH\, V_i \left(1 + \frac{\sum_{j=1}^{M} V_j \sqrt{C_j}}{V_i \sqrt{C_i}\, K y_o} - \frac{1}{K}\right)$$

The y_o measures the capacity loss due to transporter delays. This result generalizes the classical square root capacity allocation rule (Kleinrock 1964) for dealing with closed queueing network models of FMS's with transporter delays. The main insight gained here is that the optimal processing rate at machine i equals the minimal capacity THV_i, needed at this machine in order to meet the throughput goal, plus an added nonnegative increment: the percentage increment in this machine capacity is inversely proportional to the product of the visit ratio multiplied by the square root of the marginal capacity cost-rate parameter and is inversely proportional to the transporter delays in the FMS (y_o). This result also indicates that it is not optimal to set equal utilization in all machines.

Facilitating the manufacturing positioning of the firm, one has to consider the basic trade-offs among the various performance measures such as lead times, WIP, routing flexibility, variable operating costs, quality, and investments in capacity increments. The capacity targeting problem discussed by Bitran and Tirupati (1989) deals with an open multi-product queueing network with single machine stations and general arrival and process time probability distributions. The objective is to determine a least-cost schedule of capacity additions to achieve a target performance measure such as an average WIP level. In their model, capacity denotes resource availability, and a one-to-one correspondence is shown between the processing rate and capacity of each machine. Several powerful heuristics are presented for solving the resulting nonlinear program. Other related studies include Shanthikumar and Yao (1987, 1988), Bitran and Tirupati (1989), Boxma, et. al (1990), Dallery and Stecke (1990), and Vliet and Rinnooy Kan (1991).

A design optimization problem concerning the number of pallets of each type circulating in an FMS with several pallet types is considered by Solot (1990). The design objective is to maximize the throughput rate of a given product mix, subject to limiting constraints on the total investment in pallets and on the maximal turnaround times per pallet type. The heuristic solution algorithm starts with an initial set having one pallet per part type. This set is incrementally modified yielding the maximal increase of the objective function at each step. Vinod and Solberg (1985) and Dallery and Frein (1986) also consider design optimization problems in which the costs of pallets and machines must be minimized while achieving a given throughput rate.

The multitude of issues relevant to the FMS implementation and acquisition decision problems are addressed by the performance evaluation model developed by Arbel and Seidmann (1984). Their performance evaluation approach, which is carried out using the analytic hierarchy process (AHP), handles both tangible and intangible decision factors. For a number of reasons their performance evaluation approach has proved to be quite useful in the industrial applications conducted by the authors and by subsequent researchers (Canada and Sullivan (1989), Bijayananda and Chakravarty (1991)). First, it allowed the incorporation of various levels of expertise into an integrated framework. Thus, upper level management and engineering experts could contribute in their respective areas of knowledge and responsibility. This assured that all important concerns were properly addressed in the process. Second, the approach is based on combining different modules into an assessment hierarchy. This modularity property permits great flexibility in addressing different decision issues that reflect the idiosyncrasy of the particular organization. Finally, the complete process proved to be very useful in communicating ideas among the participating decision makers who identified with the focus and details of the study and the competing vendors. These external vendors were briefed on their relative proposal status and were given a chance to respond with better understanding of their customers' needs. In practice, that effort usually leads to impressive improvements in the overall cost effectiveness of the system performance.

2.3. OPERATING POLICY

We begin this section with the key observation that individual part movements are practical in FMS because of the automated material handling system. In addition, having the ability to change the machine program from one part type to another within short time intervals enables rapid capacity distribution changes among the various part types. These changes may be required as a result of varying demand patterns for the end products or as a result of machine failures. Machine failures have an immediate impact on the effective capacity and on the delivered production mix. The managerial decision options for policy selection implemented by various FMS managers vary depending on the nature of the production technology being used and on the characteristics of the major performance evaluation criteria. These decision options include *part loading, job routing, workload assignment to machines, priority scheduling, processing rates, tools set ups, shifts, overtime and batching*. Representative studies of these major choices are discussed below.

The recent study by Schweitzer, Seidmann, and Goes (1991) shows that the ability to adjust system behavior by changing operating conditions may lead to some unexpected responses: 1) increasing WIP saves money, 2) increasing capacity increases machine utilization, and 3) having unbalanced machine utilization can be beneficial. For the case reported in that paper, the authors show that the major cost savings, in order of decreasing importance, are:

1) Tool speed selection
2) Part rerouting and assignment of tasks to machines
3) Tool type selection
4) Setting multiperiod throughput goals
5) Adjusting the number of pallets.

The three level hierarchical production scheduling policy developed and implemented by Akella, Choong, and Gershwin (1984) uses available status information, past performance trends, and system flexibility to anticipate and to react to disruptive events. The hierarchical policy is designed to keep the difference between the total number of parts of type r and the total number of parts required as close as possible to zero for all r=1, 2,...R. The concept of capacity is a crucial component of that scheduling policy as the FMS capacity at any instant depends on the operational status of the machines. This status changes as machines fail, or are repaired. If parts are loaded into the FMS at a rate that violates the capacity constraints, material accumulates in buffers or in the material handling system. This results in congestion and poor performance. The status of the machines is given by the vector α, where $\alpha_i = 1$ or 0 (i=1, 2, ...M) if machine i is up or down, respectively. The *top level* of the hierarchy computes (off-line) two key decision parameters: the number of machines that type r parts visit, $A_r(\alpha)$, and the hedging point for type r part, H_r. The hedging point determines the dynamic amount of safety stock deployed to compensate for capacity shortfalls as a result of machine breakdowns. The proposed hedging point $H_r(\alpha)$ for part type r is:

$$H_{r(\alpha)} = \frac{T_r d_r(b\rho_r - a d_r) - T_f ab(\rho_r - d_r)}{(a+b)\rho_r}$$

where d_r is the demand rate for type r, T_r is the MTTR for all machines used by r, T_f is the MTBF for all machines used by r, and ρ_r is the average utilization of all machines used by r. The values of a and b represent the costs of having a surplus or a shortage of parts. As expected, the optimal hedging point increases with the MTTR and with the machine utilization, and it is inversely proportional to the machine reliability. The *middle level* of the hierarchy computes on-line the short-term production rates for each part type for each machine rate. The local objective is to compute the production rates such that the safety stock level per part type approaches and remains equal to the proposed hedging point $H_r(\alpha)$. The *lower level* of the hierarchy dispatches parts into the manufacturing system with the aim of maintaining the part-loading rate equal to the computed production rate. Extensive simulations, and an industrial application for an automated card assembly line, show that this approach can be very effective in handling capacity disturbances.

In contrast with the approach taken by Schweitzer, Seidmann, and Goes (1991), or by Akella, Choong, and Gershwin (1984), it is not likely that job routing in the FMS always follows a fixed route independent of the state of the system. The study by Yao and Buzacott (1985) is an early attempt at developing a state-dependent part-routing scheme which a'so possesses a tractable analytic structure. They consider the probabilistic shortest queue discipline (PSQ) which routes parts with the highest probability to the station which has the largest number of empty spaces, or alternatively, to the shortest input queue. In their model, part routing is defined for the part flows to the machines not for individual parts. Whenever a routing decision is to be made, an appropriate machine is identified and then a certain part is picked from the central storage area and is delivered to that machine. This scheme is essentially a randomized version of the deterministic shortest queue policy. Yao and Buzacott (1985) prove that PSQ routing leads to a time reversible process which has a product form equilibrium probability distribution.

The study by Seidmann and Tenenbaum (1991) develops two optimal routing policies and three closed-loop heuristic routing schemes for FMS similar to those studied by Yao and Buzacott (1985). One of the optimal routing policies is based on a new initiated suspension part-routing strategy. This strategy results in a reduced load on the material handling system while increasing the expected throughput (or revenue generation) rates of the machines.

Maximizing the weighted throughput rate means that the resulting queueing network model has a state-dependent arrival process and product form solutions do not hold. The alternative modeling approach developed by Seidmann and Tenenbaum (1991) is based on extending the earlier semi-Markov framework of Seidmann and Schweitzer (1984). The study by Schweitzer, Seidmann, and Goes (1991) introduces optimal static part routing procedures an FMS that is modelled as a closed network of queues. The objective of that study is to maximize the upper bound of the system capacity and to reduce the variable operating costs. The optimal routing is determined using the constrained derivatives developed for the appropriate performance measures. Significant reductions in the unit production cost are demonstrated as a result of routing jobs from heavily utilized machines to a more lightly utilized machine which is also capable of performing the same set of operations. This conclusion expects to hold for adaptive routing as well.

In many industrial applications there is a need for some job-related priority scheduling scheme in order to balance the load among the machines and to provide a preferential service to specific job classes. This scheme is relevant for capacity planning and for the operational control of existing facilities. Relevant managerial issues are, for example, the impact of the total work load and of the priority assignment policy on the relative difference between the queueing times of the high-priority and low-priority jobs, or the relationship between the priority assignment policy, the machine utilization and the product mix. The study by Shalev-Oren, Seidmann, and Schweitzer (1985) develops an analytic model of closed networks of queues with multiple product types, various non-preemptive priority service disciplines, and parallel machine stations. It incorporates the characteristic features of FMS such as robot carriers and local conveyor belts. The model was installed in a computer program named PMVA (Priority Mean Value Analysis) that was extensively tested and implemented by the authors (e.g., Seidmann, Schweitzer, and Shalev-Oren (1987)) and others. Utilizing PMVA in several operating FMS shows that throughput and flow times can be quite sensitive to the choice of priorities at the heavily loaded machines.

Schedules are sometimes implemented while assuming a given processing time for each operation on a part type using a particular machine tool. Once a throughput target is set, however, the processing times can be manipulated to reduce costs and increase tool lives (as well as to improve surface qualities) at no expense to system throughput. This interaction between machining conditions and the overall system throughput suggests that improved scheduling performance can be based on a production rate/tool wear tradeoff.

Hitomi (1976, 1977) tackles the joint problems of determining the optimal machining speeds and optimal cycle time in a deterministic multi-stage flow line. Unlimited buffer space is assumed between machines. Cost savings are obtained by slowing down noncritical machines until their cycle times match that of the bottleneck machine. McCartney and Hinds (1982) introduce a procedure to review the machining rates of parts which are first scheduled using maximum production rates. Their procedure will slow some machining rates to reduce production costs (on machines which are not on the critical path) while maintaining due-date performance. Determining buffer capacity along with optimal cutting speed and tool replacement policy in a two-machine system is discussed by Koulamas, Lambert, and Smith (1987).

Optimizing the process rates and determining changes in bottlenecks and queue lengths as the processing rates are altered is particularly intricate due to the problem of shifting bottlenecks. Watanabe and Fujii (1988) propose a heuristic control model which links the operation speed to the order tardiness. Schweitzer and Seidmann (1988, 1991) present several nonlinear queueing network optimization methodologies which determine the minimum cost processing rates given the FMS throughput target, the work-in-process level, part routes, transporter delays, and tool cost functions. Using industrial sample data, they show that a slight acceleration of the processing rates at a few economic bottleneck machines allows for significant rate reductions in others. Their studies provide for substantial gains in operating costs compared with conventional one-machine process planning models. Their results also prove that it is not necessarily optimal to: balance utilization of all machines, to balance waiting times at all machines, nor to manipulate processing times to compensate for local transporter delays.

The total number of tools required to process a set of parts on a flexible machine is usually larger than the available magazine storage capacity. As a result, a required tool may be absent from the magazine and a tool change must occur before that operation can begin. Tang and Denardo (1988a, 1988b) explore this issue for a single machine with a limited tool magazine, assuming that production requirements are known in advance. Bard and Feo (1989) address the problem of minimizing the total setup, tool replacement and machining times for individual batches subject to tool magazine capacity constraints and metal volume removal requirements.

Capacity selection decisions and the degree of heterogeneity or similarity of the jobs to be processed affect the utilization of the productive resources. These attributes have a major impact on the formation of queues with the resulting delays in the flow of jobs. Dobson, Karmarkar, and Rummel (1987) distinguish between two types of flow times for a part on the machine: *item-flow*, when a part can be delivered immediately after the machining on it is complete, and *batch-flow*, when a part must wait at the machine until the rest of the parts in its batch are completed. Karmarkar (1987), Karmarkar, Kekre and Kekre (1985, 1987) and Zipkin (1986) have examined the impact of set up times and lot sizes on manufacturing lead times and on WIP and the accumulation of safety stocks. They show that the optimal lot sizes - used for production planning with stochastic job arrival patterns - are different from those of the conventional EOQ models, which use set up costs as a device to enforce capacity constraints.

2.4. ORDERS' PROFILE

The effective use of critical manufacturing resources is achieved by carefully planning the mix of jobs and product types to be processed at any point in time. The profiles of orders and jobs have a significant impact on the overall system performance. These include decisions about the *order acceptance process, due date assignment procedures, generation of the master production schedule using rough-cut capacity planning, order release control policies, and the selection of parts and machines to be used at each planning period.* Several of the studies are presented below acknowledge the fact that in a flexible build (or produce) to order environment, the order acceptance process is the most important interface between the manufacturer and its customers. The decision model discussed by Luss and Rosenwein (1990) is particularly well suited for flexible flow lines that process numerous product types. They assume that one can generate a set h_i (i=1,2, ...n) of alternative schedules for each one of the n available orders. These alternative schedules specify the potential path and timing of tasks performed on each order. The completion date of order i in schedule $j \varepsilon h_i$ is c_{ij}. The due date requested by the customer for order i is d_i, and the weight assigned to that order is w_i. The objective is to minimize the total weighted tardiness of the existing orders:

$$\min \left\{ \sum_{i=1}^{n} \sum_{j \varepsilon h_i} [w_i x_{ij} \max\{0, c_{ij} - d_i\}] \right\},$$

where $x_{ij}=1$ if order i is assigned to schedule $j \varepsilon h_i$; zero, otherwise. One alternative schedule is selected per order, subject to capacity constraints for all resources. The results presented by the authors exploit the flexibility of the manufacturing facility and indicate significant reduction in order tardiness. These improvements are achieved by simultaneously considering a batch of jobs, thus facilitating better utilization of the most flexible bottleneck resources.

The advent of JIT has drawn attention to policies and models in which earliness, as well as tardiness, is discouraged. Seidmann and Smith (1981) introduced the due-date assignment problem for systems having randomly distributed processing times. The aggregate cost function used includes: 1) a cost that increases with lead times, 2) a cost for tardiness, and 3) an earliness cost. Several extensions of that work to a general shop operation are given by Shanthikumar and Sumita (1988). The deterministic version of the common due-date assignment problem studied by Panwalkar, Smith, and Seidmann (1982) further incorporates a flow time penalty into the cost function described above. They show that for any specified sequence, there exists an optimal value of the common due date that coincides with the completion time of one of the jobs in that sequence. Baker and Scudder (1990) provide a comprehensive review of the due-date assignment and deterministic sequencing literature. Their paper proposes some unifications and extensions to the current work in the field. Wein (1991), Glassey and Resende (1988), and Wein (1988) maintain that due-date setting and input release controls can both be thought of as *tactical design decisions*, that should be made at a *higher level* than priority sequencing and scheduling decisions.

Parts inputs and machine selection policies for FMS were studied extensively in the context of FMS operations (Gray, Seidmann, and Stecke (1990)). Stecke and Solberg (1981) present six alternative objectives for FMS loading and control. Stecke (1983) defines five FMS production management problems: the part-type selection problem, the machine grouping problem, the production ratio problem, the resource allocation problem, and the loading problem. She notes that the principal objective of the FMS is to maximize expected production rates and that surrogate criteria are typically used in order to achieve this overall objective. Mazzola, Neebe, and Dunn (1989) propose a hierarchical structure that integrates FMS production planning into a closed-loop MRP system. Other studies dealing with aggregate multi-stage capacitated scheduling of FMS are discussed by Toczylowski and Hindi (1991). The problems of determining the part types to be processed in an FMS and the selection of machines and tools are very large indeed. The algorithms available to solve them are necessarily sub optimal. Such algorithms are proposed by Whitney and Suri (1985), Stecke and Kim (1988), and Werra and Widmer (1991). These algorithms can be used together to define the most financially attractive set of parts to produce, and to ensure the feasibility of the selected set of parts with regard to tool constraints, both system-wide and at individual machines.

3. THE CHANGING NATURE OF THE PERFORMANCE MANAGEMENT FUNCTION IN FMS

The initial role of FMS was to fill the gap between high-production transfer lines and low-production job shops with numerical control machines. The volume in which complex metal products are being manufactured ranges from individual units to hundreds of thousands per order. The production mode of these products depends largely upon the volume, whether it is small, medium, or large batch production, or a mass production transfer line. The FMS was originally designed to handle medium production volume with greater flexibility and parts variety than transfer lines, and to facilitate lower variable production costs than those encountered in job shops. Such designs are in high demand, as over 50% of all complex metal parts are produced in batches of less than 50 units (Cook (1975)). The 'ideal' FMS would combine the flexibility of general purpose machine tools with the high productivity of a transfer line. This was the motive for introducing the Sundstrand line in the late sixties. It was one of the first facilities to integrate machining centers with a material handling system delivering parts and tools. It consists of eight 5-axis machining centers with two automatic multispindle drills, an automated material handling system and centralized supervisory computer control. It is capable of producing about seventy different types of housings for the speed control gears of aircraft generators. This line replaced a conventional shop that required approximately 100 conventional machine tools; the number of operators was reduced from 125 to 25; WIP was decreased by 40%. That successful concept has served as a prototype for many of the FMS that were built later. More advanced systems were designed not only for machining operations but also for welding, forming, cutting, and variety of assembly tasks. On the other hand, the idea of building unmanned factories, or of running second and third shifts unmanned, has turned out to be somewhat elusive due to the lack of sufficient exception handling capabilities in the control software and machine reliability falling short of expectations.

In recent years managers have started to realize that a manufacturing advantage that can stand up to international competition must be built upon high performance factories. These should provide rapid on-time delivery, high quality, and low delivered costs. Investing money in new automated facilities is not by itself a guarantee for performance improvement. Managed poorly, these investments can reduce the plant's performance (Hayes, Wheelwright, and Clark (1988)). Of greater significance than the new machinery is the way in which it is managed and the impact it has on the process of continued improvements in the company. For example, the author has observed several cases where FMSs are being used to support two strategic objectives: 1) reduction of time to market for new product introductions; and 2) reduction of spare parts inventories. In some automotive plants they are used by design teams to quickly produce working prototypes to be test driven and licensed. This approach has a significant impact on reducing time-to-market for new product's introductions. Similarly, in certain aerospace applications, rather than producing the anticipated demand for spares before hand, companies try and produce some of the more complex spares upon demand from their customers. Figure 2 illustrates this new role of the FMS in the product's life cycle; the next step may be expanding the role of FMS manufacturing from the 'tails to the center' to increase scope and responsiveness. It is still unclear, however, whether the present technology is sufficiently reliable and cost-effective to meet these challenges, and what is the best way to manage that transition process.

The typical FMS is involved in producing subassemblies, or components that are later incorporated with other units at the final assembly line prior to being shipped to the customers. Many of these final assembly lines are moving towards JIT production and their master production schedules no longer call for deliveries of fixed batch sizes. To meet their erratic JIT assembly schedules, these internal 'customers' want ancillary delivery contracts with some kind of upper, and maybe lower, supply bands. Another dilemma observed in industry is the aging of existing facilities. Maturing systems installed during the seventies and early eighties are coping with expansion and technological upgrade issues. It is clear that installing the latest technological systems may have many advantages. On the other hand, mixing old and new machines within the same cells will reduce compatibility (and flexibility) across all the plant resources. Finally, using a FMS eliminates major upfront investments in dedicated facilities when launching a new product to the market. For small- to mid-volume production, the variable production costs are also relatively low in FMSs. As a result, the author has observed several cases where the Strategic Business Unit manager , or the products' managers, were trying to keep and use the shared corporate FMS for relatively high production volumes. This is typically done during the ramp-up period in order to defer as much as possible the need (and the burden) for building a dedicated facility that would handle the anticipated increase in volume. In these circumstances, when one order captures a significant portion of the systems' capacity, other users face excessive leadtimes and delays. This underscores the fact that understanding the exact nature of the externalities imposed by the users is a necessary condition for achieving satisfactory performance levels in these complex and highly capable facilities.

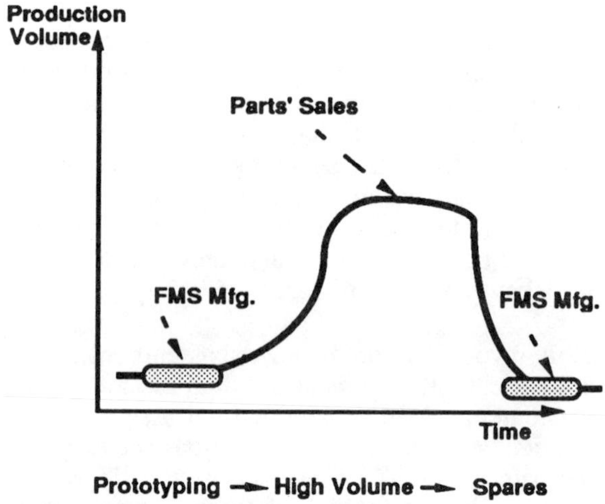

Figure 2. The FMS role in the product's life cycle.

Our analysis of the various performance issues identifies four substantial research opportunities. First, there is a growing need for richer analytical (queuing) models for dealing with the impact of special FMS performance issues: simultaneous resources, machine interference, partial machine grouping, and state-dependent part-release protocols. Second, little progress has been made so far regarding the analysis of the transient FMS behavior. Such a transient behavior will manifest itself during ramp-up production runs, adjustments in production mix, or when using dynamic routing. Third, there is a need to develop underlying theories supporting the generation of adequate master production scheduling systems. This is a key component in coordinating the FMS performance with the rest of the company. Fourth, there is a need to develop better ways to integrate conceptually the 'product design/process selection/production planning' decision phases with the key economic drivers of the manufacturing firm, rather than to squeeze them between marketing and design engineering requirements.

Acknowledgment

The author wishes to thank Professor Paul J. Schweitzer from the W. E. Simon School of Business Administration, The University of Rochester, for many useful comments and suggestions.

REFERENCES

Akella, R., Y. Choong, and S. B. Gershwin, "Performance of Hierarchical Production Scheduling Policy," *IEEE Transactions on Components, Hybrids, and Manufacturing Technology*, Vol 7, 3, (1984), 225-240.

Arbel, A. , and A. Seidmann, "Performance Evaluation of Flexible Manufacturing Systems," *IEEE Transactions on Systems, Man, and Cybernetics*, Vol 14, 4, (1984), 606-617.

Baker K.R., and G. D.Scudder, "Sequencing with Earliness and Tardiness Penalties: A Review," *Operations Research*, Vol 38, 1, (1990), 22-36.

Bard, J. and T. A. Feo, "The Cutting Path and Tool Selection Problem in Computer Aided Process Planning," *Journal of Manufacturing Systems*, Vol. 8, 1, (1989), 17-26.

Bijayananda, N. and A. Chakravarty, "Strategic Acquisition of New Technology: A review and Research Framework," Working Paper 91-DIDOM-%, A.B. Freeman School of Business, Tulane University (1991)

Bitran, G., and D. Tirupati, "Trade-offs Curves, Targeting and Balancing in Manufacturing Networks," *Operations Research*, Vol 37, 6, (1989), 547-564.

Boxma, O. J., A.H.G. Rinnooy Kan, and M. van Vliet, "Machine Allocation Problems in Manufacturing Networks," *European Journal of Operational Research*, Vol 45, 1, (1990), 47-54.

Brown, E., "IBM Combines Rapid Modeling Techniques and Simulation to Design PCB factory of the Future," *Industrial Engineering*, Vol 20, 6, (1988), 23-26.

Buffa, E. S., "Making American Manufacturing Competitive," *California Management Review*, Vol 26, 3, (1984), 29-46.

Buzacott, J. A., and J. G. Shanthikumar, *Stochastic Models of Manufacturing Systems*, Prentice Hall, Engelwood Cliffs, NJ, 1992.

Canada, J. R., and W. G. Sullivan, *Economic and Multiattribute Evaluation of Advanced Manufacturing Systems*, Prentice Hall, Engelwood Cliffs, NJ, 1989.

Cook N. H., "Computer Managed Part Manufacture," *Scientific American*, Vol 232, 2, (1975), 22-29.

Dallery, D., and Y. Frein,"An Efficient Method to Determine the Optimal Configuration of a Flexible Manufacturing System," Proceedings of the 2nd Conference on Flexible Manufacturing Systems: Operations Research Models and Applications", Ed. K. E. Stecke and R. Suri, North Holland, Amsterdam, 1986.

Dallery, Y., and K. E. Stecke, "On the Optimal Allocation of Servers and Workloads in Closed Queueing Networks," *Operations Research*, Vol 38, 4, (1990), 694-703.

Dobson, G., U. S. Karmarkar, and J. L. Rummel," Batching to Minimize Flow Times on One Machine," *Management Science*, Vol 33, 6, (1987), 784-799.

Glassey, C. R., and M.G.C. Resende, "Closed Loop Release Control for VLSI Circuit Manufacturing," *IEEE Transactions on Semiconductor Manufacturing*, Vol 1, (1988), 36-46.

Gray, A.E., A. Seidmann, and K. E. Stecke, "A Synthesis of Tool Management Issues and Decision Problems in Automated Manufacturing, " The University on Michigan Ann Arbor, School of Business Administration, Working Paper 590-b, (1990).

Hayes, R. H., S. C. Wheelwright, and K. B. Clark, *Dynamic Manufacturing: Creating the Learning Organization*, (1988), The Free Press, McMillan Inc, New York, NY.

Hitomi, K., "Analysis of Production Models; Part I: The Optimal Decision of Production Speeds; Part II: Optimization of a Multistage Production System," *AIIE Transactions*, Vol. 8, 1, (1976), 96-112.

Hitomi, K., "Optimization of Multistage Production Systems with Variable Production Times and Costs," *International Journal of Production Research*, Vol. 15, 6, (1977), 583-597.

Karmarkar, U. S., "Lot Sizes, Lead Times and In-Process Inventories," *Management Science*, Vol 33, 3, (1987), 409-418.

Karmarkar, U. S.,S. Kekre, and S. Kekre, "Lot sizing in Multi-Item Multi-Machine Job Shops," *IIE Transactions*, Vol 17, 3, (1985), 290-298.

Karmarkar, U. S.,S. Kekre, and S. Kekre, " Capacity Analysis of a Manufacturing Cell," *Journal of Manufacturing Systems*, Vol 6, 3, (1987), 165-175.

Kleinrock, L,. *Communication Nets: Stochastic Massage Flow and Delay*, McGraw-Hill, New York, 1964.

Koulamas, C. P., B. K. Lambert, and M.L. SMITH "Optimal Machining Conditions and Buffer Space Size for the Two Stage Case," *International Journal of Production Research*, Vol. 25, 3, (1987), 327-336.

Luss, H., and M. B. Rosenwein, "A Due Date Assignment Algorithm for Multiproduct Manufacturing Facilities," AT&T Bell Laboratories, Technical Report, (1990), Holmdel, NJ.

Mazzola, J.B., A.W. Neebe, and C.V.R. Dunn, "Production Planning of a Flexible Manufacturing System in a Material Requirements Planning Environment," *International Journal of Flexible Manufacturing Systems*, Vol 1, 2, (1989), 115-142.

McCartny, J., and B. K. Hinds, "Tooling Economics in Integrated Manufacturing Systems," *International Journal of Production Research*, Vol. 20, 4. ,(1982), 493-505.

Panwalkar, S.S., M.L. Smith, and A. Seidmann, "Common Due Date Assignment to Minimize Total Penalty for the One Machine Scheduling Problem," *Operations Research*, Vol 30, 2, (1982) 391-399.

Schweitzer, P. J. ,"Approximate Analysis of Multiclass Closed Networks of Queues," in *Proceedings of the International Conference on Stochastic Control and Optimization*, Free University, Amsterdam, The Netherlands, April 5-6, 1979.

Schweitzer, P. J., and A. Seidmann, "Capacity Range Analysis and Processing Rate Optimization for FMS's with Distinct Multiple Job Visits to Work Centers," Working Paper, (1988), W. E. Simon Graduate School of Business Administration, University of Rochester, Rochester, NY.

Schweitzer, P. J., and A. Seidmann, "Optimizing Processing Rates for Flexible Manufacturing Systems," *Management Science*, Vol 37, 4, (1991), 454-466.

Schweitzer, P. J., and A. Seidmann,"Bottleneck Analysis in Networks of Queues," Working Paper, (1992), W. E. Simon Graduate School of Business Administration, University of Rochester, Rochester, NY.

Schweitzer, P. J., A. Seidmann, and P. B. Goes, "Performance Management in Flexible Manufacturing Systems," *International Journal of Flexible Manufacturing Systems*, Vol 4, (1991), 17-50.

Seidmann A. and P.J. Schweitzer, "Part Selection Policy for a Flexible Manufacturing Cell Feeding Several Production Lines," *IIE Transactions*, Vol 16, 4, (1984), 355-362.

Seidmann A., P.J. Schweitzer, and S. Shalev-Oren, "Computerized Closed Queueing Network Models of Flexible Manufacturing Systems: A Comparative Evaluation," *Large Scale Systems*, Vol 12, (1987), 91-107.

Seidmann, A., and M.L. Smith,"Due Date Assignment for Production Systems," *Management Science*, Vol 27, 5, (1981), 571-581.

Seidmann, A. and A. Tenenbaum, " Throughput Maximization in Flexible Manufacturing Systems," *IIE Transactions* , (1991) (to appear).

Shalev-Oren, S., A. Seidmann, and P. J. Schweitzer, "Analysis of Flexible Manufacturing Systems with Priority Scheduling," *Annals of Operations Research*, Vol 3, (1985), 115-139.

Shanthikumar, J. G., and U. Sumita, "Approximations for the Time Spent in a Dynamic Job Shop with Applications to Due Date Assignment," *International Journal of Production Research*, Vol 26, (1988), 1329-1352.

Shanthikumar, J. G., and D. D. Yao,"On Server Allocation in Multiple Center Manufacturing Systems," *Operations Research*, Vol 36, 2, (1988), 333-342.

Shanthikumar, J. G., and D. D. Yao,"Optimal Server Allocation in a System of Multi-Server Stations," *Management Science*, Vol 33, 9, (1987), 1173-1180.

Stecke, K.E.,"Formulations and Solutions of Nonlinear Integer Production Planning Problems for Flexible Manufacturing Systems," *Management Science*, Vol 29, 3, (1983), 273-288.

Stecke, K.E., and J.J. Solberg, "Loading and Control Policies for a Flexible Manufacturing System," *International Journal of Production Research*, Vol 19, 5, (1981), 481-490.

Solot, P. ,"A Heuristic Method to Determine the Number of Pallets in a Flexible Manufacturing System with Several Pallet Types," International Journal of Flexible Manufacturing Systems, Vol 2, 2, (1990), 191-216.

Stecke, K.E., and I. Kim, "A Study of FMS Part Type Selection Approaches for Short-Term Production Planning," *International Journal of Flexible Manufacturing Systems*," Vol 1, (1988), 7-29.

Suri, R., and C.K. Whitney, "Algorithms for Parts and Machine Selection in Flexible Manufacturing Systems," *Annals of Operations Research*, Vol 3, (1985), 239-262.

Tang C., and E. V. Denardo "Models Arising from a Flexible Manufacturing Machine, Part I: Minimization of the Number of Tool Switches," *Operations Research*, Vol. 36, 5, (1988a), 767-777.

Tang C., and E. V. Denardo, "Models Arising from a Flexible Manufacturing Machine, Part II: Minimization of the Number of Switching Instances," *Operation Research* ,Vol. 36, 5, (1988b), 778-784.

Toczylowski, E., and K. S. Hindi, "Aggregate Capacitated Lot-Size Scheduling for a Class of Flexible Machining and Assembly Systems," *IIE Transactions*, Vol 23, 3, (1991), 259-266.

van Vliet, M. and A. G. H. Rinnooy Kan, "Machine Allocation Algorithms for Job Shop Manufacturing," *Journal of Intelligent Manufacturing*, Vol 2, 1, (1991), 83-94.

Vinod, B. and J. J. Solberg, "The Optimal Design of Flexible Manufacturing Systems," *International Journal of Production Research*, Vol 23, 6, (1985), 1141-1151

Watanabe, T. , and Fujii, R., "Determining the Job Operation Speed and Schedule for Machine Tools in FMS by Forecasting using a Simulator and a System Performance Index," *ASME Proceedings of the USA-Japan Symposium on Flexible Automation*, (1988),Minneapolis, MN, 751-756.

Wein, L. M."Scheduling Semiconductor Wafer Fabrication," *IEEE Transactions on Semiconductor Manufacturing*, Vol 1, (1988), 115-130.

Wein, L. M. ,"Capacity Allocation in Generalized Jackson Networks, "*Operations Research Letters*, Vol. 8, 3, (1989), 143-146.

Wein, L. M. "Due-Date Setting and Priority Sequencing in a Multiclass M/G/1 Queue," *Management Science*, Vol 37, 7, (1991), 834-850.

Werra, D, and M. Widmer, "Loading Problems with Tool Management in FMS's: A few Integer Programming Models," *International Journal of Flexible Manufacturing Systems*, Vol 3, 2, (1991), 71-82.

Yao, D.D., and J. A. Buzacott," Modelling a Class of State-Dependent Routing in Flexible Manufacturing Systems," *Annals of Operations Research*, Vol 3, (1985), 153-167.

Zipkin, P. H.,"Models for Design and Control of Stochastic Multi-Item Batch Production Systems," *Operations Research*, Vol 34, 1, (1986), 91-104.

MODELS FOR TIGHTLY-COUPLED PRODUCTION SYSTEMS

Kenneth R. Baker
Amos Tuck School of Business Administration
Dartmouth College
Hanover, NH, USA

INTRODUCTION

Shorter lead times and lower inventories have become major goals in the operation of contemporary manufacturing systems. In the past, inventories were used to decouple adjacent stages of production and to provide some autonomy to individual stages of the process. The longer lead times that resulted were not viewed as a disadvantage in traditional markets. Today, in the new paradigm for world-class manufacturing, such inventories prevent a firm from fulfilling its potential on response time, product quality, and indirect cost, thus diminishing competitiveness. As a consequence, there is much more concern now with removing inventories and operating tightly-coupled production systems.

Tight coupling is epitomized by the notion of just-in-time material flow. Ideally, the operations in a just-in-time process all take the same amount of time, and material moves from station to station at regular time intervals in synchronized fashion. Under these conditions the output potential of the system can be realized without any provisions for in-process inventories. However, this ideal is difficult to attain in environments characterized by constant innovation in products and processes. In reality, managers of production systems strive for just-in-time operation without assurances that all of its preconditions will exist. In particular, production systems are designed for perfect synchronization but must operate in the presence of considerable variability.

The complexity of system design, in a context of zero-inventory aspirations and unavoidable variability, might not be such a great problem if there existed certain well-understood performance laws governing the behavior of complex production systems, analogous to the law of gravity or the laws of thermodynamics. Lacking these, it would be useful if designers of production systems could rely on design rules that could provide some broad guidelines and useful heuristics. However, the science of designing production lines is still in its infancy, and there is little more than intuition and experience to guide designers. For example, intuition may tell us that a production line should be evenly balanced or that there is no difference in alternative loading mechanisms for assembly. As we shall see, such intuition is not always reliable.

Generic models of tightly-coupled production systems provide a scientific basis for discovering guidelines and heuristics for the design of actual systems. In this paper we review a portion of the literature on tightly-coupled systems, emphasizing the effects of process-time variability. Tight coupling is problematic when operation times are constant, as in a transfer line, but random breakdowns occur. It is also problematic in the absence of breakdowns, when operation times vary unpredictably. In fact, there are essentially two branches of the research literature corresponding to these two issues—one focusing on the effects of breakdowns, the other on the effects of process-time variability. (For convenience, we use the term "variability" to mean process-time variability in the remainder of the paper.) This survey deals with the effects of variability, which is an important factor in light of two current trends. One is the trend to reduce downtimes with improved maintenance policies. This trend promises to make breakdowns a less significant operational problem. The other trend is to reduce predictability with more flexible systems. This trend promises to make variability a more significant operational problem.

A generic model for a tightly-coupled system can be viewed as representing an automated production line that handles a diverse set of related products. In other words, the diversity within a general product family leads to small production lots and to variability in the requirements at each work station. Alternatively, the model can be viewed as representing a traditional manual line, with variability induced by human performance of repetitive tasks.

This paper is organized into two main sections, one dealing with serial lines and the other with assembly systems, in which we look at several common themes. The simplest models have two stages: in serial lines, this means an initial station and a final station; in assembly systems, it means a component stage (with parallel stations) and an assembly stage. We first analyze these simple models; then we examine the effects of enlarging the model. We also look specifically at the role of variability in the system, and we address questions of allocating the work among stations. Our basic operating assumptions are as follows:

- Input material is always available
- Operation times are independent random variables
- Breakdowns never occur
- Defective parts are never produced

Variability in operation times gives rise to line inefficiency due to starving or blocking. A station is *starved* when it is ready to work but no part is available;

a station is *blocked* when it completes a part but cannot pass it on until work is completed at some subsequent station. Blocking and starving each entail some idleness and hence inefficiency.

The design criterion is *throughput*, defined as the long-run average output per unit time. For a given description of a line, we are interested in measuring its throughput; when there are design choices, we are interested in maximizing the throughput. For simplicity in notation, let T denote throughput. Additionally, let n denote the number of stations in the system and X_i the random variable for the operation time at station i.

We say that a line is *balanced* if the average workload at each station is the same. Specifically, if the X_i values all have the same mean, then the line is balanced. A balanced line is also said to have *identical stations* if the X_i all have identical distributions. Finally, a line is *synchronized* if each part is held at its work station until all parts are ready to proceed to their next operation. (A transfer line is essentially a synchronized line.) However, in an *unsynchronized* line, parts proceed as soon after completion as the subsequent work station is free. In this paper we deal with the most extreme case of tight coupling, when there is no buffer of any kind between work stations. Therefore, all of the cases we address are unbuffered systems.

SERIAL LINES

Two-Station Case

Perhaps the simplest tightly-coupled model is the case of two stations in series. Jobs flow through the system, first to station 1, then to station 2, with operation times that are independent and random at each station. We use X_1 and X_2 to denote the operation times at stations 1 and 2, respectively. In a *synchronized* system, jobs wait until all operations are complete, and then all work proceeds simultaneously. In the two-station case, this means that the job at station 2 leaves the system, the job at station 1 moves to station 2, and a new job enters the system at station 1, all at the same time. Let CT denote the cycle time, which can be defined as the time between successive departures from the system. For the synchronized system, the cycle time is $CT = \max(X_1, X_2)$. The expected cycle time, ECT, is simply the expectation of this quantity; and the throughput is defined by:

$$T = [ECT]^{-1} = \mathrm{E}[\max(X_1, X_2)]^{-1} \tag{1}$$

Closed-form expressions for T can be derived for a number of common distribution forms.

Now consider the *asynchronous* version of the same model, where jobs proceed as soon as they can. In the two-station case, the job at station 2 leaves as soon as it finished; the job at machine 1 moves ahead at its completion time or the completion time at station 2, whichever is later. Although the individual cycle times are different in synchronous and asynchronous systems, both systems have the same throughput, at least in the two-station case. (To see this, note that jobs enter the system at the same points in time in both systems.) Thus, we can use (1) to find the throughput for an asynchronous two-station system. The exponential case, originally due to Hunt (1956), can be written as follows.

$$T = \frac{ab(a+b)}{a^2 + ab + b^2} \qquad (1a)$$

where a and b denote the mean processing rates at stations 1 and 2, respectively. See Rao (1975a, 1976a), Muth (1977), and Mishra et al. (1985) for some more complicated cases.

In a *balanced* system, the mean operation time is the same at each station, and we shall assume that the model is scaled so that this mean is equal to 1.0. For the most part, the throughput for a balanced two-station line is calculated as a special case of (1). For example, in the exponential case, the expression in (1a) reduces to $T = 2/3$, and in the uniform case where the mean is equal to the range, the expression in (1) reduces to $T = 6/7$.

Three-Station Case

For three stations, the analysis of the synchronous case is easily generalized. In fact, for an n-station system the throughput is written

$$T = \mathrm{E}[\max(X_1, X_2, X_3, \ldots, X_n)]^{-1} \qquad (2)$$

When the stations are identical and each has an exponential operation time, the special form of (2) becomes:

$$T = (1 + \frac{1}{2} + \frac{1}{3} + \cdots + \frac{1}{n})^{-1} \qquad (2a)$$

Similarly, when the stations are identical and each has a uniform operation time with maximum b (and minimum $2 - b$), the special form of (2) becomes:

$$T = \frac{n+1}{2 + b(n-1)} \tag{2b}$$

The three-station, asynchronous case is much more difficult to analyze, and closed-form expressions are harder to obtain. In the case of the exponential distribution, however, a formula exists:

$$T = \{\frac{1}{a} + \frac{1}{b} + \frac{1}{c} - \frac{(a+c)(a+b+c)^3}{(a+b)(a+c)[(a+c)^2(a+b+c)-abc]}\}^{-1} \tag{3}$$

This expression was originally derived by Makino (1964) and later presented by Muth (1984) and by Muth and Alkaff (1987b). In the balanced case ($a = b = c$), the expression in (3) reduces to $T = 22/39$. Numerical approaches to determining throughput in the Erlang case were introduced by Hillier and Boling (1967) and Muth (1977), and an analytic approach was later provided by Muth (1984). Neither approach leads to closed form expressions; however, in the case of the uniform distribution, Muth (1977) gave the following formula for the balanced case:

$$T = (1 + 0.8037cv)^{-1} \tag{4}$$

where cv denotes the coefficient of variation of the operation time distribution.

Later, Muth (1987) proposed a general approximation for balanced lines, in the form shown below:

$$T = \frac{1 + 0.0775cv}{1 + 0.9125cv} \tag{5}$$

This formula is significant as a distribution-free result for the three-station case with identical stations. No simple approximation has been offered for nonidentical stations, although a method due to Muth and Alkaff (1987b) can be used under certain broad restrictions on the family of distributions assumed. Recently, however, Baker, Powell and Pyke (1990c) proposed an approximation procedure. In essence, this procedure asserts that any three-station line can be viewed as a weighted combination of simpler three-station lines in which the operation times are either zero or one. These lines, in turn, are amenable to direct analysis, or to at least a very close approximation; the weighted combination thus gives a good approximation to the throughput of the given line. Tests indicate this approximation to be within about 4% of the theoretical figure for throughput.

Little progress has been made in analyzing n-station lines for $n>3$. Hillier and Boling (1967) gave some results for the exponential distribution and the Erlang distribution, but these results apply only to balanced lines of length $n \leq 6$. Muth (1987) developed an approximate formula for the n-station balanced line, which was specified in correct form by Blumenfeld (1990):

$$T = [1 + \frac{1.67cv(n-1)}{1+n+0.31cv}]^{-1} \qquad (6)$$

We can conclude that the exact analysis of other cases, even for relatively small line lengths, is difficult to perform. Three-station lines represent the current analytic frontier for general results. Nevertheless, the analysis of three-station models may be useful in approximating throughput for larger systems if we think in terms of decomposing those systems into a front end, a middle, and a back end, which, in turn, can be approximately viewed as nonidentical stations in a three-station line.

Effect of Line Length

For an n-station system, the throughput decreases as n increases. The precise relationship is difficult to determine, however, because the throughput cannot be calculated except in very special cases. An approximate relationship can be determined by simulating various cases to estimate throughput and then fitting a regression model to the data. This type of methodology is exemplified by the work of Anderson and Moodie (1969). For the balanced exponential case, the Anderson-Moodie equation takes the following form:

$$T = 1 - \frac{2.07 - 2.36n^{-1.16}}{3.74 - 1.10n^{-0.75}} \qquad (7)$$

Although such curves depend on the operation time distribution, it appears to be the coefficient of variation that largely accounts for throughput. This observation has persuaded most analysts to use cv along with n as the key independent variables. More recent developments in this spirit have been presented by De Kok (1990) and Blumenfeld (1990).

It is interesting to ask what asymptote the curve approaches. Conway, Maxwell, McClain, and Thomas (1988), for example, addressed this question directly. In the context of their work, the range of variability encountered in practice is covered by a family of uniform distributions with $cv \leq 0.3$. They concluded that the asymptote lies at about 0.75 for the largest amount of variability in this range. Using the limit of (2b), we can compute a lower bound on this figure of about 0.66.

The researchers who developed regression models based on simulations did not address this question explicitly, but it is often possible to extrapolate from their results and calculate the asymptote. In (7), for example, the limiting value for the throughput is about 0.447. A brief summary of such results is listed in Table 1. Included in this table is the perspective of Magazine and Silver (1978). They dealt with the case of exponential distributions at each station and developed an approximation for the throughput as a function of n. The limit of their approximation value, as n grows large, is equal to the complement of the golden mean, or roughly 0.382. The simulations of the exponential case reported by El-Rayah (1979), Conway et al. (1988), and Baker, Powell and Pyke (1990a), are consistent with this limit. (These simulations yielded throughputs between 0.45 and 0.42 for $8 \leq n \leq 12$.)

Exponential Case (cv = 1.0)		
Basu (1977)	0.333	Approximate model
Panwalkar-Smith (1979)	0.367	Regression on Hillier-Boling
Muth (1987)	0.375	Approximation scheme
Magazine-Silver (1978)	0.382	Approximation scheme
Knott (1970)	0.400	Regression fit to simulation
Anderson-Moodie (1969)	0.447	Regression fit to simulation
Normal Case with cv = 0.3		
Muth (1987)	0.666	Approximation scheme
Anderson-Moodie (1969)	0.728	Regression fit to simulation
Knott (1970)	0.747	Regression fit to simulation
Buzacott (1967)	0.771	(Erlang case)
Uniform Case with cv = 0.3		
Muth (1987)	0.666	Approximation scheme
Knott (1970)	0.727	Regression fit to simulation
Conway et al. (1988)	0.75	Extrapolation of simulation

Table 1. Asymptotic throughput estimates for three distributions.

Two theoretical results provide additional perspective on the asymptote for the exponential case. First, consider the behavior of a synchronous system. From (2a), it follows that as n approaches infinity, the throughput goes to zero. This result is not general: for example, if the operation times at each station follow a uniform distribution with a maximum of $b = 2$, then, from (2b), the asymptotic throughput is just 0.5. Thus, the exponential distribution may represent a boundary case in which throughput under synchronous operation eventually vanishes, but throughput under asynchronous operation does not.

Secondly, in the case of systems with *communications blocking* (where production cannot start at a station until the downstream station is free), Liggett (1975) has shown that the asymptotic throughput is 0.25. Obviously, communications blocking represents a less efficient mode of operation than what we encounter in production systems; asymptotically, the throughput under production blocking appears to be about 50% greater than under communications blocking.

Effect of Variability

Muth (1973) asserted that throughput in an n-station line is essentially a function of the first two moments of the operation time distributions. We refer to this as the *two-moment property*. An equivalent statement for the special case of balanced lines with unit mean is that throughput is a function of the *cv*. The two-moment property is not rigorously true, however. Rao (1975b) found some two-station cases where different distributions with the same *cv* produce slightly different throughput. Lau (1986) amplified this result by finding two-station examples in which higher moments of the distribution influence throughput. But the two-moment property still appears to have considerable practical validity when viewed from the perspective of the entire literature on this subject.

Muth's observation about the two-moment property was made in the context of three-station systems. Because three-station models are so difficult to analyze theoretically, there has been no analytic demonstration of similar results for systems with more than two stations. However, in an extensive simulation study, Baker et al. (1990a) provided additional support for the two-moment property. They simulated a nine-station system and tested several different distributions. They found that that there are slight differences due to distribution form, but that the *cv* is, for all practical purposes, the dominant factor in determining throughput.

The Bowl Phenomenon

Hillier and Boling (1966) studied the effects of reallocating work among the stations along a serial line. Their base case is a serial line with identical exponential distributions at each station, with time scaled so that the mean operation time at each station is one time unit. As an alternative, they considered a model in which the mean times at the three stations are variables u_i.

They imposed the requirement that the total amount of work in the system is held constant (that is, $\Sigma u_i = n$), and they asked whether throughput could be raised by "unbalancing" the work (that is, by taking $u_i \neq 1$). They discovered that the optimal allocation of work (in terms of mean operation times) is indeed unbalanced—high at the ends of the line and low in the middle of the line. A graph of the optimal work allocation as a function of station number is thus bowl-shaped, and their result has therefore been called the *bowl phenomenon*. The optimal allocations u_i for various line lengths are shown in Table 2.

Station:	*1*	*2*	*3*	*4*	*5*	*6*
$n = 3$	1.086	0.828	1.086			
$n = 4$	1.137	0.863	0.863	1.137		
$n = 5$	1.169	0.898	0.866	0.898	1.169	
$n = 6$	1.192	0.923	0.885	0.885	0.923	1.192

Table 2. Optimal allocations of work for serial lines.

The bowl phenomenon has intrigued researchers because it is, for the most part, counterintuitive. Although people may claim after the fact that the bowl phenomenon is indeed common sense, it is difficult to build a compelling intuitive justification for it. For researchers, a natural question is whether the bowl phenomenon is robust. After all, the Hillier-Boling analysis originally relied on the exponential distribution, which is a one-parameter distribution and which may not be representative of other distribution forms. This question persisted after the sequel by Hillier and Boling (1979), which described the bowl phenomenon for lines in which the Erlang distribution holds at each station.

Rao (1976a) analyzed a three-station serial line in which both end stations have exponential times and the middle station has constant times. The optimal unbalancing for this line is not bowl shaped. On the contrary, throughput is maximized by allocating more work to the middle station than to either of the end stations. We might speculate that the bowl phenomenon might not occur if there are different operation time distributions along the line, particularly including one with no variability at all as in Rao's model.

A set of simulation studies, exemplified by Carnall and Wild (1976) and El-Rayah (1979b), indicated that there is a "variance effect" distinct from a "mean effect" in the bowl phenomenon. Specifically, this line of work suggested that, other things being equal, lower variances should be placed in the middle of the line and higher variances should be placed at the ends. The existence of a variance effect raises questions about what drove the original Hillier-Boling results. Since the exponential distribution has mean equal to standard deviation, the unbalancing of means carries with it the unbalancing of variances. In the optimal designs of Table 2, the lines have smaller variances at the middle stations and larger variances at the end stations. One might wonder whether it is the reallocation of means or the reallocation of variances that primarily accounts for the improvement in throughput.

Muth and Alkaff (1987a) analyzed a family of three-station serial lines and were able to separate the mean effect from the variance effect, demonstrating that both can exist. In addition, other studies used simulation to examine longer lines and reinforced the conclusion that maximum throughput can be attained with an unbalanced line. For example, El-Rayah (1979a) and So (1989) found bowl phenomena in lines with more than three stations. However, it is clear that the bowl phenomenon diminishes as the level of variability drops: in the deterministic limit, where all variances are zero, a balanced allocation is optimal.

From a practical perspective, the bowl phenomenon calls into question the way we think about line balancing and about bottlenecks. The classical line balancing problem takes a deterministic view and asks how work should be allocated to stations along a serial line. The objective is typically to allocate work so that it comes as close as possible to equal balance. However, the bowl phenomenon reflects a nondeterministic view, leading to the conclusion that equal balance is not necessarily ideal. In principle, if the line balancing model were used as a deterministic proxy for a stochastic line, it makes more logical sense to set some bowl-shaped work allocation as the ideal.

The traditional thinking about bottlenecks likewise suffers from the limitations of a deterministic perspective. Suppose we were operating a three-station line with exponential distributions at each station. Suppose further that the line were unbalanced with the mean times of 1.04, 0.90, and 1.06 at stations 1, 2, and 3, respectively. In this setting, which station would we consider a bottleneck? The conventional notion of a bottleneck, of course, is the location of the heaviest mean workload, or station 3 in this example. Few people would immediately pick station 2 as the bottleneck. Yet, compared to the optimal three-station allocation, station 3 is underloaded, and station 2 is overloaded. At the margin, changes in the design of the line should treat station 2 as the bottleneck, even though its load is the lightest. Knowledge of the bowl phenomenon thus gives us a perspective on bottleneck analysis that differs from conventional thinking.

As provocative as the bowl phenomenon is, the potential improvement achieved by optimal unbalancing is usually quite small. In the three-station exponential line, the optimal allocation is only 0.55% better than the balanced allocation. In fact, the largest potential improvement in the cases studied by Hillier and Boling was 1.4%, for a six-station exponential line. Magazine and Silver (1978) used their approximation to compare optimal and balanced designs with respect to throughput. They found that the maximum potential improvement for exponential lines occurs at about 10 stations and is no higher than about 1.6%.

One explanation for the small improvement due to unbalancing in the exponential case relates to the special relation between mean and variance. As the mean operation times are unbalanced, the total variance in the entire system necessarily increases. Since greater variability tends to reduce throughput, the "pure" beneficial effect of unbalancing the means may be greater than the roughly 1% observed in the Hillier-Boling analyses. However, simulation studies using other distributions, such as El-Rayah (1979a), and the limited analysis due to Muth and Alkaff (1987a) have also estimated the potential improvement from unbalancing to be roughly 1%. Larger improvements seem to occur only in cases where the $cv \geq 2.0$.

ASSEMBLY SYSTEMS

Simultaneous and Independent Loading

The structure of a serial line captures some of the important material flow features of many manufacturing systems found in practice. Another structure of interest is an assembly system, in which component parts are produced in separate feeder lines and brought together at a final assembly station. The simplest assembly system is the three-station case, where the feeder lines each consist of one station and there are two feeder lines. In general, when the feeder lines may be longer than one station, we assume that the first station in each feeder line is never starved.

A fundamental hypothesis about assembly systems is that their throughput performance can be evaluated by "unwrapping" them. In the three-station case, we unwrap the model by treating the assembly station as the middle station in a serial line and the two feeder stations as the end stations of the serial line. Then, this hypothesis continues, we can analyze the resulting serial line to determine how the assembly system will perform. The unwrapping hypothesis holds theoretically in at least one model structure, a specialized kanban system described by Hopp and Simon (1989).

In assembly systems there is an important distinction between two mechanisms for loading the assembly operation. Consider the three-station system and suppose that the feeder stations are busy and that the assembly station has just finished an operation. What happens when one of the feeder stations completes work on a component? Under simultaneous loading, movement to assembly is synchronized. This means that the assembly station remains idle until both components are ready. The assembly station thus functions as if it is looking for two ready signals, one at each feeder station, in order to initiate the movement of components into the assembly process. By contrast, under independent loading, movement to assembly is asynchronous. When the first component is ready, it moves to the assembly station and waits in an assembly workspace until its mate is ready. This means that the assembly station remains idle until both components are ready, but while it is waiting for the second component there is no blocking of the station that completed the first component. We might think of the independent loading system as having a half-buffer at the assembly station.

The distinction between the two loading systems for assembly operation is important for two reasons. First, as a matter of design efficiency, independent loading is better. Other things equal, an independent system will have higher throughput than a simultaneous system unless there is no variability at all. Second, the distinction shows the limitation of the unwrapping hypothesis: since the independent mode and the simultaneous mode differ, a single serial line model cannot serve as a proxy for both of them. In fact, the throughput for a three-station assembly system operating with simultaneous loading is equivalent to the throughput of a *synchronous* three-station serial line. This equivalence generalizes only in the following sense: an assembly system with simultaneous loading and feeder lines of length k has throughput equal to that of a synchronous serial line with $(2k + 1)$ stations if the loading at assembly is simultaneous. For example, the three-station exponential case has throughput equal to $T = 0.545$, from (2a).

With independent loading, however, no such equivalence exists. As Baker, Powell and Pyke (1990b) observed, the throughput for a three-station assembly system operating with independent loading is not equivalent to the throughput of an asynchronous three-station serial line. For example, the throughput for the three-station exponential case is $T = 0.578$ (Hendricks, Moodie and Williams, 1988). Some other throughput values (mostly from simulation runs) are shown in Table 3. These results suggest that the throughput of an assembly system with two feeder lines is greater than that of the serial line obtained by unwrapping, but lower than that of the next smaller serial line.

Effect of Variability

As in the case of serial lines, the cv has a major influence on throughput in assembly systems. Throughput decreases with an increase in cv, and the coefficient of variation explains nearly all of the variation in the throughput. In other words, the two-moment property, as a practical if not rigorous phenomenon, carries over to assembly systems.

Distribution	*cv*	*Serial* *n = 4*	*Assembly* *n = 5*	*Serial* *n = 5*
Exponential	1.000	0.515	0.493	0.486
Uniform	0.289	0.789	0.784	0.777
Normal	0.118	0.903	0.900	0.897
Uniform	0.115	0.903	0.901	0.897

Distribution	*cv*	*Serial* *n = 6*	*Assembly* *n = 7*	*Serial* *n = 7*
Exponential	1.000	0.467	0.460	0.453
Uniform	0.289	0.771	0.770	0.766
Normal	0.118	0.892	0.891	0.889
Uniform	0.115	0.894	0.893	0.891

Table 3. Balanced assembly systems and balanced serial lines.

Effect of Unbalancing

In the balanced three-station assembly system, the mean operation time is 1.0, for the assembly station as well as for the feeder stations. As in the case of serial lines, we can ask whether a reallocation of the workload will lead to an improvement in throughput. In essence, this question asks whether there is an analogy of the bowl phenomenon for assembly systems.

To begin addressing this question, notice where the unwrapping hypothesis would lead us: it would suggest that the optimal work allocations for a three-station serial line would be optimal as well for the corresponding three-station assembly system. In particular, some unbalancing would always be desirable.

Baker, Powell and Pyke (1990d) point out that a balanced system is optimal for the three-station assembly system under simultaneous loading. On the other hand, under independent loading, an improvement is possible. For the case of exponential operation times, it is not difficult to analyze the imbedded Markov chain in order to relate throughput to workload allocation. Using this approach, it is possible to show that the optimal allocation calls for a mean workload of 1.15 at the feeder stations and only 0.70 at the assembly station. This is further evidence of the limitations of the unwrapping hypothesis, for if we had taken the three-station serial line as a proxy for the assembly system, the corresponding reallocation parameters would have been 1.085 and 0.83, respectively.

Therefore, the optimal allocation of work is more unbalanced for the assembly system than for its corresponding serial line. Moreover, the optimal throughput is 1.56% greater than that of the balanced assembly system, as compared to an improvement of 0.55% in the serial line model.

Baker et al. used numerical methods and simulation to explore the robustness of this result. For two cases of the uniform distribution, in which there is less variability than exhibited by the exponential distribution, the results show that the optimal allocation is unbalanced, with less work at the assembly station than at the feeder stations, but the effect is not as severe for the uniform case as for the exponential case. Stated another way, it appears that the severity of optimal unbalancing declines with the *cv*. This result mirrors the findings of Hillier and Boling for the Erlang case in serial lines.

The study also examined two other structural variations of the basic assembly system. In one, the system consists of two feeder lines, each composed of two-station serial lines. In this case, the optimal allocation calls for approximately 0.77 at assembly and achieves an improvement of 1.47% over that of a balanced five-station system. Both of these figures are less dramatic than the comparable figures for a three-station assembly system, suggesting that the severity of optimal unbalancing declines as the line length increases.

A second variation involves greater parallelism. In each case the feeder lines consist of just one station, but the assembly station is fed by k parallel feeders. In this model, the unbalancing results are more dramatic than for the basic three-station model (k = 2), as exemplified by the exponential case in Table 4. In particular, the optimal portion of the total workload in assembly drops quickly with k, and essentially reaches zero for k near 8. Meanwhile, the improvement due to optimal unbalancing increases with k, reaching almost 16% for k=20. Here, the implication is that the severity of optimal unbalancing grows as the extent of the parallelism increases.

Parallelism	*Throughput*		*Assembly*	*Optimal*
k	*Balanced*	*Optimal*	*Allocation*	*Improvement*
2	0.5780	0.5810	0.71	1.6%
3	0.5329	0.5530	0.48	3.8%
4	0.5010	0.5303	0.30	5.9%
5	0.4788	0.5175	0.18	8.1%
10	0.4209	0.4815	0.00	14.4%
20	0.3709	0.4309	0.00	15.6%

Table 4. Optimal allocation of work for different degrees of parallelism

CONCLUSIONS

Models of tightly-coupled systems provide helpful tools in the design of production systems. In the family of models surveyed in this paper there is a serial line of asynchronous operations, carried out at stations that never break down, possibly culminating with an assembly operation. Furthermore, no disruptions occur at the boundary with the external environment; hence, there is no starving at initial operations nor blocking at final operations. The key design problem is to relate the system's capacity, measured as throughput, to the configuration of the system and the variability of operation times. At the design phase, throughput is likely to be the critical measure of performance.

These models are helpful from two perspectives, but there are limitations in each case. First, to the extent that the assumptions are realistic, these models provide precise estimates for throughput. The rigorous analysis of tightly-coupled systems, however, barely extends beyond three-station cases. For larger systems, we must impose further assumptions about the probability distribution of operation times and about the balance among stations. It appears that tightly-coupled systems with more than three stations will largely be the subject of approximation techniques as research in this area develops.

The second perspective relates to the value of insights rather than tangible estimates, especially when the assumptions are restrictive. In this view, the lessons gained from the analysis of simple models are likely to be valuable even in larger and more complicated systems. For example, increments to the size of a large serial line have a negligible impact on throughput; but in small systems, an addition of a station or two may cause a significant degradation. Additionally, Table 3 suggests that the performance of certain assembly systems can be predicted from the performance of analogous serial lines.

Examining the effect of variability, we see the key role played by the coefficient of variation, apart from the mean, in affecting system throughput. For example, when the coefficient of variation is about 0.3, variability alone can account for a degradation in throughput of as much as 25% from nominal throughput.

Finally, models for the reallocation of work content challenge our simplistic notions of line balancing and bottleneck behavior. In serial systems, the bowl phenomenon is a guide to robust design choices. In asynchronous assembly systems, similar results point toward the desirability of fast assembly operations, particularly when several parallel processes are feeding assembly. Such insights broaden the understanding of system designers and help set the agenda for future research in the design of tightly-coupled production systems.

REFERENCES

Anderson, D.R., and C.L. Moodie 1969. Optimal Buffer Storage Capacity in Production Line Systems. *International Journal of Production Research* 7, 233-240.

Baker, K.R., S.G. Powell and D.F. Pyke 1990a. The Performance of Push and Pull Systems: A Corrected Analysis. *International Journal of Production Research* 28, 1731-1736.

Baker, K.R., S.G. Powell and D.F. Pyke 1990b. Buffered and Unbuffered Assembly Systems with Variable Processing Times. *Journal of Manufacturing and Operations Management* 3, 200-223.

Baker, K.R., S.G. Powell and D.F. Pyke 1990c. A Predictive Model for the Throughput of Unbuffered Three Station Serial Lines. Working Paper No. 257, Amos Tuck School of Business Administration, Hanover, NH.

Baker, K.R., S.G. Powell and D.F. Pyke 1990d. Optimal Allocation of Work in Assembly Systems. *Management Science* (to appear).

Basu, R.N. 1977. The Interstage Buffer Storage Capacity of Nonpowered Assembly Lines: A Simple Mathematical Approach, *International Journal of Production Research* 15, 365-385.

Buzacott, J.A. 1968. Prediction of the Efficiency of Production Systems without Internal Storage. *International Journal of Production Research* 6, 173-188.

Carnall, C. and R. Wild 1976. The Location of Variable Workstations and the Performance of Production Flow Lines. *International Journal of Production Research* 14, 703-710.

Conway, R.W., W.L. Maxwell, J.O. McClain and L.J. Thomas 1988. The Role of Work-In-Process Inventory in Serial Production Lines. *Operations Research* 36, 229-241.

De Kok, A.G. 1990. Computationally Efficient Approximations for Balanced Flowlines with Finite Intermediate Buffers. *International Journal of Production Research* 28, 401-419.

El-Rayah, T.E. 1979a. The Efficiency of Balanced and Unbalanced Production Lines. *International Journal of Production Research* 17, 61-76.

El-Rayah, T.E. 1979b. The Effect of Inequality of Interstage Buffer Capacities and Operation Time Variability on the Efficiency of Production Line Systems. *International Journal of Production Research* 17, 77-89.

Hendricks, K., D. Moodie and K. Williams 1988. The Effect on Throughput and Flowtime of the Configuration of a System of Identical Machines without Buffers. Working Paper 88-09, Johnson Graduate School of Management, Cornell University.

Hillier, F. and R. Boling 1966. The Effect of Some Design Factors on the Efficiency of Production Lines with Variable Operation Times. *Journal of Industrial Engineering* 17, 651-658.

Hillier, F. and R. Boling 1967. Finite Queues in Series with Exponential or Erlang Service Times: A Numerical Approach. *Operations Research* 15, 286-303.

Hillier, F. and R. Boling 1979. On the Optimal Allocation of Work in Symmetrically Unbalanced Production Line Systems with Variable Operation Times. *Management Science* 25, 721-728.

Hopp, W.J. and J.T. Simon 1989. Bounds and Heuristics for Assembly-like Queues. *Queueing Systems* 4, 137-156.

Hunt, G.C. 1956. Sequential Arrays of Waiting Lines. *Operations Research* 4, 674-683.

Knott, A.D. 1970. The Inefficiency of a Series of Work Stations: A Simple Formula. *International Journal of Production Research* 8, 109-119.

Lau, H.-S. 1986. A Directly-Coupled Two-Stage Unpaced Line. *IIE Transactions* 18, 304-312.

Liggett, T.M. 1975. Ergodic theorems for the asymmetric simple exclusion process. *Transactions of the American Mathematical Society* 213, 237-261.

Makino, T. 1964. On The Mean Passage Time Concerning Some Queueing Problems of the Tandem Type. *Journal of the Operations Research Society of Japan* 7, 17-47.

Magazine, M.J. and G.L. Silver 1978. Heuristics for Determining Output and Work Allocations in Series Flow Lines. *International Journal of Production Research* 16, 169-181.

Mishra, A., D. Acharya, N.P. Rao and G.P. Sastry 1985. Composite State Effects In Unbalancing of Series Production Systems. *International Journal of Production Research* 23, 1-20.

Muth, E.J. 1973. The Production Rate of a Series of Work Stations with Variable Service Times. *International Journal of Production Research* 11, 155-169.

Muth, E.J. 1977. Numerical Methods Applicable to a Production Line with Stochastic Servers. *Algorithmic Methods in Probability*, M.F. Neuts (ed.), Studies in the Management Sciences, 7, 143-159.

Muth, E.J. 1979. The Reversibility Property of Production Lines. *Management Science* 25, 152-158.

Muth, E.J. 1984. Stochastic Processes and their Network Representations Associated with a Production Line Queueing Model. *European Journal of Operational Research* 15, 63-83.

Muth, E.J. 1987. An Update on Analytical Models of Serial Transfer Lines. Research Report No. 87-15, Department of Industrial and Systems Engineering, University of Florida, Gainesville.

Muth, E. J. and A. Alkaff 1987a. The Bowl Phenomenon Revisited. *International Journal of Production Research* 25, 161-173.

Muth, E. J. and A. Alkaff 1987b. The Throughput Rate of Three-Station Production Lines: A Unifying Solution. *International Journal of Production Research* 25, 1405-1413.

Panwalkar, S.S. and M.L. Smith 1979. A Predictive Equation for Average Output of K Stage Series Systems with Finite Interstage Queues. *AIIE Transactions* 11, 136-139.

Rao, N.P. 1975a. On the Mean Production Rate of a Two-stage Production System of the Tandem Type. *International Journal of Production Research* 13, 207-217.

Rao, N.P. 1975b. Two Stage Production Systems with Internal Storage. *AIIE Transactions* 7, 414-421.

Rao, N.P. 1976a. A Generalization of the Bowl Phenomenon in Series Production Systems. *International Journal of Production Research* 14, 437.

Rao, N.P. 1976b. A Viable Alternative to the Method of Stages Solution of Series Production Systems with Erlang Service Times. *International Journal of Production Research* 14, 699-702.

So, K.C. 1989. On the Efficiency of Unbalancing Production Lines. *International Journal of Production Research* 27, 717-729.

DYNAMIC MAINTENANCE OF A DETERIORATING SYSTEM UNDER UNCERTAINTY

John B. Larsen
College of Business
Florida State University
Tallahassee, Florida, USA

James S. Dyer
Graduate School of Business
The University of Texas at Austin
Austin, Texas, USA

I.INTRODUCTION

The problem of maintaining productive systems is gaining more and more attention of late. As productive machinery becomes more sophisticated and complex (and expensive), it behooves a manager to ensure that the system functions as often as possible, and, if it does break down, to repair it as quickly as possible. In modern Just-In-Time-based systems, preventive maintenance is considered a vital production function. Generally, a part of each worker's day is set aside to perform such duties on the machinery and equipment that he or she operates. The extent to which productive systems are properly maintained affects both productivity and quality.

In this paper, we consider the problem of controlling a deteriorating, complex physical system. More specifically, we concentrate on discrete time models developed to support this task, with a particular emphasis on the role of information, or more precisely, the lack of information available to the decision maker. In particular, we examine the effect of the availability of and the quality of information regarding the actual state of the system to be maintained. A more wide-ranging review of maintenance and control models can be found in

Pierskalla and Voelker(1976). We intend to address a subset of that literature, going into more depth and considering more recent works in our chosen arena.

The outline of the paper is as follows. In Section II we describe the basic mechanics and characteristics of a discrete-time maintenance/control model and introduce some common notation. Then we consider the role of state information in such models. In Section III we present a short review of past work when perfect state information is available, with special emphasis on the form of the optimal control policy. Section IV then considers the incomplete or imperfect information case. Again, we survey past work in this area, considering both solution forms and solution techniques. As will be seen, the current ability to model the imperfect information case far outstrips the ability to solve such problems. Section V then describes some of our own recent work in algorithm development for the imperfect information case, pointing out its advantages and shortcomings. A brief summary of empirical studies based on our approach is also presented. Section VI concludes with a discussion of future research in the area, specifically concentrating on extensions of our own work

II. A DISCRETE-TIME MAINTENANCE MODEL

The discrete-time maintenance model describes a system that progresses through a series of states of disrepair, finally ending up as a failed system if it is not repaired in time. This set of states is often denoted by the set of integers {0,1,2,...L} where the state represented by 0 is a new, perfectly operating system (defect-free) and the state represented by L is a failed system. This deterioration process is described by a Markov Process with known transition probabilities. These are denoted by p_{ij}^{t} and represent the probability of the system moving from state i to state j at time t. The decision maker has three basic control options - inspect, repair or replace, or do nothing, which can be performed at discrete time points, t={0,1,2,3,4,...}. The inspect option gives information about the state of the system. Obviously, there are crucial differences in this option between the perfect and imperfect information cases. The repair or replace options change the state of the system to a "better" state than the one in which it is at the time of this action. Some articles will treat repair and replace as different options, but we will not make that distinction. For our current purposes, the important consideration is that both of these options somehow improve the system. The last option, do nothing, means no maintenance or control functions are performed - merely let the system under consideration do that which it is designed to do; i.e., produce a product if it is a manufacturing system or live and breathe if it is a human body system.

III. PERFECT INFORMATION

In many models, it is assumed that the decision maker has access to or can acquire through inspections, complete and perfect information about the state of the system. That is, if $X_t \in \{0,1,2,...,L\}$ represents the state of the system at time t, the decision maker has direct knowledge of X_t

Much of this work is based on Derman's early models (1962) and (1970). In these works, he showed that, under some conditions on the values in the transition matrices, optimal policies are stationary and nonrandomized, and have a control-limit form. That is, there exists a system state i* such that if, at any time t, $X_t \le i^*$ do nothing, but if $X_t > i^*$ replace the machine.

Works by Kolesar(1966) and Ross (1969) generalized Derman's models to include an additional cost of occupying a state and a continuous state space, respectively, and still found control-limit type policies to be optimal. Kao (1973) then modelled the deterioration process as a Semi-Markov process, and also found similar types of policies to be optimal.

Assaf and Shanthikumar (1987) consider the problem of maintaining a group of machines under either continuous or periodic inspection policies. They too determined that control-limit policies are optimal. In particular, they defined the system state N_t as the number of failed machines in the group and showed that there is a critical number n^* such that if $N_t \ge n^*$ then all failed machines should be repaired, otherwise allow the system to continue operating.

As can be seen from the above discussion, most models and analyses (with perfect state information) have fairly simple optimal control-limit type policies. This would seem to be a desirable goal- to be able to provide simple, easily implemented control policies. In fact, this is likely to be critical to the success of "real-time" control.

In the next section, we examine some of the models developed with partial information. For some simple problems, control-limit policies can be proven to be optimal. But general results for more realistic problems seem to be unobtainable. Therefore, the emphasis turns to solution algorithms.

Note that the underlying structure of the modeled systems remains basically the same as above - a Markov Process. Thus, these new problems are referred to as Partially Observable Markov Decision Processes (POMDP). A survey of research on POMDP's can be found in Monahan (1982). Part A below follows the same conceptual format as Monahan, adding very little to his work. It is included here for completeness. Part B differs somewhat from Monahan. We consider here some more recent work in the field, some of which raises concerns over the tractability of the most popular solution algorithms.

IV. PARTIAL OR IMPERFECT INFORMATION

A. Special Models Initial treatments of the partially-observed problem involved small, simple problems. The most common effort was to attempt to find control-limit type solutions. Girshick and Rubin (1952) developed a model with four system states (two "good" states and two "under-repair" states). The allowable actions are: produce, inspect and repair. In their model, inspection gives perfect information. The partially observable aspect of their model comes from the fact that inspections are assumed to be expensive so that they are not always called for. Thus, the system state is hidden by a lack of constant inspection. They conclude that a control limit solution exists which then determines how many units should be skipped between inspections. The control function and the control limits depend on the conditional distribution of output quality and the probability of transition from the "best" state to the "next-best" state. Their conclusions were proven false by counter-example by Taylor (1966), but their model is still referred to as being perhaps the first "partially-observable" one.

Ross (1971) then considered a problem with two possible system states ("good" and "bad"). Allowable actions again are produce, inspect, and repair. As in the Girshick and Rubin case, inspection reveals the state of the system with certainty and repair puts the system into the best state with certainty. He proves that there exists a control limit solution to this problem. The control variable is the probability that the system is in the "bad" state. Under general conditions, the solution will have four control regions defined by three critical numbers, a, b, and c, such that $0 \leq a \leq b \leq c \leq 1$. Denoting the control variable as p, the general solution is: if $0 \leq p < a$ produce, $a \leq p < b$ inspect, $b \leq p < c$ produce, and $p \geq c$ repair.

White (1977) generalizes Ross' work, allowing for imperfect information in the produce option. First, he formulates the general control problem as a Dynamic Program. The state descriptor is the vector of probabilities associated with each possible system state. He shows that the cost function is concave in all control regions (without really defining what a control region is or describing how to determine such a thing).

White then considers the special case of a two-state problem (a la Ross) and finds that, if perfect information from the inspect option is still available, there will never be more than four control regions, and they will take the form described by Ross. He also presents conditions in which only three regions are present in the optimal control policy, and conditions in which the inspect option is never part of the optimal policy.

Rosenfield (1976) considers a similar problem with more than two states. He develops conditions under which a similar four region control policy is optimal. The conditions are that the repair and operating costs increase as the system deteriorates, the difference between repair and operating costs decrease as the system deteriorates, the system has a single ergodic chain, and the transition matrices are upper-triangular and totally positive of order 2. The state variable is an ordered pair (i,k) which implies that k time units ago, the operator knew that the process was in state i. Note here also that inspections are again assumed to provide perfect information, so that the partially observed aspect arises from less than 100% inspection.

The above models attack well-defined problems with special characteristics (small state spaces, perfect inspection) which allow for characterization of the form of the optimal solution. Pate-Cornell, Lee, and Tagaras (1987) take a somewhat different approach. They consider a four state system with imperfect inspections and imperfect repairs. Instead of trying to find optimal solutions, they consider four reasonable classes of maintenance policies, finding the "optimal" solution within each class. A test problem is devised. Some of the parameters are allowed to vary, and the performance of these policies is then compared for the various permutations. Thus, their approach is not geared to finding the optimal solution, but to comparing some common, easily implemented policies, paying attention to their robustness with respect to various task dimensions.

The next section discusses a branch of the POMDP literature which deals with general solution approaches. Due to the increased generality, the form of the optimal solution is indeterminate. Instead, the emphasis is on the development of algorithms which can then be applied to specific problems. It should be noted, perhaps, that an algorithm may determine that the optimal solution does in fact have the characteristics of a control limit solution, but, *a priori*, we have no notion of how many critical numbers or how many control regions there might be in the optimal solution.

B. General Solution Approaches Eckles (1968) formulates a general problem in which inspections are not necessarily perfect. Repairs always return the system to the "good" state. He formulates the problem as a Dynamic Program with the state variable being a pair consisting of: a) the posterior distribution (in the Bayes' sense) over the possible levels of system deterioration, and b) the age of the system (measured from the last repair action). His solution technique then makes use of the perfect nature of the repair option. If the repair option is chosen η time periods after the previous repair, the dynamic programming recursion will has the same value at the beginning and ending time periods. By choosing and then iteratively adapting this boundary value, Eckles can approach its actual value (by standard dynamic programming - based Markov Decision Process techniques), thus solving the problem.

Sondik (1971) and Smallwood and Sondik (1973) developed the most commonly cited method for attacking general POMDP problems. They also formulate the general problem as a Dynamic Program without restrictions on the quality of information from inspections or on the effectiveness of repairs. They show that a vector of posterior probabilities of the underlying system states is sufficient to describe the POMDP. In this way the POMDP can be formulated as a MDP with a state description consisting of the vector probabilities rather than the (unobservable) system state. They also show that the response surface is piecewise linear and convex in this observer state space. They exploit this property to develop a linear programming based algorithm which finds the hyperplanes that partition the observer state space into regions corresponding to the different linear pieces of the response surface. When the dividing hyperplanes are found, the response surface can be characterized quantitatively, thus determining the optimal control alternative. This procedure is applied to the observer state space at each time period in the problem horizon.

Satia and Lave (1973) also considered a general POMDP problem formulated in much the same way as Sondik's. The state of the system is described by a deterioration measure. They also use the posterior probability of the system being in the various states as the variable in their dynamic programming functional equation. In contrast to the Sondik approach, though, they use an extensive form analysis rather than a normal form of analysis.[1] They then use an implicit enumeration procedure (Branch and Bound). They show that bounds can be obtained by 1) solving the (fully-observed) MDP for each system state, and then 2) multiplying those results by the appropriate probabilities associated with each system state.

The Sondik algorithm is widely credited as the most efficient approach to the POMDP control problem. Unfortunately, Papadimitriou and Tsitsiklis (1987) examined the computational aspects of this type of problem and showed, under this approach, it is PSPACE-HARD. The notion of PSPACE HARD-ness is related to, but more discouraging than, the more well-known notion of NP-Completeness. A PSPACE -HARD problem is not only difficult to solve, but it is also likely that the solution to such a problem is too large to be stored in

---

[1] The major difference between these two approaches is that the extensive form assumes a particular prior state vector. We will discuss some other computational and operational considerations which stem from this distinction later. For a more complete "philosophical" discussion, see Raiffa (1970).

a reasonable amount of computer memory.

Lovejoy (1987) does not consider a specific model and makes no attempt to "solve" a maintenance problem. Instead, he tries to characterize some properties of solutions to a subset of POMDP's. In particular he determines that when certain conditions are satisfied by the problem parameters, the optimal value function is monotone with respect to the probability distribution on the underlying system states. That is, if the state space of probability vectors is ordered by likelihood ratios, the optimal value function is monotone on this ordering. The parametric assumptions relate to the forms of the reward function, the observation matrices, and the system-state transition matrices.

Later, Lovejoy (1991) considers the bigger problem of solving general POMDP's. His approach involves dividing the continuous state space represented by the posterior probability distributions by a flexible gridding technique and then finding upper and lower bounds over each region. The coarseness of the grid determines the quality of the solution so obtained, but this can be adjusted to allow as good an approximation as desired. It seems that his approach represents an attempt at a hybrid strategy between the normal and extensive solution forms. The ability to obtain a high-quality solution (near optimal) will be affected by the size of the state space. As the number of system states increases, the dimension of the state space grows, so that a fine grid will result in a vast number of regions, increasing the computational effort.

Parkan and Hollands (1990) describe a novel hybrid between the extensive and normal forms. They consider the problem of maintaining a sick human patient. Their solution procedure is linear programming based and shares some common characteristics with Sondik's. The major difference is that they introduce the decision maker's uncertainty concerning the prior probability state vector explicitly into the formulation in the form of a constraint. The major difficulty with this algorithmic approach is that it seems to require the explicit enumeration of all possible strategies. Thus, it seems feasible for only very small problems.

All of the above partially observed algorithms experience computational difficulties when the problem size grows. As a result, we were led to consider an alternative solution strategy.

V. EXTENSIVE FORM ALGORITHMS

Larsen (1989) and Larsen and Dyer (1990) developed an extensive form algorithm which exhibits great promise in solving POMDP's of realistic size. We now describe this approach, and briefly summarize the empirical results that have been obtained.

A. Deteriorating Physical System Larsen (1989) and Larsen and Dyer (1990) consider a POMDP model for the solution of a problem proposed by Kleinmuntz (1985). There is a deteriorating physical system, abstractly defined. The "quality measure" (the opposite of the degree of deterioration) is assumed to be a noisy linear process on the range of 0 to 100. The expected reading declines linearly if the system is not repaired, or improves linearly if it is. The actual reading is drawn from a normal distribution about this mean. If the reading is 0 or less, the system has failed. This failure is assumed to have catastrophic consequences, so that the objective is to minimize the probability of its occurrence. The rates of deterioration and improvement, the noisiness of the observation process, and the initial expected quality are all assumed to be known.

The system has one of a known set of defects, but the decision maker does not know which one. To diagnose the defect, the decision maker has a set of tests that can be tried. These are imperfectly diagnostic, with known conditional probabilities. Similarly, there is a set of known repair actions that can be applied. These are also imperfect in that they will only succeed with some probability less than one. Also, they are assumed to be quite risky in the sense that they may be cause immediate failure of the system with a known, non-zero probability.

The decision maker receives two types of feedback- the results of any test just applied (imperfect information about the defect) and an observed quality measure (imperfect information about the status of the system; i.e., is it repaired or not?) It is easy to see that the state space for this problem can be enormous. As an example, consider a problem with three possible defects, four diagnostic tests, and three corrective actions. Obviously, at the beginning of the problem (time 0), there are only three possible system states corresponding to a defective system with one of three possible defects. At the end of the first decision epoch, there are 33 possible system states. At the end of the second, there are well over 200 possible system states.

This general problem structure seems appropriate for the dynamic maintenance of a number of real world systems. Examples that come to mind include maintaining the operation of a unit of a nuclear power plant, maintaining the operation of a continuous refining or chemical process in the petrochemical industry, maintaining the operation of a key piece of machinery on an automated assembly line, or even maintaining the human body through medical interventions.

In the control room of a unit of a nuclear power plant, a large number of instruments are used to monitor the status of the reactor on a real-time basis. However, only a few of these instruments, such as the one that measures the temperature of the reactor core, are really key indicators of the overall viability of the operating system. For example, a systematic trend of increasing temperature in the reactor core could lead to a complete shutdown of the unit, which can cost several million dollars in both direct and opportunity costs. If the operators of the unit detect a trend of increasing temperature, they will be highly motivated to diagnose the cause of this increase, perhaps by running a series of tests, in order to identify an appropriate remedial action (or "repair") that can stop or reverse this trend before more drastic, and costly, options must be considered. It should be clear that this dynamic maintenance problem could be modeled within the general structure of the POMDP that we have described. By analogy, the other examples that we have suggested above should lend themselves to analysis with similar models.

This general model has some important advantages over models in the above POMDP literature. The bottom line is that it is in many ways more realistic.

The defect which is present in the system is unknown. In all others, either the nature of the defect was unimportant, or it was assumed that it would be easily discoverable. A related consideration is that the system is allowed to have more than a few possible states. The relevant system state is a combination of two factors: the nature of the defect and the level of deterioration of the system. This two dimensional aspect greatly increases the size of the state space.

Unfortunately, as with any other modelling-algorithm situation, increased generality in the model often places greatly increased strains on the solution algorithms. In particular, the increased state space of our model makes it unlikely that the algorithms described above would be able to solve it in an acceptable amount of computing time.

Therefore, we took a slightly different tack. Instead of attempting a "normal form" analysis, we used an "extensive form" analysis. The major difference between the two approaches is that the extensive form assumes the decision maker knows or can estimate the initial (prior) probability of the various possible system states. The problem then is conceptualized as a decision tree, with Bayesian processing of the feedback from the test results and the quality level observations.

The normal form on the other hand attempts to solve the problem as a function of the prior probability state space. This is essentially the approach used by Sondik (1971). As mentioned above, it has been shown that the computational complexity of this approach makes it intractable for realistic sized problems. The amount of computational effort depends directly on the number of response surfaces which will obviously increase as the dimension of the observer state space increases.

The extensive form, on the other hand, is not affected by the dimension of the observer state space. A related observation is that stationarity of the transition matrices (conditional probabilities) is not an issue. Strictly speaking, the stationarity property is not required in Sondik's algorithm either. But, practically speaking, it does ensure that the system will approach a steady-state, so that the number of distinct regions in the observer state-space will tend to level off as the time horizon increases.

The extensive form algorithm will have its own computational problems, depending on the number of possible alternatives, outcomes, and decision epochs in the time horizon. In particular, the number of arcs and nodes in a decision tree is an exponential function of these quantities. Our solution approach, then, solves the decision tree without having to consider the whole tree.

B. AO*-Based Algorithms Our solution procedure is based on the AO* algorithm taken from the Artificial Intelligence literature (Nilsson(1980), e.g.) on heuristic search. The decision tree is a special case of a graph structure called an AND/OR tree. That is, at some nodes, called OR nodes, the decision maker chooses an arc, while at the other nodes, AND nodes, the decision maker must consider all of the arcs. Obviously, the decision nodes are OR nodes and the outcome nodes are AND nodes. AO* is an implicit enumeration procedure for searching an AND/OR graph for an optimal solution. It uses heuristic estimates to guide the search through the graph. It has been shown(Nilsson(1980)) that if the estimates satisfy a bounding property that defines them as "admissible", the AO* is guaranteed to lead to an optimal solution. If the estimates satisfy this bounding property, they are said to be "admissible". The correspondence between AO* and the Branch and Bound (B&B) solution strategy has been verified by Nau, Kumar, and Kanal (1984).

We basically used the AO* algorithm "as is" to find optimal solutions to the problems. Very loose bounds were obtained based on the structure of our model formulation. (A general bounding approach can be found in Satia and Lave(1973).)

This optimal algorithm was successful in solving many of our test problems, but not all. In order to "solve" the bigger ones, we modified the algorithm slightly. In particular, we allowed it to search for ϵ - optimal solutions, where ϵ is a predetermined tolerance. We developed two versions of this modified AO* algorithm. One version would find an appropriate "complete" solution, where a complete solution is one which specifies the decision to choose at each decision node in the solution tree-that is, it specifies a policy. The other version finds a "first-period" solution-that is, it specifies the appropriate decision to make at the first decision node.

There are a few points that should be made clear about these modified AO* algorithms. First, the estimates used to guide the search need not be bounds. In fact, some of the mechanisms used to reduce computational effort will cause them to lose their bounding properties; that is, they become inadmissible. But, as long as we can bound the extent of this inadmissibility, we can bound the extent to which the solution deviates from the optimal solution.

Second, the first-period algorithm is more useful than it may sound at first consideration. To use it in a real-time control setting, obviously, one would need to solve the first-period problem, observe the outcome, set up a new problem starting at the second period, and solve again in a timely manner. It should be noted, though, that the second problem is significantly smaller than the first due to a shortened time horizon.

Also, the first-period algorithm does provide an ϵ -optimal solution value. Thus, if our major emphasis is on the solution value rather than on the form of the solution (the policy), the first-period solution is sufficient.

As we can see from the discussion of the mechanics of the heuristic search-based AO* algorithms, a distinguishing feature is that they employ information about the problem to reduce computational effort. This we see as a particular advantage of the extensive form of analysis. In particular, we have the bounding information provided by the particular parameters of the problem and the problem structure. Also, it may well be possible to use some of the "special-case" structural results described earlier to help lead the search for a solution to a more general problem. As an example, we may determine some control limits similar to Ross(1971) and White(1977), which, though not necessarily optimal for the problem at hand, may be valuable in an algorithmic sense as a guide for directing the search in a profitable direction. This latter idea seems promising, but we have not yet implemented a corresponding algorithm.

C. Summary of Computational Results In order to evaluate the heuristic search-based AO* algorithm, we generated a set of forty POMDP problems based on the structure described above, but with randomly chosen numbers of possible defects, diagnostic tests, corrective actions, and decision epochs (problem horizon). Similarly, the conditional probabilities representing test diagnosticty and corrective action effectiveness were also randomly generated within reasonable ranges. A complete description of these problems and the computational details are contained in Larsen (1989) and Larsen and Dyer (1990) . Here we briefly highlight the most significant findings of this study.

An attempt was made to solve the test problems to optimality by two methods: 1) generating and evaluating the entire decision tree associated with the problem, and 2) using the standard AO* algorithm. Within the computational time and memory constraints (15M storage and 1000 CPU seconds on the IBM 3081 at the University of Texas at Austin), only six of the forty problems could be solved by the first method, while twenty-one were solved by the second. Table 1 shows the relative computational effort in terms of the number of arcs generated for the six problems solved by both methods. The AO* provided significant computational savings, generating, on average, 1% of the arcs in the full tree.

The test problems were also solved to ϵ - optimality where, as with other problem parameters, a unique ϵ was randomly chosen from the range of 0.005 to 0.05 for each problem. This corresponded to a percent of optimality of 0.6% to 9.2%. The test problems were solved with the modified AO* algorithm for the complete solution and with the modified AO* algorithm for the first-period solution. The enormous computational advantages associated with the combination of an ϵ -optimal, first-period solution are also illustrated in Table 1. For the entire set of test problems, the modified AO* algorithm required, on average, only 16% as many arcs for the first-period solutions as it did for the complete solutions, with a low of 1.7% and a high of 46.7%.

In order to test the effects of the choice of ϵ, the acceptable deviation from optimality, on computational effort, a base problem was constructed and the value of ϵ was then varied systematically. Figure 1 shows the results of this experiment graphically for the complete and first-period solution cases. The vertical axis is the number of arcs generated relative to the number generated by the AO* algorithm to find the optimal solution and the horizontal axis is ϵ The results show that there is a very steep drop in computational effort for very small ϵ's. As an example, a first period solution with $\epsilon=0.005$, an acceptable deviation of less than 1%, requires less than 21% as much effort as an optimal solution obtained via AO*. As an additional point of reference, compared to the full tree, the same solution requires less than 0.1% as many arcs.

VI. IMPLICATIONS FOR FUTURE RESEARCH

We believe this algorithmic approach will provide researchers with a valuable new tool for attacking these troublesome problems. We envision future research efforts being directed along two general lines - development of real-time control policies and designing systems for controllability.

Problem Number	Problem Size				Number of Arcs		
	D	C	T	TH	Full Tree	AO*	Modified AO* (ϵ - optimal, first-period)
1	4	4	2	5	250034	4084	429
11	3	4	3	5	734512	13742	4836
12	3	3	5	4	1166722	7876	1334
14	2	4	4	5	2022708	12334	1582
21	2	4	4	5	1013772	16779	625
25	3	4	3	5	771376	7668	184

Legend: D - Number of Possible Defects
C - Number of Corrective Actions
T - Number of Tests
TH- Length of Time Horizon

Table 1. Optimal and ϵ - Optimal Solutions

<u>A. Control Policies</u> Optimal control policies are likely to be problem-specific. Simple, easily implementable, and optimal control-limit policies are likely to be difficult or impossible to obtain for actual industrial applications. Instead, if optimal policies are required, one of the algorithms described above will need to be applied to the specific instance. As discussed previously, though, all candidates are computationally expensive, and the optimal policies are likely to be complicated and expensive in terms of computer storage.

An alternative is to search for simple, effective heuristic policies. The actual form of these heuristic policies is as yet an open question. It may be that they will bear some resemblance to the simple control limit policies described earlier. A potential difficulty with this approach is that the logical control

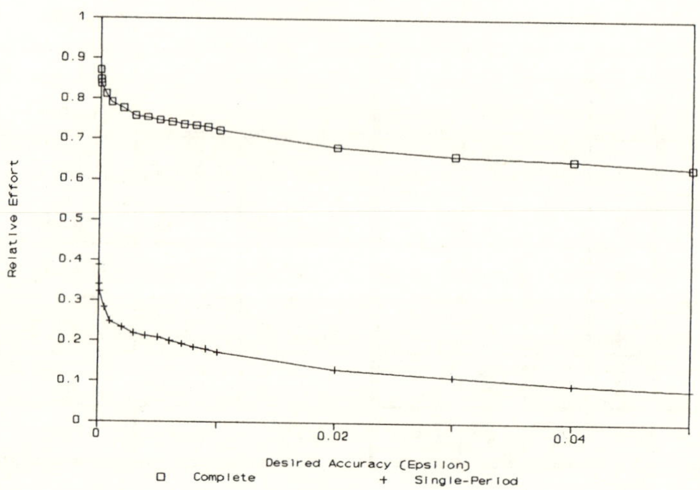

Figure 1. Effect of ϵ on Computational Effort

variable will be the posterior probability vector associated with the possible system states. These calculations may be more than some decision makers want to perform. It might also force the decision maker to feel that he is placing a great deal of faith in the estimates of the problem parameters - the prior probabilities of the system states and the conditional probabilities of the inspection and repair options.

With that caveat in mind, we discuss below some possible questions that would need to be answered in the development of heuristic policies, and show how the modified AO* algorithms can be useful in a research setting.

1. When seeking state information, which diagnostic test should be attempted? Some research has suggested that information which attempts to disconfirm the most likely state will be most valuable (Hogarth(1980)). Are

some test more "disconfirming" than others? Can we develop decision rules for determining which available test is most informative?

2.The obvious corollary to this question in a multi-defect and multi-repair context is "Which corrective action should be applied?" This answers may depend on riskiness, time pressure, state probabilities and the conditional probabilities of the possible repair actions. The answers to these questions are likely to be quite complicated. Developing effective heuristic decision rules will require a great deal of creativity - especially developing general rules to be used in various specific problems.

3.The two questions above reflect an implicit assumption that the maintenance problem has two distinct phases - defect diagnosis and system repair. It seems clear that this would be a helpful structure for heuristic policies to follow. A valuable research stream would then investigate the validity of this structure. How effective are such policies? This should not only look at their average performance, but also at their worst-case performance since it is easy to imagine a situation in which a policy has mistakenly identified the defect and, after attempting ineffective repairs, a manager feels the need to do more testing. Can we devise policies that allow for such deviations from the structure or is the manager then flying by the seat of his/her pants?

A related question concerns when the switch from the information-seeking stage to the repair stage should occur. When is enough state information enough? The answer to this question will depend on many factors including the number and diagnosticity of available tests, the extent to which feedback from corrective actions relates to defect diagnosis, and the degree of time pressure.

4.How do we know that the system is defective (as assumed). One reasonable answer is that we may have process control charts to signal that something is wrong. Note that such devices do not signal what is wrong - only that something is likely to be wrong. How does it signal? Three sigma limits? Two sigma limits? A trend of readings toward a limit? The strength of the signal depends on our criteria for concluding an "out-of-control" situation exists. That choice of criteria may then also determine the initial quality level of the system when we start the solution model.

B. Design Issues How do we design productive systems for maintainability. This question actually divides into two parts - the design of the productive system itself and the design of its associated maintenance system. With regard to the design of the system itself, we may be able to study the effects of redundancy and component robustness. With regard to the design of the maintenance system, we can explore questions of the following sort: Which is better: A system with many cheap tests (i.e., ones whose diagnosticity is weak), or a system with a smaller number of more sensitive and expensive tests? Which is better: A system with powerful, but more risky corrective actions, or one with less powerful and less risky corrective actions?

An interesting question here concerns the interaction between the system design and the control policies. It seems clear that the available options (design) will constrain the possible actions (control policies). How can these interactions be included in the decision rules?

The design and control questions posed above are of obvious interests to managers responsible for ensuring the effective operation of an operating system. The AO* -based algorithms described above have the potential to allow research in these areas to proceed. In particular, using the first-period solution techniques, we can quickly obtain an ϵ-optimal solution value that can be compared to the values of set of possible heuristic maintenance policies, thus identifying effective and efficient policies.

VII. BIBLIOGRAPHY

Assaf, D. and J. Shanthikumar (1987). Optimal Group Maintenance Policies with Continuous and Periodic Inspections. Management Science, 33, 1440-1452.

Derman, C. (1962). On Sequential Decisions and Markov Chains. Management Science, 9, 16-24.

Derman, C. (1970). Finite State Markovian Decision Processes. Academic Press, New York.

Eckles, J. (1968). Optimum Maintenance with Incomplete Information. Operations Research, 16, 1058-1067.

Girshick, M. and H. Rubin. (1952). A Bayes Approach to a Quality Control Model. Annals of Mathematical Statistics, 23, 114-125.

Hogarth, R. (1980). Judgement and Choice. John Wiley and Sons, New York.

Kao, E. (1973). Optimal Replacement Rules. Operations Research, 21, 1231-1249.

Kleinmuntz, D. (1985). Cognitive Heuristics and Feedback in a Dynamic Decision Environment. Management Science, 31, 680-702.

Kolesar, P. (1966). Minimum Cost Replacement Under Markovian Deterioration. Management Science, 12, 694-706.

Larsen, J. (1989). A Decision Tree Approach to Maintaining a Deteriorating Physical System. Unpublished Ph.D. Dissertation. The University of Texas. Austin, Texas.

Larsen, J. and J. Dyer. (1990). Using Extensive Form Analysis to Solve Partially Observable Markov Decision Processes. Working Paper 90/91-3-1, The University of Texas, Graduate School of Business. Austin, Texas.

Lovejoy, W. (1987). Some Monotonicity Results for Partially Observed Markov Decision Processes. Operations Research, 35, 736-743.

Lovejoy, W. (1991). Computationally Feasible Bounds for Partially Observable Markov Decision Processes. Operations Research, 39, 162-175.

Monahan, G. (1982). A Survey of Partially Observable Markov Decision Processes: Theory, Models, and Algorithms, Management Science, 28, 1-16.

Nau, D., V. Kumar, and L. Kanal. (1984). General Branch and Bound, and Its Relation to A* and AO*. Artificial Intelligence, 23, 29-58.

Nilsson, N. (1980). Principles of Artificial Intelligence. Tioga Publishing Company, Palo Alto, California.

Papadimitriou, C. and J. Tsitsiklis (1987). The Complexity of Markov Decision Processes. Mathematics of Operations Research, 12, 441-450.

Parkan, C. and L. Hollands (1990). The Use of Efficiency Linear programs for Sensitivity Analysis in Medical Decision Making. Medical Decision Making, 10, 116-125.

Pate-Cornell, M., H. Lee, and G. Tagaras. (1987). Warnings of Malfunction: The Decision to Inspect and Maintain production Processes on Schedule or on Demand. Management Science, 33, 1277-1290.

Pierskalla, W. and J. Voelker. (1976). A Survey of Maintenance Models: The Control and Surveillance of Deteriorating Systems. Naval Research Logistics Quarterly, 23, 353-388.

Raiffa, H. (1970). Decision Analysis: Introductory Lectures on Choice Under Uncertainty. Addison - Wesley, Reading, Mass.

Rosenfield, D. (1976). Markovian Deterioration with Uncertain Information. Operations Research, 24, 141-155.

Ross, S. (1969). A Markovian Replacement Model with a Generalization to Include Stocking. Management Science, 15, 702-715.

Ross, S. (1971). Quality Control Under Markovian Deterioration. Management Science, 17, 587-596.

Satia, J. and R. Lave (1973). Markovian Decision Processes with Probabilistic Observation of States. Management Science, 20, 1-13.

Smallwood, R. and E. Sondik. (1973). The Optimal Control of Partially Observable Markov Decision Processes Over a Finite Horizon. Operations Research, 21, 1074-1088.

Sondik, E. (1971). The Optimal Control of Partially Observable Markov Decision Processes. Unpublished Ph.D. Dissertation. Stanford University, Palo Alto, California.

Taylor, H. (1966). Markovian Sequential Replacement Processes. Annals of Mathematical Statistics, 38, 871-890.

White, C. (1977). A Markov Quality Control Process Subject to Partial Observation. Management Science, 23, 843-852.

THE ACCURACY OF AGGREGATE LP PRODUCTION PLANNING MODELS

Harvey M. Wagner
Vincent A. Vargas
Narinder N. Kathuria
The Kenan-Flagler Business School
University of North Carolina
Chapel Hill, NC 27599-3490

ABSTRACT

This paper provides valuable evidence concerning the use of small, simple models to represent complex systems. The focus is on linear programming planning models. The starting point is a large, detailed model which serves as a proxy for reality. From the large model, we derive a smaller one through aggregation. Through a process of calibration, we develop estimation models for determining the value of the detailed model objective function value from the aggregate model objective function value and a few additional explanatory variables. This overall procedure proves to be remarkably accurate.

1. INTRODUCTION

Research in aggregation addresses the issue of what degree of aggregation produces meaningful models. (See [8] for a survey of aggregation research.) Smaller models reduce the task of data collection, and are easier to understand, study, and explain to managers. Since larger models are more complex, and thus have "more moving parts," they are more likely to obscure insights and be subject to egregious errors.

Research in production planning models has produced bigger and better models. Departing from this trend, we pursue the theme: How far can a modeling structure be simplified and adequately represent a complex one? [9,10] We deal specifically with the use of an aggregate linear programming model (LP) to estimate the optimal objective function of its corresponding detailed model. This paper builds on earlier work [7] and provides evidence that the results can be extended to other applications.

LP planning models can flexibly represent multi-plant, multi-product, multi-market situations subject to constraints on labor, raw materials, production capacities, yields, transportation, and inventory levels. Models of this sort are prevalent in industry [2].

Aggregation in the context of LP models involves combining "similar items into product groups, different machines into machine centers, different labor skills into labor skill centers, and individual customers into market regions" [5]. The algebraic process of aggregation is weighted row and column aggregation. Decision variables in the original problem are grouped and replaced by a single decision variable in the aggregate model. Similarly, constraints of the original problem are grouped and replaced by a single constraint in the aggregate model.

Dembo and Zipkin use the following matrix algebra notation to represent aggregation [3]. The original problem in canonical form is denoted as

$$\max c^T x$$

subject to

$$Ax \leq b$$

$$x \geq 0,$$

where A is an $m \times n$ matrix, x and c are n-dimensional vectors, and b is an m-dimensional vector. The corresponding aggregate problem is

$$\max \bar{c}^T y$$

subject to

$$\bar{A}y \leq \bar{b}$$

$$y \geq 0,$$

where

$$\bar{A} = W_R\, A\, W_C$$

$$\bar{c}^T = c^T\, W_C$$

$$\bar{b} = W_R\, b,$$

for W_R and W_C non-negative matrices with dimensions p×m and n×q, where, typically, $p \ll m$ and $q \ll n$. Under weighted column and row aggregation, where each decision variable (column of A) and each constraint (row of A) is combined into only one aggregate decision variable or constraint, the matrices W_C and W_R have a special structure. Each row of W_C has only one non-zero entry and each column of W_R has only one non-zero entry.

Column aggregation has a restrictive effect on the optimal objective function value. Under column aggregation, the aggregate decision variables are taken to be fixed linear combinations of the decision variables in the original problem. The restrictive effect occurs because in the disaggregation of an aggregate model optimal solution, the original decision variables must occur in fixed proportions.

Row aggregation is a relaxation on the optimal objective function value. Suppose two constraints in the original model are combined to form an aggregate constraint, and that in an optimal solution to the original problem only one of these constraints is binding. Then, in the aggregate model, the constraint would also have slack for the corresponding detailed model optimal solution. Thus (except in the case of degeneracy), the aggregate model's solution can be improved upon.

When both variables and constraints are aggregated, the impact on the objective function of the aggregate model is unclear. The issue, then, is how "close" is the optimal objective function value of the aggregate model to that of the detailed model.

Research in LP aggregation has established conditions for "perfect" aggregation. Unfortunately these conditions are of little practical use. In order to achieve perfect aggregation, rows must be combined in proportion to optimal values of the dual variables, and columns must be combined in proportion to optimal values of the primal variables. From a practical point of view, if we already know these proportions, there is no need to aggregate. In applications, however, weights used for aggregation are often dictated by marketing or production concerns.

Zipkin showed that the error due to aggregation of LP models is bounded, and that these bounds can be estimated [12,13,14]. We assess the use of the aggregate model to estimate the detailed model's objective function.

We develop reliable estimation models for a detailed model's objective function value (DMOBJ) using its aggregate model's objective function value (AMOBJ) and some additional explanatory variables. Such estimation provides top management a tool for making strategic decisions. To evaluate the impact of changing conditions in the firm's external environment or the impact of proposed strategic decisions relating to management policy or capacity expansion, top management can manipulate the smaller, aggregate model and obtain reliable estimates of the effect on DMOBJ. Since the process of calibration lends itself to automation, the development of software tools to implement this approach is promising.

By establishing the accuracy of aggregate LP planning models, this research fosters their adoption as a strategic planning tool useful to top manage-ment. Further, these results suggest that it is reasonable to expect that the calibration process can be extended to other constrained optimization applications.

2. RESEARCH DESIGN

Overview

The research design is a two-stage process. First, we generate 120 "synthetic" LP planning models from which five are chosen for further study. Second, these five synthetic LPs and one from Kathuria [7] are used to study the robustness of regression models for estimating DMOBJ. We denote these six selected LPs as "base-case" LPs. We apply the calibration methodology to each LP, and conduct a reliability analysis of the estimation models.

The 120 synthetic LPs are constructed so as to be "purposefully malevolent." That is to say, the model coefficients are generated randomly and in a fashion as to introduce aggregation error. No a priori assessment is made to ensure that the variables (columns) to be combined are proportional (in fact, the opposite is likely to occur). The products to be aggregated do not share similar demand seasonalities or variable production costs, criteria that are cited by Dembo and Zipkin [3] as well as Hax and Candea [4] for minimal aggregation error and "reasonable" aggregation. Five of these synthetic models are selected using a stratified sampling technique according to percent aggregation error. We employ this method to ensure an adequate representation of the parameter space. See Wagner, Vargas and Kathuria [11] for the complete details concerning the generation of the 120 synthetic LPs and the subsequent selection of the five base-case models.

The second stage of the research design has two phases, calibration and validation. In the calibration phase, we consider seven different operating assumptions called "scenarios." The calibration process involves solving a set of LPs that are variants of a base case. There is a calibration set of LPs for each of the scenarios. For each LP in a set, the detailed model and its corresponding aggregate model are solved. This information is used to develop models for estimating DMOBJ from AMOBJ and other limited data. For each of the six base-case LPs, 100 calibration cases are run under each scenario for a total of 700 calibration cases.

In the validation phase, we assess the reliability of an estimation model using a separate set of LPs. Seven hundred additional LPs are generated for each base-case LP. Both the associated aggregate and detailed models are solved. We examine the error estimating DMOBJ for each model for validation cases resembling the cases under which it was calibrated. In addition, for one of the estimation models, the accuracy of extrapolative estimation is investigated.

The basic structural LP

We describe the basic structural LP and its aggregate before describing the calibration procedure. We use a hypothetical 12-product, 4-period production planning LP structurally identical to an LP in Kathuria [7].

Define the decision variables for product i, where i = 1,...,N, and for period t, where t = 1,...,T, as

$SALES_{it}$ = sales,
$PRDN_{it}$ = production,
$PRCH_{it}$ = purchases from an outside vendor,
$INVN_{it}$ = inventory at the end of the period.

Define the decision variables for the workforce as

$NBWRK_t$ = workers on the payroll at the beginning of the period (after hiring and firing),
$NBSTF_t$ = staff members on the payroll at the beginning of the period (after hiring and firing),
$FIREW_t$ = workers fired at the start of the period,
$FIRES_t$ = staff members fired at the start of the period,
$HIREW_t$ = workers hired at the start of the period,
$HIRES_t$ = staff members hired at the start of the period,
$REGHRW_t$ = regular-time worker hours used for production,
$REGHRS_t$ = regular-time staff-member hours used for production,
$OTHRW_t$ = overtime worker hours used for production,
$OTHRS_t$ = overtime staff-member hours used for production.

Define the following parameters for each unit of product i as

rev_i = sales price,
cv_i = variable production cost,
ch_i = per period holding cost,
cp_i = purchase cost from an outside vendor,
mch_i = machining requirement,
fin_i = finishing requirement,
str_i = storage requirement,
$cprd_i$ = direct manufacturing cost,
$wrhr_i$ = worker-hour production requirement,
$sthr_i$ = staff-member-hour production requirement.

Define the sales and inventory parameters for product i and period t as

d_i = annual forecast demand,
a_{it} = seasonality fraction applied to annual forecast demand,
ub_{it} = fraction by which sales can exceed forecast demand, $0 \le ub_{it} \le 0.30$,
lb_{it} = fraction of maximum demand determining minimum sales restriction, $0 \le lb_{it} \le 1$,
ss_i = fraction determining safety stock level,
$supper_{it}$ = $(1 + ub_{it}) \times d_i \times a_{it}$, upper bound for sales,
$slower_{it}$ = $lb_{it} \times supper_{it}$, lower bound for sales,
$sstok_{it}$ = $ss_i \times a_{i,t+1} \times d_i$, safety stock level.

Define the capacity parameters as

capmch = capacity limit on machining,
capfin = capacity limit on finishing,
capstr = capacity limit on storage,
maxinv$ = limit on value of inventory,
maxwrk = limit on number of workers,
maxstf = limit on number of staff.

Define the workforce parameters as

cww = per period wage per worker,
cws = per period wage per staff member,
cfw = firing cost per worker,
cfs = firing cost per staff member,
chw = hiring cost per worker,
chs = hiring cost per staff member,
cow = hourly overtime cost per worker,
cos = hourly overtime cost per staff member,
xw = regular-time hours available per worker per period,
xs = regular-time hours available per staff member per period,
yw = maximum allowable worker overtime expressed as a fraction of worker regular-time hours,
ys = maximum allowable staff-member overtime expressed as a fraction of staff-member regular-time hours.

The constraints of the LP can then be expressed as follows. Sales must be between specified upper and lower bounds

(2.1) $\text{SALES}_{it} \leq \text{supper}_{it}$ for $i = 1,...,N,\ t = 1,...,T$

(2.2) $\text{SALES}_{it} \geq \text{slower}_{it}$ for $i = 1,...,N,\ t = 1,...,T$.

Inventory is constrained to be at least the prespecified safety stock levels

(2.3) $\text{INVN}_{it} \geq \text{sstok}_{it}$ for $i = 1,...,N,\ t = 1,...,T$.

Sales are satisfied from production, outside purchase, and inventory

(2.4) $\text{SALES}_{it} + \text{INVN}_{it} - \text{INVN}_{i,t-1} - \text{PRDN}_{it} - \text{PRCH}_{it} = 0$
for $i = 1,...,N,\ t = 1,...,T$.

Steady-state constraints are imposed on inventory levels

(2.5) $\text{INVN}_{iT} \equiv \text{INVN}_{i0}$ for $i = 1,...,N$.

Production is subject to constraints on two types of machine capacity

(2.6) $\sum_{i=1}^{N} \text{mch}_i\ \text{PRDN}_{it} \leq \text{capmch}$ for $t = 1,...,T$

(2.7) $\sum_{i=1}^{N} \text{fin}_i\ \text{PRDN}_{it} \leq \text{capfin}$ for $t = 1,...,T$.

Limits are imposed on the physical storage and dollar value of inventory

(2.8) $\sum_{i=1}^{N} \text{str}_i\ \text{INVN}_{it} - \text{BUY\_STR}_t \leq \text{capstr}$ for $t = 1,...,T$

(2.9) $\sum_{i=1}^{N} \text{cprd}_i\ \text{INVN}_{it} - \text{XS\_INV}_t \leq \text{maxinv\$}$ for $t = 1,...,T$,

where the variable BUY_STR_t represents the amount of additional storage leased from an outside vendor and XS_INV_t indicates the value of inven-tory on hand in excess of maximum inventory value limitations. We assign penalty costs to these variables. The cost of leasing storage space (cbs) is 10% of the average product value per unit of storage required, weighted by the product forecast demand. That is,

$$\text{cbs} = (.1)\left[\sum_{j=1}^{N}\left\{\frac{\text{cprd}_j}{\text{str}_j}\left(\frac{d_j}{\sum_{i=1}^{N} d_i}\right)\right\}\right].$$

The excess dollar value of inventory is penalized at 5% per period for an annual rate of 20% in addition to ordinary holding costs. The purpose of this penalty cost structure is to ensure that the model is always feasible with respect to the storage and investment requirements as well as to suggest when it may be desirable to obtain additional warehouse space or an increased line of credit.

Production is subject to constraints on two categories of labor

$$(2.10) \quad \sum_{i=1}^{N} \text{wrhr}_i \, \text{PRDN}_{it} = \text{REGHRW}_t + \text{OTHRW}_t \text{ for } t = 1,\ldots,T$$

$$(2.11) \quad \sum_{i=1}^{N} \text{sthr}_i \, \text{PRDN}_{it} = \text{REGHRS}_t + \text{OTHRS}_t \text{ for } t = 1,\ldots,T.$$

Regular-time hours available from workers and staff members cannot exceed a fixed proportion times the number of employees on the payroll

$$(2.12) \quad \text{REGHRW}_t \leq \text{xw NBWRK}_t \text{ for } t = 1,\ldots,T$$

$$(2.13) \quad \text{REGHRS}_t \leq \text{xs NBSTF}_t \text{ for } t = 1,\ldots,T.$$

Overtime hours available from workers and staff members cannot exceed a fixed fraction of available regular time hours

$$(2.14) \quad \text{OTHRW}_t < \text{yw REGHRW}_t \text{ for } t = 1,\ldots T$$

$$(2.15) \quad \text{OTHRS}_t \leq \text{ys REGHRS}_t \text{ for } t = 1,\ldots T.$$

The numbers of workers and staff members on the payroll in any given period are determined from the prior period workforce level and current hiring and firing decisions

(2.16) $NBWRK_t - NBWRK_{t-1} - HIREW_t + FIREW_t = 0$ for $t = 1,...,T$

(2.17) $NBSTF_t - NBSTF_{t-1} - HIRES_t + FIRES_t = 0$ for $t = 1,...,T$.

Workforce level for workers and staff members may not exceed a prespecified maximum

(2.18) $NBWRK_t \leq maxwrk$ for $t = 1,...,T$

(2.19) $NBSTF_t \leq maxstf$ for $t = 1,...,T$.

All variables in the model are constrained to be non-negative.

The objective is to maximize contribution to profit. Thus the problem can be stated as

$$\max \sum_{i=1}^{N} \sum_{t=1}^{T} (rev_i\, SALES_{it} - cv_i\, PRDN_{it} - ch_i\, INVN_{it} - cp_i\, PRCH_{it})$$

$$- \sum_{t=1}^{T} (cww\, NBWRK_t + cfw\, FIREW_t + chw\, HIREW_t + cow\, OTHRW_t)$$

$$- \sum_{t=1}^{T} (cws\, NBSTF_t + cfs\, FIRES_t + chs\, HIRES_t + cos\, OTHRS_t)$$

$$- \sum_{t=1}^{T} (cbs\, BUY\_STR_t + .05\, XS\_INV_t),$$

subject to constraints (2.1) - (2.19).

<u>The aggregate model</u>

The aggregate model shares the structure of the detailed model and is derived from it using weighted column aggregation and simple row addition. Let K denote the number of aggregate product groups. Let S_k denote the set of products to be aggregated to form product group k. Let n_k denote the number of elements in the set $\{S_k\}$, that is, the number of products that comprise product group k. Assume that the twelve products are to be aggregated into three product groups consisting of five, four and three products, respectively. In this case, $S_1 = \{1,2,3,4,5\}$, $S_2 = \{6,7,8,9\}$, and $S_3 = \{10,11,12\}$. We define the aggregate decision variables as follows for product group k, where $k = 1,...,K$, and period t, where $t = 1,...,T$,

$$ASALES_{kt} = \sum_{i \in S_k} SALES_{it}$$

$$APRDN_{kt} = \sum_{i \in S_k} PRDN_{it}$$

$$APRCH_{kt} = \sum_{i \in S_k} PRCH_{it}$$

$$AINVN_{kt} = \sum_{i \in S_k} INVN_{it}$$

The column weights for product i in group S_k are fixed as

$$w_i = \frac{\sum_{t=1}^{T} supper_{it}}{\sum_{t=1}^{T} \sum_{i \in S_k} supper_{it}}.$$

Define the following parameters for aggregate product group k from the parameters of the original model and the column weights w_i,

$$arev_k = \sum_{i \in S_k} w_i \, rev_i$$

$$acv_k = \sum_{i \in S_k} w_i \, cv_i$$

$$ach_k = \sum_{i \in S_k} w_i \, ch_i$$

$$acp_k = \sum_{i \in S_k} w_i \min(cp_i, rev_i)$$

$$amch_k = \sum_{i \in S_k} w_i \, mch_i$$

$$\text{afin}_k = \sum_{i \in S_k} w_i \, \text{fin}_i$$

$$\text{astr}_k = \sum_{i \in S_k} w_i \, \text{str}_i$$

$$\text{awrhr}_k = \sum_{i \in S_k} w_i \, \text{wrhr}_i$$

$$\text{asthr}_k = \sum_{i \in S_k} w_i \, \text{sthr}_i.$$

The workforce parameters and capacity limits remain as in the detailed model.

The aggregate constraints on sales become

(2.1a) $\text{ASALES}_{kt} \leq \text{asupper}_{kt}$ for $k = 1,\ldots,K,\ t = 1,\ldots,T$

(2.2a) $\text{ASALES}_{kt} \geq \text{aslower}_{kt}$ for $k = 1,\ldots,K,\ t = 1,\ldots,T,$

where $\text{asupper}_{kt} = \sum_{i \in S_k} \text{supper}_{it}$ and $\text{aslower}_{kt} = \sum_{i \in S_k} \text{slower}_{it}$.

The aggregate constraints that inventories be at least safety stock levels become

(2.3a) $\text{AINVN}_{kt} \geq \text{asstok}_{kt}$ for $k = 1,\ldots,K,\ t = 1,\ldots,T,$

where $\text{asstok}_{kt} = \sum_{i \in S_k} \text{sstok}_{it}$.

The aggregate materials balance constraints become

(2.4a) $\text{ASALES}_{kt} + \text{AINVN}_{kt} - \text{AINVN}_{k,t-1} - \text{APRDN}_{kt} - \text{APRCH}_{kt} = 0$

for $k = 1,\ldots,K,\ t = 1,\ldots,T.$

The steady-state inventory constraints become

(2.5a) $\text{AINVN}_{kT} = \text{AINVN}_{k0}$ for $k = 1,\ldots,K.$

The production capacity constraints become

(2.6a) $$\sum_{k=1}^{K} amch_k \; APRDN_{kt} \leq capmch \text{ for } t = 1,\ldots,T$$

(2.7a) $$\sum_{k=1}^{K} afin_k \; APRDN_{kt} \leq capfin \text{ for } t = 1,\ldots,T.$$

The storage and dollar-value limits on inventory become

(2.8a) $$\sum_{k=1}^{K} astr_k \; AINVN_{kt} - BUY\_STR_t \leq capstr \text{ for } t = 1,\ldots,T$$

(2.9a) $$\sum_{k=1}^{K} acprd_k \; AINVN_{kt} - XS\_INV_t \leq maxinv\$ \text{ for } t = 1,\ldots,T.$$

The labor availability constraints become

(2.10a) $$\sum_{k=1}^{K} awrhr_k \; APRDN_{kt} = REGHRW_t + OTHRW_t \text{ for } t = 1,\ldots,T$$

(2.11a) $$\sum_{k=1}^{K} asthr_k \; APRDN_{kt} = REGHRS_t + OTHRS_t \text{ for } t = 1,\ldots,T.$$

Since the workforce variables and parameters are unaffected by product aggregation, the constraints of the form (2.12) to (2.19) are unchanged.

The objective is to maximize contribution to profit. The aggregate problem can be expressed as

$$\max \sum_{k=1}^{K} \sum_{t=1}^{T} (arev_k \; ASALES_{kt} - acv_k \; APRDN_{kt} - ach_k \; AINVN_{kt}$$

$$- acp_k \; APRCH_{kt})$$

$$- \sum_{t=1}^{T} (cww \; NBWRK_t + cfw \; FIREW_t$$

$$+ \text{chw HIREW}_t + \text{cow OTHRW}_t)$$

$$- \sum_{t=1}^{T} (\text{cws NBSTF}_t + \text{cfs FIRES}_t + \text{chs HIRES}_t$$

$$+ \text{cos OTHRS}_t)$$

$$- \sum_{t=1}^{T} (\text{cbs BUY\_STR}_t + \text{cxs XS\_INV}_t),$$

subject to constraints (2.1a) to (2.11a) and (2.12) to (2.19). The detailed model has 240 variables, 108 constraints, and 152 upper and lower bounds, whereas the aggregate model has 96 variables, 63 constraints, and 44 upper and lower bounds, excluding slack variables.

Calibration

Overview

For expository convenience, the model parameters are categorized into six groups.

Group 1:	Sales and inventory parameters a_{it}, ub_{it}, lb_{it}, ss_i, ch_i, where $ch_i = 0.5 \times cprd_i$.
Group 2:	Annual forecast demand d_i.
Group 3:	Price/cost parameters rev_i, cv_i, and cp_i.
Group 4:	Technological parameters mch_i, fin_i, str_i, $wrhr_i$, and $sthr_i$.
Group 5:	Labor-related parameters cfw, cfs, chw, chs, cww, cws, xw, xs, yw, ys, otratew, and otrates.
Group 6:	Capacity parameters capmch, capfin, capstr, maxinv\$, maxwrk, and maxstf.

Values for the group 1 parameters are selected for each calibration case under all scenarios. In addition, for scenario 1 calibration cases, we vary group 2 parameters moderately (±25%); for scenario 2 calibration cases, we vary group 2 parameters at greater levels than in scenario 1 (±50%). For scenarios 3, 4, 5, and 6, we vary the group 2 parameters as in scenario 1 together with parameters in one of the groups 3, 4, 5, and 6, respectively. In scenario 7, we vary all parameters simultaneously, and group 2 parameters as in scenario 2.

We develop 100 calibration cases under each scenario (for a total of 700 calibration cases) for each of the six base-case LPs. The procedure for generating the calibration cases is explained in detail next.

Changes in group 1 parameters

For each calibration case, new values for the group 1 parameters are selected.

The demand seasonality parameters a_{it} take on values from the set {.15, .20, .25,..., .40} such that $\sum_{t=1}^{T} a_{it} = 1$. The parameters a_{it} are selected according to the following method. Let u, v, and w be random variables, each drawn from the following sets with equal probability

$$u \in \{.15, .20, .25, .30, .35, .40\}$$

$$v \in \{.25, .30, .35, .40, .45, .49\}$$

$$w \in \{.43, .45, .50, .55, .60, .65\}.$$

The variable u is assigned to a_{it}, where t is chosen randomly from {1, 2, 3, 4}. When t = 1, the values for the other three quarters are determined as follows

$$a_{i,t+1} = (1 - a_{it})\ v, \text{ rounded to the nearest .05,}$$

$$a_{i,t+2} = (1 - a_{it} - a_{i,t+1})\ w, \text{ rounded to the nearest .05,}$$

$$a_{i,t+3} = (1 - a_{it} - a_{i,t+1} - a_{i,t+2}).$$

The analogous approach is used for other values of t.

The parameters ub_{it} and lb_{it} represent the proportion by which sales can exceed the demand forecast and the lower bound on sales as a fraction of the upper bound, respectively. The parameter values of ub_{it} are selected at random from 0(.05).30. The parameter values of lb_{it} are selected from the set {0(.5), .25(.1), .50(.1), .75(.1), 1(.2)}, where the associated probability is shown in parentheses. A value of $lb_{it} = 0$ implies no lower bound, whereas a value of $lb_{it} = 1$ implies that sales must equal the upper bound.

The parameter ss_i, which determines the safety stock level for product i as a fraction of next period's demand, is determined as follows. Let $z = \frac{\sum_{t=1}^{T} ub_{it}}{4}$, rounded to the nearest .05 value. Then ss_i is taken to be max{0, z - .05} with probability .25, z with probability .5, and z + .05 with probability .25.

Experimental cases for scenarios 1 and 2

All parameters in groups 3 through 6 are kept at their base-case levels. For a given base case, let d_{ij} denote the annual forecast demand for product i in calibration case j. Determine d_{ij} from

(2.20) $d_{ij} = d_i \times DVF_j \times MIX_{ij}$,

where d_i denotes the annual forecast demand for product i in the base case. DVF_j is the demand variation factor for case j and MIX_{ij} is the mul-tiplicative factor for product i in case j. For scenario 1 calibration cases, the values of DVF_j are selected from the distribution uniform[.75,1.25]. For scenario 2 calibration cases, the values of DVF_j are selected from the distribution uniform[.50,1.50]. The values of MIX_{ij} are selected from the distribution uniform[.75,1.25].

So, for scenario 1 cases, overall demands vary within 25% of the base case, but individual product demands vary between 56% to 156% of the base-case values. For scenario 2 calibration cases, overall demands vary within 50% of the base case, but individual product demands vary between 38% to 188% of base-case values.

<u>Experimental cases for scenarios 3, 4, 5, and 6</u>

For each case in scenarios 3, 4, 5, and 6, we vary the group 2 parameters in the same manner as in scenario 1.

For scenario 3 calibration cases, let $basecost_{hi}$ denote the base values of rev_i, cv_i, and cp_i for h = 1,2,3, respectively. Let $cost_{hij}$ denote the perturbed values of rev_i, cv_i, and cp_i in case j, where

$$(2.21) \quad cost_{hij} = basecost_{hi} \times PVF_j \times IPPF_{hij},$$

and PVF_j denotes the price variation factor for case j, and is drawn at random from uniform[.8,1.4]. $IPPF_{hij}$ denotes the individual product price factor and is chosen from uniform[.75,1.25] for each hij. Overall price/cost variations are between 80% to 140% of the base-case values and individual price/costs variations are between 60% to 175% of base-case values.

For scenario 4 calibration cases, the parameters mch_i, fin_i, str_i, $wrhr_i$, and $sthr_i$ are assigned values that are less than or equal to base-case values. For g = 1,...,5, let $basetech_{gi}$ and $tech_{gij}$ denote the values for mch_i, fin_i, str_i, $wrhr_i$, and $sthr_i$, where $basetech_{gi}$ represents the parameter value in the base case and $tech_{gij}$ denotes the perturbed value for case j. Let

$$(2.22) \quad tech_{gij} = basetech_{gi} \times TIF_{gij},$$

where TIF_{gij} represents the technological improvement factor and is drawn independently for each parameter according to the distribution shown in the first column of table 2.1. Note that lower variable production costs cv_i may be associated with technological improvements.

For scenario 5 calibration cases, let $baselabor_f$ and $labor_{fj}$ denote the values of the labor-related parameters cfw, cfs, chw, chs, cww, cws, xw, xs, yw, ys, otratew, and otrates, where f = 1,...,12, in the base case and case j, respectively. Let

$$(2.23) \quad labor_{fj} = baselabor_f \times ILPF_{fj},$$

where $ILPF_{fij}$ denotes the individual labor parameter factor and is drawn independently from the distribution shown in the first column of table 2.2.

Table 2.1 Technological Improvement Factors TIF_{gij}

Probability	mch_i	fin_i	str_i	$wrhr_i$	$sthr_i$	cv_i
0.2	0.5	0.7	0.8	0.4	0.6	0.7
0.2	0.8	0.75	0.85	0.6	0.7	0.8
0.2	0.9	0.8	0.9	0.8	0.8	0.9
0.4	1.0	1.0	1.0	1.0	1.0	1.0

Table 2.2 Individual Labor Parameter Factors $ILPF_{fij}$

Probability	cfw	chw	cww	xw	yw	otratew
0.2	0.8	0.7	0.9	0.9	1.25	0.8
0.2	1.2	1.1	1.1	0.95	0.5	1.2
0.2	1.4	1.2	1.2	1.05	0.75	1.3
0.4	1.0	1.0	1.0	1.0	1.0	1.0
	cfs	chs	cws	xs	ys	otrates
0.2	0.5	0.85	0.9	1.05	1.5	1.2
0.2	0.8	1.15	1.1	1.1	2.0	1.4
0.2	1.2	1.2	1.2	0.95	2.5	1.6
0.4	1.0	1.0	1.0	1.0	1.0	1.0

For the scenario 6 calibration cases for e = 1,...,6, let $basecaps_e$ and $caps_{ej}$ denote the values of the capacity parameters capfin, capmch, capstr, maxinv$, maxwrk, and maxstf in the base case and case j, respectively. Let

(2.24) $caps_{ej} = basecaps_e \times (CAF_j + IRCV_{ej})$,

where CAF_j denotes the average capacity addition factor drawn from uniform[1.2,1.5] and $IRCV_{ej}$ denotes the individual resource capacity variance factor drawn from uniform[-0.25,0.25]. For calibration cases representing capacity changes, overall capacity varies from 120% to 160% of base-case levels while individual capacities vary from 95% to 185% of base-case levels.

Experimental cases for scenario 7

We construct scenario 7 calibration cases by allowing the group 2, 3, 4, 5, and 6 parameters to vary simultaneously. The demand parameters (group 2) are varied according to the ±50% case. In generating the scenario 7 calibration cases, for changes to the groups 3, 4, 5, and 6 parameters, we use the same calibration factors as developed for scenarios 3, 4, 5, and 6. Note that the parameter cv_i reflects the net effect of cost increases and technological improvements; for any given case, cv_i can be above or below the base-case level.

Table 2.3 summarizes parameter variation across all scenarios. An "X" entry means that the parameters vary for the indicated scenario. The calibration analyses are reported in section 3.

Validation

In order to assess the accuracy of the estimators developed for each scenario, we use cases other than those involved in calibration. The procedures described above are repeated to obtain an independent set of 700 validation cases for each of the six base-case LPs.

In summary, for a given base-case LP, for each scenario, we develop 100 calibration and 100 validation cases. We generate the validation cases in the same manner as we generate the calibration cases; the validation cases can be viewed as a holdout sample by which we assess the accuracy of the estimation models developed for each scenario.

3. RESULTS AND ANALYSIS

Introduction

In this chapter we develop and analyze models for estimating DMOBJ for each of the six base-case LPs and seven scenarios. We designate the five synthetic LPs as A, B, C, D, and E, respectively, and the LP from Kathuria [7] as Z.

For each LP and scenario combination, we consider three alternative approaches. First, we use AMOBJ as the estimate of DMOBJ. We designate this as the naïve approach. Second, we estimate DMOBJ by the regression

(3.1) $\quad \text{DMOBJ} = b_0 + b_1 \times \text{AMOBJ} + \text{error}.$

Table 2.3

Parameter	Count[1]	Group Number	Scenario Number 1	2	3	4	5	6	7
a_{lt}, ub_{lt}, lb_{lt}	NT×3	1	X	X	X	X	X	X	X
ss_l	N×1	1	X	X	X	X	X	X	X
d_l	N×1	2	X	X[1]	X	X	X	X	X[2]
rev_l, cv_l, cp_l, ch_l[4]	N×3	3			X				X
mch_l, fin_l, str_l, $wrhr_l$, $sthr_l$, ch_l[4]	N×5	4				X			X
cfw, cfs, chw, chs, cww, cws, xw, xs, yw, ys, otratew, otrates, ch_l[4]	1×12	5					X		X
capmch, capfin, capstr, maxinv$, maxwrk, maxstf	1×6	6						X	X
cbs,cxs[3]	2	--							

1 N = 12 and T = 4.

2 represents greater demand variation (±50%).

3 these parameters represent the cost to buy storage and the cost to carry excess inventory and are invariant over all scenarios.

4 ch_l is 5% of direct manufacturing cost which is a function of group 3, 4, and 5 parameters, i.e., $ch_i = .05\times\{cv_i + wrhr_i(\frac{cww}{xw}) + sthr_i(\frac{cws}{xs})\}$.

And finally, we estimate DMOBJ by the regression

$$(3.2) \qquad \text{DMOBJ} = b_0 + b_1 \times \text{AMOBJ} + b_2 \times \text{POTCON} + b_3 \times \text{CONRATIO} + b_4 \times \text{PSRATIO} + \text{error},$$

where POTCON is defined as the potential contribution if all sales in all periods are met through regular production, CONRATIO is the ratio of AMOBJ to POTCON, and PSRATIO is the ratio of aggregate production to total potential sales. That is,

$$\text{POTCON} = \sum_{i=1}^{N} \sum_{t=1}^{T} \text{supper}_{it} \left[\text{rev}_i - \{\text{wrhr}_i\,(\text{cww}/\text{xw}) + \text{sthr}_i\,(\text{cws}/\text{xs}) + \text{cv}_i\}\right]$$

$$\text{CONRATIO} = \frac{\text{AMOBJ}}{\text{POTCON}}$$

$$\text{PSRATIO} = \frac{\sum_{t=1}^{T} \sum_{k=1}^{K} \text{APRDN}_{kt}}{\sum_{t=1}^{T} \sum_{i=1}^{N} \text{supper}_{it}}.$$

The values of the explanatory variables are easily determined from either the parameters of the detailed model or the aggregate model optimal solution. POTCON represents the theoretical upper bound on DMOBJ. CONRATIO represents the degree to which this upper bound is realized in the optimal solution to the aggregate LP. PSRATIO measures the fraction of potential sales met through production in the aggregate model. Intuitively, the closer this ratio is to 1, the less constrained is the aggregate LP, and the greater DMOBJ should be.

Accuracy

In table 3.1 we present error measures associated with the calibration set of LPs for the naïve approach and for regressions (3.1) and (3.2). Table 3.2

presents analogous results when the estimators are applied to an independent set of validation LPs.

AMOBJ is a biased predictor of DMOBJ. Of the 4200 calibration cases (100 calibration cases per scenario × 7 scenarios per LP × 6 LPs = 4200), AMOBJ is less than DMOBJ in 3756 instances or 89.4% of the time. The "bias" column in table 3.1 gives a breakdown of the estimation bias by LP and scenario, indicating the number of times AMOBJ is less than DMOBJ.

Regressions (3.1) correct for bias and provide better estimates of DMOBJ than the naïve approach. Regressions (3.2), however, represent a significant improvement over regressions (3.1). The average percent reduction in standard error is 43.1%, where % reduction =

$$\frac{\text{STDERR}_{(3.1)} - \text{STDERR}_{(3.2)}}{\text{STDERR}_{(3.1)}} \times 100\%.$$

From table 3.1 we see that the R^2 values for the regressions (3.2) under scenarios 1, 2, 3, 6, and 7, are greater than .980 for 26 out of 30 instances. None of the 30 have an R^2 less than .967. For the regressions (3.2) fit under scenario 4, the R^2 values are at least .948. For those fit under scenario 5, the R^2 values are at least .929.

In order to be certain that a regression relationship beyond (3.1) exists for all base-case LPs and all scenarios, we test the null hypothesis that for (3.2), $b_2 = b_3 = b_4 = 0$, using an appropriate F test. For 41 of the 42 models we reject the null hypothesis with reported p-value less than .0001. For scenario 5, LP Z, the evidence for rejecting the null hypothesis is strong, but not as convincing as in the other 41 cases. In that instance, the p-value is .0529.

As a further test of accuracy, the regression models (3.1) and (3.2) are applied to an independent set of validation cases. The associated accuracy measures compiled in table 3.2 are "mock R^2" and root mean squared error (RMSE). These measures closely parallel the calibration statistics R^2 and standard error. Mock $R^2 = 1 - \frac{\text{SSE}}{\text{SSTO}}$, where SSE denotes the error sum of squares for the estimated values of DMOBJ and SSTO is the corrected sum of squares of the DMOBJ values, $\text{SSTO} = \sum_{j=1}^{100} (\text{DMOBJ}_j - \overline{\text{DMOBJ}})^2$.

The results shown in table 3.2 are comparable to those in table 3.1. Regressions (3.2) perform significantly better than regressions (3.1). The average percent reduction in RMSE is 42.8%, where

$$\% \text{ reduction} = \frac{RMSE_{(3.1)} - RMSE_{(3.2)}}{RMSE_{(3.1)}} \times 100\%.$$

From table 3.2 we see that the mock R^2 values for the regressions (3.2) under scenarios 1, 2, 3, 6, and 7, are greater than .980 for 27 out of 30 instances. None of the 30 have mock R^2 less than .969. For the regres-sions fit under scenario 4, mock R^2 values are at least .949. For those fit under scenario 5, the mock R^2 values are at least .914.

We note that accuracy varies by scenario. This seems reasonable given the differences in the nature of the parameters varied and, for scenarios 1 and 2, the range of parameter variation. Among the regressions (3.2) developed under scenarios 1, 2, 3, 4, 5, and 6, those developed under scenarios 1, 2, and 6 have the highest R^2 values and the lowest standard errors. These scenarios correspond to varying the vector of right-hand-side parameters b in the canonical representation of the LP given in section 1. From sensitivity analysis, we know that for certain ranges of variation of right-hand-side parameters, the impact on the solution can be readily assessed. On the other hand, sensitivity analysis tells us that changes to any basic column of the matrix A affect the solution in a complex way. This is in accordance with the lower R^2 values and higher standard errors for scenario 4 cases, where the technological parameters vary. In scenario 5, where labor parameters vary, both the vector of cost coefficients and the vector of right-hand-side parameters vary. (By changing the amount of hours available per worker/staff member and amount of permissible overtime, the net effect is to alter the available labor capacity.) For this reason, it is not surprising that the models developed under scenario 5 have the lowest R^2 values among those developed under scenarios 1 through 6.

It is particularly noteworthy that for the regressions (3.2) developed under scenario 7 all have R^2 of .995 and above. While this can be attributed, in part, to high correlation between DMOBJ and AMOBJ, regressions (3.2) represent a significant improvement over regressions (3.1). Under scenario 7, the standard errors for (3.2) are 56.5% less than those for (3.1) on average.

Table 3.1
Error Measures for Calibration Cases

Base LP	Scenario	DMOBJ=AMOBJ RMSE	BIAS	Reg. (3.1) R^2	STDERR	Reg. (3.2) R^2	STDERR	DMOBJ Mean	S.D.
A	1	60938	89	.872	37519	.980	15175	1013292	104298
B	1	62833	96	.924	30741	.989	11827	1016486	110794
C	1	85588	99	.927	27427	.978	15199	1005174	101160
D	1	107554	100	.947	23553	.985	12675	1011796	101769
E	1	120469	100	.941	21987	.970	15823	996219	89808
Z	1	98453	100	.968	21459	.997	6499	1009535	119029
A	2	71846	94	.964	42172	.993	18894	1013728	221774
B	2	71372	93	.982	32272	.996	14480	1021233	237320
C	2	89422	95	.979	28102	.989	21081	994420	194616
D	2	119430	94	.980	27919	.995	14538	1006117	196142
E	2	115238	90	.978	27824	.991	17668	975786	186386
Z	2	84521	92	.986	27248	.998	9536	991333	233247
A	3	103160	80	.966	74938	.984	52700	1244877	405039
B	3	97242	90	.976	60403	.989	41388	1201055	387033
C	3	126170	90	.962	76213	.985	49075	1240250	390508
D	3	159973	97	.953	85173	.980	55766	1231334	389585
E	3	131787	97	.973	62113	.982	51420	1193694	373564
Z	3	149179	98	.965	69646	.986	44686	1199761	369909
A	4	120129	92	.836	69973	.955	37161	1404367	171764
B	4	112911	96	.902	51196	.959	33785	1337087	162674
C	4	144055	97	.861	71300	.949	43868	1507754	190354
D	4	154991	96	.804	75029	.949	38802	1497117	168412
E	4	137557	96	.860	60393	.951	36087	1407497	160509
Z	4	141821	99	.859	54572	.948	33682	1353040	144780
A	5	60785	82	.882	42700	.929	33499	938932	123538
B	5	61885	90	.931	34766	.960	26852	956017	131996
C	5	87226	99	.942	30848	.955	27613	929076	127455
D	5	100724	99	.900	39399	.945	29548	943682	123755
E	5	113621	100	.936	29486	.957	24618	918648	116380
Z	5	87500	100	.974	22198	.976	21656	936801	136307
A	6	68812	78	.898	49256	.985	19260	1085304	153440
B	6	60709	81	.939	39972	.985	19868	1078984	160978
C	6	68749	80	.921	41048	.982	19793	1079611	145507
D	6	86360	90	.887	49937	.967	27319	1115277	147600
E	6	98033	92	.869	51070	.978	21204	1094959	140168
Z	6	81997	92	.945	38062	.995	12075	1036667	161358
A	7	102842	68	.983	83275	.995	43949	1518355	628792
B	7	104328	67	.981	77530	.998	28480	1431710	563453
C	7	146308	64	.975	105402	.997	38049	1644894	663989
D	7	123669	75	.984	87336	.995	48481	1643356	679386
E	7	110404	60	.979	90461	.997	33130	1534377	627506
Z	7	130306	69	.979	86771	.996	37583	1485793	600241

Table 3.2
Error Measures for Validation Cases

Base LP	Scenario	Reg. (3.1) Mock R^2	Reg. (3.1) RMSE	Reg. (3.2) Mock R^2	Reg. (3.2) RMSE	DMOBJ Mean	DMOBJ S.D.
A	1	.870	44171	.986	14677	994441	122917
B	1	.935	35159	.993	11904	998908	138343
C	1	.935	29019	.977	17241	992378	114008
D	1	.958	22909	.985	13510	999418	111982
E	1	.937	23669	.969	16613	978542	94700
Z	1	.969	23544	.997	7127	987386	134510
A	2	.962	42674	.994	16710	986637	220717
B	2	.981	32475	.996	14228	985171	237061
C	2	.981	27717	.991	19475	978348	203769
D	2	.985	24956	.995	13749	993440	201573
E	2	.974	28335	.984	22274	961898	177723
Z	2	.990	23280	.999	8210	969424	237918
A	3	.968	64365	.985	44179	1199884	363511
B	3	.967	59167	.987	36940	1151700	326175
C	3	.959	76193	.980	52919	1204415	379450
D	3	.961	72026	.980	51985	1199350	365574
E	3	.977	54989	.983	47828	1162783	364266
Z	3	.971	57329	.987	38591	1158684	340059
A	4	.904	61018	.958	40219	1355713	198215
B	4	.918	55003	.975	30524	1297847	193172
C	4	.932	55550	.976	33245	1452134	214014
D	4	.895	66195	.969	35984	1447417	205404
E	4	.899	53355	.949	37715	1354439	168414
Z	4	.906	48468	.972	26672	1296543	158825
A	5	.894	37283	.914	33514	970643	115101
B	5	.932	31639	.959	24550	983406	122250
C	5	.932	29965	.957	23936	954222	115902
D	5	.894	36946	.931	29774	971123	114016
E	5	.916	29461	.934	26022	943712	102084
Z	5	.962	22838	.968	21103	974265	118264
A	6	.886	44621	.977	19935	1083504	132973
B	6	.920	39115	.980	19397	1071785	138898
C	6	.908	40799	.979	19414	1085261	135373
D	6	.842	51706	.969	23053	1113888	130677
E	6	.830	52269	.983	16712	1095779	127239
Z	6	.931	37923	.994	11121	1033875	144879
A	7	.973	95682	.995	40616	1627347	588447
B	7	.978	80878	.997	31221	1540679	544319
C	7	.968	111952	.997	33219	1722460	624822
D	7	.972	106224	.996	41564	1757524	637443
E	7	.969	107933	.995	42471	1652181	615475
Z	7	.965	108881	.992	52854	1593998	582300

We note that a pattern arises when comparisons are made between tables 3.1 and 3.2. Within a given scenario, the differences for any base-case LP between the mean DMOBJ value in table 3.1 and the corresponding mean DMOBJ value in table 3.2 are roughly equal. This pattern occurs as a result of the experimental design. In general, the group 1 parameters and the factors listed in table 3.3 have values that vary with the scenario (1,...,7), the case number (1,...,100), and the case type (whether it is a calibration or validation case). They do not, however, vary with the base-case LP (A,...,D,Z). That is, they have the same values for all base cases.

Among the six LPs, LP Z is the presumably the best candidate for aggregation. So, it is reasonable to expect the estimation models developed for LP Z to perform best vis à vis standard error and R^2. From tables 3.1 and 3.2 we see that while this does appear to be true for scenarios 1 and 2, the difference is less pronounced for scenarios 5 and 6 and does not exist for the other scenarios.

Extrapolative Performance

Extrapolative performance refers to the application of an estimator to cases that are dissimilar to its calibration set. Our analysis concerns the estimators developed under scenario 7 as applied to scenario 3, 4, 5, and 6 validation cases.

When estimators (3.2) developed under scenario 7 are applied to scenario 3, 4, 5, and 6 validation cases, we observe a substantial bias. We conclude that scenario 7 estimators should not be used to estimate DMOBJ for the other scenarios.

Inferential Analyses

The error term in estimator (3.2) results from the inherently approximative nature of the mathematical form that fits the four variables to DMOBJ. Inspection of the numerical values for the error reveals that they appear sufficiently close to normal that standard statistical inferential techniques may be usefully applied.

To illustrate, table 3.4 shows the accuracy of confidence intervals applied to the validation cases. Specifically, for each base-case LP and scenario, the standard error shown in table 3.1 is used to compute a confidence interval

Table 3.3
Parameters and Factors Used to Generate Calibration and Validation Cases for the Indicated Scenarios

Reference	Parameters or Factors	Scenarios
group 1 parameters	a_{it}, ub_{it}, lb_{it}, ss_i	all
(2.20)	DVF_j, MIX_{ij}	all
(2.21)	PVF_j, $IPPV_{ij}$	3 and 7
(2.22)	TIF_{gij}	4 and 7
(2.23)	$ILPF_{fj}$	5 and 7
(2.24)	CAF_j, $IRCV_{ej}$	6 and 7

Table 3.4 Normality Test for Confidence Intervals*

			Scenario						
k	Normal	All	1	2	3	4	5	6	7
1.5	86.6	87.6	84.0	86.8	91.0	89.5	87.5	88.3	86.0
2.0	95.4	94.6	94.2	93.3	96.0	95.8	95.7	95.2	92.0
2.5	98.8	97.2	97.8	96.0	97.5	97.8	99.2	97.7	94.5
3.0	99.7	98.7	98.8	97.3	99.3	99.5	100.0	99.2	96.7

*Each entry in the scenario columns represents the percentage of confidence intervals covering DMOBJ, taken over all base LPs (600 cases).

for DMOBJ, namely, estimator (3.2) + k × (standard error). These confidence intervals are calculated for each validation case, and table 3.4 indicates the percentages of correct inferences as compared to the theoretical percentages that occur when errors are normally distributed. In a practical situation, calculating a confidence interval assuming normality of the error term would be a reasonable approximation.

As a corollary, making an inferential judgment about an unobserved value of DMOBJ in one case vis à vis another on the basis of the corresponding estimated values from (3.2) gives type 1 and 2 errors roughly comparable to regression results when the error term is normally distributed with a comparable estimated standard error.

Further Research Directions and Concluding Remarks

We have demonstrated the robustness of the calibration process under a set of circumstances designed to be purposefully malevolent vis à vis aggregation. Thus, we believe this area is promising for continued research. As suggested by Zipkin in a personal communication, what remains to be discovered are the underlying principles by which this approach works [15]. The work of Dembo and Zipkin [3] suggests two promising explanatory variables that could be used when this method is applied to <u>any</u> LP aggregation. These variables measure the primal and dual infeasibility of the disaggregated aggregate model optimal solution. They embody the degree of overestimation (primal infeasibility) and under-estimation (dual infeasibility) in the disaggregated solution. To calculate their values, it will be necessary to have constructed the full detailed model.

REFERENCES

[1] Bitran, G. R., E. A. Haas and A. C. Hax, "Hierarchical Production Planning: A Two-Stage System," *Operations Research*, 29 (1981) 717-743.

[2] Buffa, E. S., and R. K. Sarin, *Modern Production / Operations Management*, 8th edition, Wiley, New York, 1987.

[3] Dembo, R. and P. Zipkin, "Construction and Evaluation of Compact Refinery Models," *Energy Models and Studies*, B. Lev (editor), North-Holland, New York, 1983.

[4] Hax, A. C. and D. Candea, *Production and Inventory Management*, Prentice-Hall, Englewood Cliffs, N.J., 1984.

[5] Hax, A. C. and H. C. Meal, "Hierarchical Integration of Production Planning and Scheduling," M. Geisler (editor), *TIMS Studies in Management Science, Vol.1, Logistics*, North-Holland/American Elsevier, New York, 1975.

[6] Holt, C. C., F. Modigliani, J. F. Muth and H. A. Simon, *Planning Production Inventories and Work Force*, Prentice Hall, Englewood Cliffs, N.J., 1960.

[7] Kathuria, N., The Predictive Accuracy of Aggregate Planning LP Models, Ph.D. Thesis, UNC-Chapel Hill, Chapel Hill, North Carolina (1988).

[8] Rogers, D. F., R. D. Plante, R. T. Wong and J. R. Evans, "Aggregation Techniques and Methodology in Optimization," *Operations Research*, 39 (1991) 553-582.

[9] Wagner, H. M., "Research Portfolio for Inventory Management and Production Planning Systems," *Operations Research*, 28 (1980) 445-475.

[10] Wagner, H. M., M. H. Rothkopf, C. J. Thomas, and H. J. Miser, "The Next Decade in Operations Research: Comments on the CONDOR Report," *Operations Research*, 37 (1989) 664-672.

[11] Wagner, H. M., V. A. Vargas, N. N. Kathuria, Use of Aggregate Production Planning Models for Estimation, Technical Report No. 1, Kenan-Flagler Business School, UNC-Chapel Hill (1991).

[12] Zipkin, P., Aggregation in Linear Programming, Ph.D. Thesis, Yale University, New Haven, Connecticut (1977).

[13] ______, "Bounds on the Effects of Aggregating Variables in Linear Programs," *Operations Research*, 28 (1980) 445-476.

[14] ______, "Bounds for Row Aggregation in Linear Programming," *Operations Research*, 28 (1980) 903-916.

[15] ______, personal communication with H.M. Wagner, August 26, 1988.

MATHEMATICAL MODELS IN INTEGRATED-CIRCUIT MANUFACTURING: A REVIEW

Christopher S. Tang
Anderson Graduate School of Management
University of California, Los Angeles
Los Angeles, CA 90024

and

Lieven Demeester
Anderson Graduate School of Management
University of California, Los Angeles
Los Angeles, CA 90024

1. Introduction

In this paper, we review mathematical models that have been developed for improving the performance of a wafer fab that faces yield uncertainty and system uncertainty such as machine failures. Specifically, we focus on models that deal with the design and operational issues arising from a wafer fab. We classify these models into two main categories: yield models and wafer-design models. These models address different strategic and planning issues that related to yield uncertainty and system uncertainty in IC manufacturing, and can be used to predict system performance of a specific production plan or system design. Besides the predictive power, these models can be used for optimization.

The goals of this paper are: (1) to review existing mathematical models that help IC manufacturers to deal with yield uncertainty and system uncertainty; (2) to provide a list of references covering various types of mathematical models; and (3) to identify potential research areas.

The organization of this paper is as follows. Section 2 presents a description of the fabrication process for integrated circuits. The reader is referred to Hamilton and Howard (1975) and Reinhard (1987) for details. In section 3, we describe the various types of defects that cause yield uncertainty in IC manufacturing, and review some mathematical models (yield models) for predicting yield and monitoring the yield process. Section 4 presents some mathematical models for wafer design. We conclude this section with some comments on the existing models and remaining design issues for future research. Section 5 provides some concluding remarks.

2. **Wafer Fabrication Process**

In IC manufacturing, many identical ICs are formed on a single (silicon) wafer, where each IC consists of multiple layers of circuits. To form one layer of circuit on the wafer, several types of processes are needed. The sequence and the types of the processes depend on the characteristics of the circuit to be formed. Burman et al. (1986) and Chen et al. (1988) have grouped the processes into seven types, as follows:

(1) **Clean**: This operation removes particulate matter on the wafer before a new layer is formed. To prevent contamination, cleaned wafers must go to the next step within a prespecified time.

(2) **Deposition**: This operation is essentially a "batch" operation in which a thin layer of material (such as silicon oxide or metal) is grown on the wafer. In this operation, the wafer is heated to a temperature of approximately 1100°C and then the wafer is exposed to oxygen or steam.

(3) **Lithography**: This operation consists of two procedures: (i) photoresist and (ii) masking. In the photoresist procedure, a light-sensitive coating (photoresist) is placed on top of the oxide layer (as shown in figure a). Then in the masking procedure, the coated wafer is exposed to ultraviolet light through a mask that contains the pattern of the oxide to be removed (as shown in figure b).

(4) **Etching:** In this operation, the "unexposed" part is dissolved in a chemical solvent, leaving an opening "window" in the coating, as shown in figure c. Then the silicon oxide at those opening locations are etched away, as shown in figure d.

(5) **Ion Implant**: This operation changes the electrical properties of the surface not protected by photoresist. In this operation, the wafer is exposed to accelerated ions which are implanted to a predetermined depth.

(6) **Photoresist Strip**: This operation removes the remaining photoresist on the wafer, leaving the printed circuit on the silicon wafer, as shown in figure e.

(7) **Inspection and Measure**: After a layer of circuit is formed on the wafer, this operation inspects and determines defects of this particular layer of circuit. Failed wafers are either scrapped or reworked.

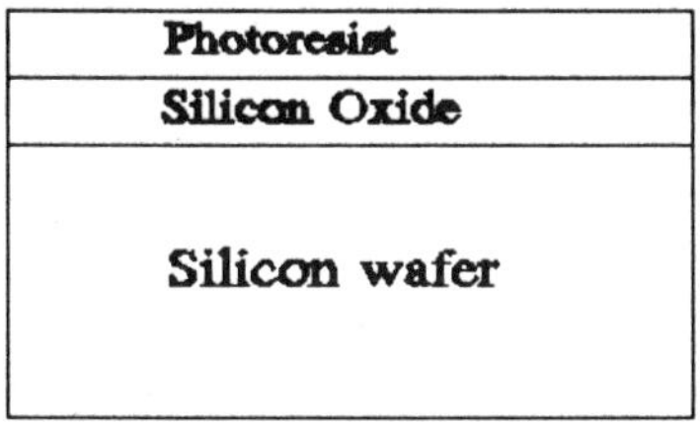

(a) photoresist coating

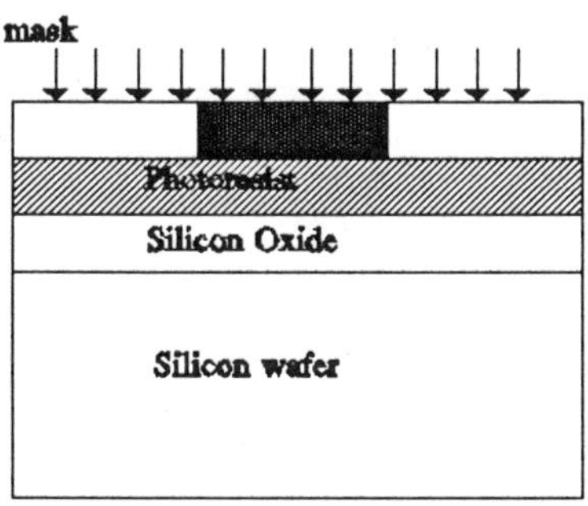

(b) Lithography

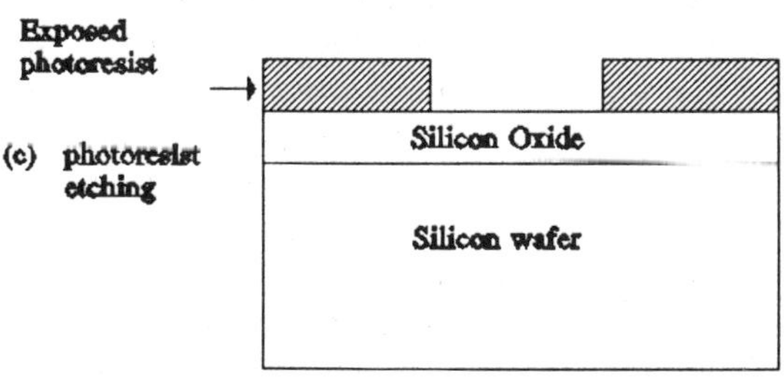

(c) photoresist etching

(d) Oxide etching

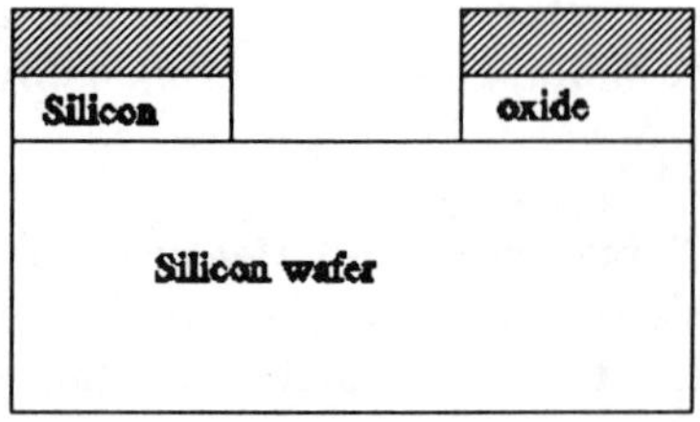

(e) after strip

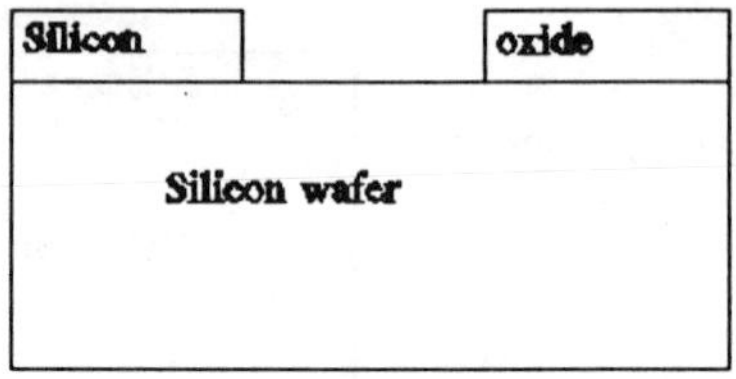

Operations (1) through (7) form a sequence of operations for forming a layer of circuit on a wafer. In IC manufacturing, multiple layers of circuits are formed on a wafer. Each layer of circuit has its own sequence of operations. Therefore, each type of wafer has it own "process plan" that specifies the required sequence of operations. In most IC fabrication facilities, various types of inspections and measurements are taken at different stages of the process. Any wafer that is defective is either scrapped or reworked. The uncertain defective rate (or <u>yield</u> <u>uncertainty</u>) has been a primary source of difficulties in managing wafer-fabrication facilities.

3. Models for Yield Management

In this section we address some issues that are related to yield management in a wafer fab. Specifically, we discuss the importance of yield models in yield management. The wafer-fabrication process, as described in the previous section, requires a very high level of accuracy at every step of the process. It is extremely sensitive to particle contamination. Real throughput (in terms of non-defective chips) is only a fraction of the capacity of the released wafers. This fraction is called the **yield** and typically ranges from 10 to 95%, depending on the size and complexity of the produced circuits and the maturity of the processing technology and circuit design.

In order to improve the yield, it is necessary to understand the causes of yield loss. To do so, many IC manufacturers are now equipped with inspection stations and monitoring systems for obtaining direct measures of the yield of a certain operation or group of operations. This yield management tool is called yield monitoring. Yield monitoring not only is an indispensable step towards yield improvement, but it also provides necessary information for short and long term planning. For instance, process engineers use yield monitoring to respond quickly to process errors; and design engineers use the feedback they get to design circuits with higher yields. Production planners need yield information to determine the real capacity of the wafer fab and to adapt their production plan, while managers use it to decide when to switch to a new circuit design or new process technology with higher circuit complexity.

In section 3.1, we present a rough classification of the sources for yield loss and describe some yield monitoring techniques. Then in section 3.2, we review models that predicting yield. Finally, in section 3.3, we present some potential research areas in which these yield models can be used for optimization.

3.1 **Yield Monitoring**

Generally speaking, there are 3 kinds of defects on a wafer. These defects are:

(1) General 'area' defects These defects affect the entire wafer or areas that are large compared to the chip size. Examples are diffusion or etching errors, masking errors, general contamination, ... etc.

(2) 'Line' defects These defects generally result from mishandling or breakage by operators and do not constitute a significant part of overall losses prior to chip test in a well controlled process.

(3) Highly localized 'point' defects These defects constitute the predominant type of defect causing losses. Especially dust particles, diffusion pipes, oxide pinholes and other leakage defects, introduced at the various masking steps, are common and troublesome in large-area devices.

Yield monitoring tries to detect the defects as soon as possible in order to identify the sources and take appropriate action to correct the process. There are two ways to detect defects on wafer circuitry: electrical-property testing, and visual inspection through optical or electron microscopy. Electrical-property testing is performed mostly after deposition of a metal layer, which can be part of the final circuit or which is deposited only for testing purpose. Optical visual inspection stations are most useful after the photolithography or metallization process to check that the geometric pattern has been well aligned and that no parts are missing. Electron microscopy is used to detect oxide pinholes and other invisible defects. The readers are referred to Stapper (1982) for an extensive review of yield-monitoring techniques.

The inspection and monitoring systems are often imperfect and most often designed to detect only specific kinds of defects. Hence, there is always a certain degree of uncertainty about the source of a defect once it is detected. The inspection methods also are time consuming, labor intensive, and, therefore, expensive. Small sample sizes seem to be a good compromise, but they also introduce additional uncertainty about the yield at certain stages. These uncertainties make yield monitoring a challenging task. Given the yield-monitoring information, the next step is finding ways to predict the yield when changes are made to the process, the technology, or to the circuit design. One would like to be able to answer questions like: How much would the overall yield improve if tolerances in the etching process are reduced by 20%, if operator errors in the photolithography stage can be reduced by 50%, or if a new design reduces the sensitivity defects by 30% ?

The question of how to organize the huge amount of information of yield monitoring into models for predicting yield has been addressed by several authors. These yield models proposed in the literature tend to focus on the behavior of the 'point' defects on each wafer, mainly for three reasons. First, the 'point' defects constitute the predominant cause for failures. Second, the point defects behave strangely: they have a tendency to cluster, causing the defect densities to vary from chip to chip and from wafer to wafer. Third, some design optimizations are possible by the very nature of point defect distribution. The models for predicting yield are now discussed.

3.2 Yield models

A <u>yield</u> <u>model</u> is defined as a set of mathematical relationships that predicts the effects of various defects encountered in the manufacturing process on yield. The first to report on the practical use of such models were Murphy (1964), Lawson (1964), and Dennard (1966). In this section we first describe yield that are based on the aggregate defect density. Then we show how these models can be made more accurate and better applicable when the different defect types are treated separately.

3.2.1 Basic yield models

Generally, the yield rate of a wafer can be expressed in the form:

(3.1) $$Y = Y_0 Y_1(D_0, A, \alpha) ,$$

where Y is the ratio of good (or non-defective) chips per wafer to the number of chip sites per wafer. Notice that $(1 - Y_0)$ is the fraction of chip sites that yield defective chips either because of general area, line or sensitivity defects and that $(1 - Y_1)$ is the fraction of the remaining chip sites that yield defective chips, mostly because of point defects. In addition, Y_1 is a function of D_0, A, and α, where D_0 is the average density of point defects per unit area, A is the area (or the size) of the chip, and α is a parameter that is uniquely defined for each yield model. In the yield model, Y_1 is especially interesting because it relates the yield Y to the chip size A.

In order to find the form of $Y_1(D_0, A, \alpha)$, one has to make an assumption about the distribution of point defects. It seems reasonable to assume that the number of point defects in the area A of one chip has a Poisson distribution, or in other words, every infinitesimal piece of wafer surface is equally likely to contain a defect. Thus, Y_1 is equal to the probability that no point defect occurs in the area A, where

(3.2) $$Y_1 = \exp(-AD_0), \quad \text{with}$$

$$AD_0 = \text{average number of point defects per chip}$$

The Poisson model has been found to give poor yield predictions. The reason is that point defects have a strong tendency to cluster. In 1964, B. T. Murphy noted that: "The surface density of defects which result in bad components varies from chip to chip and from slice to slice, resulting in an effective correlation between the yields of separate components on the same chip."

The clustering of point defects results in more chips with either no defect or many defects, and fewer chips with only a few defects compared to the Poisson model defined in (3.2). The strong gap between the model and the reality also became clear when comparing the ratio of the variance to the mean of the number of defects on one chip. For the Poisson model this ratio is equal to one, while the empirical results have indicated that these ratios tend to range from 5 to 15. These observations raised the need for a more realistic distribution function for the point defects on a chip.

Murphy (1964) proposed a compound Poisson model, which is similar to the Poisson model except that the defect density D can vary from chip to chip as if D is drawn from a second, compounding, distribution. In this case D_0 is the expected value of the random defect density D and Y_1 takes on the following form.

(3.3)
$$Y_1 = \int_0^{\infty} e^{-AD} dF(D)$$

where F(D) is the compounding distribution function. Notice that Y_1 is equal to the moment-generating function of the compounding distribution evaluated at -A.

Seeds (1967) , Ansley (1968), and Moore (1970) continued on the same path to extend Murphy's work. Warner (1974) took a slightly different approach to explain defect clustering and hypothesized that during the manufacturing process, IC's are subject to two defect mechanisms, one that distributes point defects randomly throughout the wafer, and one that leads to high defect densities in certain areas.

Stapper (1985, 1986, and 1987) tried out several compounding distributions and reported a best fit of empirical data when D has a gamma distribution with parameters α and D_0/α, where α can be considered as the 'clustering' parameter ranging from 0 to $+\infty$, and D_0 is the expected value for the defect density D. Note that α is small when the defect density D shows a great variance, and α is large when the variance of D is small. Using the moment-generating function of the gamma distribution it can be easily checked from (3.3) that Y_1 now has the following form :

(3.4)
$$Y_1 = (1 + D_0 A/\alpha)^{-\alpha},$$

Note that Y_1 tends to $\exp(-AD_0)$ (i.e., the Poisson model) as α tends to $+\infty$. Note that as the chip area A increases, the probability of no defect in that area becomes small very fast. This is one of the reasons why the manufacturers are interested in reducing chip size.

Stapper's effort to fit the empirical data to the gamma-compounded Poisson distribution has shown that the clustering parameter α depends on the chip area A. Meyer and Pradhan (1989) have offered an explanation as follows: When the area is small, it might tend to be contained within a cluster of defects and therefore does not reflect the extent of clustering. This is because there are nearby defects that belong to the cluster but are not included in the count (since they occur in neighboring areas). As the area is increased, it is more likely that an entire cluster would fit neatly in an area. Hence, the clustering observed is maximized, yielding a distribution with a lower clustering parameter α. As area is increased, some clusters may fall within the area together with much area free of defects, so the clustering is less apparent, and α starts to rise. This would indicate that the clusters are not uniformly distributed.

Another phenomenon reported by Yanagawa (1972) and Stapper (1982) is that certain types of point defects show a strong "radial dependence" in the frequency of their occurrence. This relationship holds especially for cosmetic defects, such as handling damage, misalignment errors, photoresist residue, ... etc. The radial variation in defect density can be represented by the expression:

(3.5) $$D(r) = D_0 + D_R \exp((r- R)/L)$$

where D_0 is the defect density associated with the center of the wafer, D_R is the increase in defect density at the edge of the wafer, R is the radius of the wafer, r is the radial coordinate $0 \le r \le R$, and L is a parameter associated with the edge-related defects. This radial dependence has motivated IBM (Stapper, 1982) to partition a wafer into different ring- formed sections, which can be used to predict yield.

More sophisticated models are still being examined. For example, Meyer and Pradhan (1989) propose a center-satellite model to capture both clustering and radial distribution, but have no empirical data to prove the usefulness of their model.

3.2.2 Multi-type defects Models

In practice the yield for each wafer is affected by many types of defects. Each type of defect affects different circuits to a different extent. To make the model more general and tractable, it is necessary to treat each type of defect separately. Suppose there are N types of defects, and, further, that each defect type n, where $n \in \{1,..,N\}$, is characterized by its mean density D_{n0}, its clustering parameter α_n, and the portion of the total chip area A_n that is susceptible to that particular type of defect. Making use of Stapper's gamma-compounded Poisson model (3.4), the yield for each defect type n can be expressed as:

$$Y_n = (1 + \frac{A_n D_{n0}}{\alpha_n})^{-\alpha_n}$$

Since the yields associated with different defect types are assumed to be independent, the overall yield is the product of the yield for each known type of defect. In this case, the overall yield, denoted by Y, has the form:

(3.6) $$Y = \prod_{n=1}^{N} Y_n = \prod_{n=1}^{N} (1 + \frac{A_n D_{n0}}{\alpha_n})^{-\alpha_n}$$

For a given processing technology, the D_{n0}'s are constant, the α_n's are functions of the size of the circuit (c.f., Stapper(1986)), and the A_n's can be determined from the circuit design.

Suppose that we would like to predict the yield for two different circuits to be processed by the same technology. Since the D_{n0}'s only depend on the processing technology, we only have to determine the A_n's and α_n's for each circuit in order to predict the yield (when the defects are independent).

3.3 **Potential Research Topics**

We now present potential research areas that are directly related to the yield models discussed in this section.

3.3.1 Optimal chip partitioning The optimal size of a chip, with respect to the number of circuit functions is a compromise among several competing considerations: partitioning of the circuit system, the expected yield of good circuits, packaging and system- assembly cost, and the overall reliability of the complete system. For example, if a complex circuit system is divided into a number of (smaller) ICs with simpler circuit systems, then the yield of these simpler circuits may be very high, due to the fact that the yield is inversely proportional to the chip size. However, the total cost of packaging a large number of these smaller chips and assembling them on a circuit board may be greater than the cost of manufacturing a smaller number of larger chips with complex circuit at a lower yield, and packaging and assembling these larger chips on a circuit board. To divide the complete circuit system into an optimum number of ICs, we need to evaluate the cost of an IC as a function of circuit size, and combine that information with the size-yield relationship from the yield models. Murphy (1964) derives a generalized cost-size relationship for a circuit consisting of N homogeneous circuits, accounting for the variations in component density, yield, and assembly costs. In order to design an optimal partition of a complex circuit system consisting N heterogeneous circuits, one needs to develop a more sophisticated model to evaluate the tradeoffs between the higher yield of producing smaller and less complicated chips and the cost of assembling a larger number of smaller chips.

3.3.2 Optimal inspection plan Whenever defect data are encountered in industrial quality-control applications, the Poisson distribution is generally assumed to be the underlying distribution. It follows from our discussion in section 3.2, that this assumption is invalid in the case of defects on IC wafers. Albin and Friedman (1989) use the Neyman distribution, a Poisson-compounded Poisson distribution, to design acceptance-sampling plans for various stages of the manufacturing system. They find that the sample size required for a product with a Neyman defect-distribution is larger than the sample size required for a product with a Poisson distribution by a factor equal to variance/mean. This result may provide a technique to determine the sample sizes used for yield monitoring in IC manufacturing. Specifically, their model may provide a foundation for developing an optimal inspection plan that accounts for the tradeoff between yield information and inspection cost.[1]

3.3.3 Optimal Redundancy Level Many large-area VLSI circuits are designed with redundant circuitry, which can be used to replace defective circuit elements. The replacement of a defective circuit is usually accomplished by using "fusible" links, which can be fused by using laser or other techniques. To apply this technique it is very important to know the spatial dispersion of defects on individual wafers. Redundancy at the lowest levels -- occurring in close spatial proximity -- may be defeated by clustered defects. Meyer and Pradhan (1989) provide examples to demonstrate that inaccurate yield models could lead to the wrong design decisions. This phenomenon further illustrates the value of yield monitoring and accurate yield models. In addition, this leads us to believe that a yield model can be used to determine the optimal level of redundancy in VLSI design.

3.3.4. Optimal timing for technology introduction

Due to process improvement, the yield tends to improve over time. However, the process technology and the chip design can become obsolete. In that case, manufacturers may introduce a new design or a new technology. This raises an interesting question regarding the introduction of a new design or a new technology from R&D to manufacturing. Clearly, there is a trade-off between producing the "old" chips and the "new" chips. For instance, the new chips are probably smaller and more sophisticated than the older ones, but their initial yield may be low and the time it takes to improve the yield of the new chips to an acceptable level is very difficult to predict. Although there exist mathematical models on technology introduction (see Balcer and Lippman (1984)), we are not aware of any model that addresses the issue of technology introduction in IC manufacturing.

The yield-monitoring system has provided meaningful information for process improvement and yield improvement. However, yield uncertainty remains to be a chronic problem especially when the manufacturer initiates

---

[1] Tang (1991) has developed a model for determining optimal inspection plan for a serial production line.

the production of new IC chips with a higher complexity. This observation has motivated some researchers to investigate the issue of wafer design as a way to improve the throughput rate. The issue of wafer design is discussed in the following section.

4. Models for Wafer Design

In wafer fabrication, each wafer is partitioned into hundreds of grids (or chip sites), where each chip is formed at a site. Traditionally, all chips to be formed on a wafer are of the same type. Such wafers are called single-type wafers. However, some IC manufacturers are considering new technologies (such as electronic beam lithography, and X-ray lithography) to produce different types of chips simultaneously on a wafer.[2] (See Wilson (1986) and Shimkunas (1984).) Such wafers are called multi-type wafers. This new development has motivated the following questions: (1) What is the economic benefit of producing multi-type wafers? (2) How should the chip sites be allocated to different types of chips? These two questions are easy to ask but difficult to answer. In this section, we shall present model that justifies the economical value of multi-type wafers in 4.1. Then in 4.2, we discuss models for designing multi-type wafers. Finally, we comment on these models in section 4.3.

4.1. Models for comparing single-type wafers and multi-type wafers

To our knowledge, Avram and Wein (1990) is the only paper that analyzes the economic value of multi-type wafers. In their study, they consider a situation in which the customer orders the chips in sets, where a set of chips consists of a specified number of chips for each of several chip types. This situation is rather common, especially when customers order multiple types of chips for printed circuit-board production. Due to the fact that the changeover cost[3] is extremely high when the system changes from processing a specific type of multi-type wafer to a different type of multi-types wafer (with a different chip allocation), Avram and Wein (1990) restrict their

---

[2] For instance, according to an article reported in Los Angeles Times, May 27, 1991, AT&T Bell Laboratories have installed an X-ray lithography system designed by Hampshire Instruments of Rochester.

[3] When using the traditional ultra-violet lithography process, the changeover cost includes the cost of making different mask sets for different multi-type wafers with different chip allocation.

attention to static design (i.e., the chip allocation on the wafer does not vary over time). This model is described below.

Consider a wafer that consists of S chip sites, indexed by $s = 1,\ldots,S$, and a set of chips that is comprised of n_k type k chips, $k = 1,\ldots, K$. It is assumed that $S > K$. Avram and Wein (1990) considered three types of yield uncertainties: defective rate of a lot, defective rate of a wafer, and defective rate of a chip. For each non-defective wafer, let p_{sk} be the average number of non-defective chips produced of type k when we assign chip type k to chip site s.

Avram and Wein (1990) present two types of analyses: deterministic analysis and stochastic analysis. The deterministic analysis focuses on the long-run average throughput rates, while the stochastic analysis concentrates on the throughput up to time t. We shall discuss both analyses in turn. In order to make the model tractable, the fabrication facility is modeled as an open quasi-reversible queueing network that possesses a stationary distribution. In addition, assume that the multi-type wafers arrive at the system according to a Poisson process with rate λ, and that the single-type wafers of type k arrive at the system according to independent Poisson process with rate $r_k\lambda$, where r_k is the fraction of wafers that are of type k.

4.1.1 Deterministic Analysis

In single-type wafer fabrication, the "average" number of non-defective chips of type k on a non-defective (single-type) wafer of type k is denoted by p_k. Since it takes n_k/p_k non-defective type k wafers to form a set on average, it takes $\Sigma_k (n_k/p_k)$ non-defective wafers of all types to form a set. Therefore, the average ratio of type k wafers should be equal to r_k, where

(4.1) $$r_k = (n_k/p_k) \;/\; \Sigma_j (n_j/p_j).$$

In multi-type wafer fabrication, let x_{sk} be the decision variable that specifies the "fraction" of site s to be allocated to type k chips. Hence, we have $\Sigma_k x_{sk} = 1$ for each s. In this case, the total average number of non-defective type k chips is equal to $\Sigma_s p_{sk} x_{sk}$. Since n_k chips are required to form a set, the average number of sets to be formed from a non-defective multi-type wafer is equal to $\text{Min}_k \{(\Sigma_s p_{sk} x_{sk})/n_k\}$. Hence, it is desirable to determine a chip allocation so that the average number of sets formed is maximized. Thus, one can formulate the chip-allocation problem as the following program:

(P)	$\underset{x_{sk}}{\text{Max}}$ $\underset{k}{\text{Min}}$ $\{ (\Sigma_s p_{sk} x_{sk})/n_k \}$		
(4.2)	$\Sigma_k x_{sk} = 1$	for each s, and
(4.3)	$x_{sk} \geq 0$	for each s and each k.

Consider the case when we set $x_{sk} = r_k$ for all s and k. Clearly, this is a feasible solution to problem (P). Since $\Sigma_s p_{sk} = p_k$, the objective value associated with this feasible solution is equal to $1/\{\Sigma_k (n_k/p_k)\}$. This implies that the number of non-defective multi-type wafers that is needed to form a set is equal to $\Sigma_k (n_k/p_k)$, which is the same for the case of single-type wafer fabrication. Since this objective value is associated with a feasible solution to problem (P), we can conclude that the optimal objective value to problem (P) is greater than or equal to $1/\{\Sigma_k (n_k/p_k)\}$. In other words, the number of sets formed by a non-defective multi-type wafer is greater than or equal to that of a non-defective single-type wafer.

4.1.2. Stochastic Analysis

Let N(t) and F(t) be the number of sets of non-defective chips that have departed from the system during the time interval [0, t] associated with single-type wafer fabrication and multi-type wafer fabrication, respectively. Clearly, F(t) is dependent on the value of x_{sk}, and the problem is to choose x_{sk} to maximize $\lim_{t\to\infty} (t^{-1} E(F(t)))$. However, it is very difficult to determine the optimal chip allocation x_{sk}. For this reason, the authors consider a specific chip allocation that has $x_{sk} = r_k$. When the yield rate is "separable"; i.e., $p_{sk} = c_s p_k$ with $\Sigma_s c_s = 1$, the authors showed that the average throughput rate for multi-type wafer fabrication is greater than or equal to that of single-type wafer fabrication. More formally, the authors have shown that

$$\lim_{t\to\infty} E(F^*(t))/\surd t \geq \lim_{t\to\infty} E(N(t))/\surd t,$$

where $F^*(t)$ is the throughput rate associated with a specific chip allocation by setting $x_{sk} = r_k$ for all s and k.

4.2. **Models for allocating chip types to chip sites**

The model presented in 4.1 has justified the economical value of multi-type wafers. This observation motivates researchers to develop models for designing multi-type wafers. In this section, we shall review models that focus on the issue of allocating chip types to chip sites. It appears that the work of Singh et al. (1990) is the first such study. In that paper, the authors assumed that the chip sites are homogeneous, and that p_{sk} depends on the chip type k only. This implies that the yield y_k depends only on the chip type k, where the yield is defined as the probability of a chip of type k being non-defective. Let x_k be the decision variable that represents the number of chips of type k to be allocated to a wafer that has S sites. In this case, a feasible chip allocation must satisfy the following constraints:

(4.4) $\Sigma_k x_k \leq S$ for $k = 1,..., K$, and

(4.5) $x_k \geq 1$, integer for $k = 1,..., K$.

Let z_k be the number of non-defective chips of type k generated from a multi-type wafer. In this case it is assumed that z_k has a Binomial distribution with parameters (x_k, y_k). Clearly, z_k depends on x_k.

The objective is to determine an optimal chip allocation to maximize the probability that the non-defective chips generated from a multi-type wafer forms at least one set is maximized; i.e., Max $P(z_1 \geq n_1)*P(z_2 \geq n_2)*...*P(z_K \geq n_K)$. The authors propose a heuristic for determining a near-optimal chip allocation with respect to this performance measure. That heuristic is based on the following steps: (a) relax the integer requirement of x_k; (b) extend the probability function $P(z_k \geq n_k)$ (defined on integer value of x_k) to a different function that is defined on real numbers; (c) determine an optimal continuous solution; and (d) round the continuous optimal solution to obtain an integer solution.

This observation has motivated Tang (1990) to reformulate the problem as a dynamic programming problem that can be solved optimally within a polynomial number of operations. Besides this performance measure, Tang (1990) has considered additional performance measures that can be described as follows:

(a) Target Production Let t_k be the "target" number of non-defective chips of type k generated from a multi-type wafer. Then it is desirable to determine chip allocations that either minimize the total expected squared sum of deviations; i.e., Min $E\{\Sigma(z_k - t_k)^2)\}$, or to minimize the maximum expected squared deviation; i.e., Min $\text{Max}_k E\{(z_k - t_k)^2\}$. Tang (1990) has formulated these two problems as dynamic programs that can be solved within polynomial time.

(b) Set Production Besides the performance measure of maximizing the probability of forming a set from a multi-type wafer, Tang (1990) has considered another measure that corresponds to the number of sets generated from a non-defective multi-type wafer. The objective function can be written as Max $E\{ \text{Min}_k \lfloor z_k/n_k \rfloor \}$, where the term $\text{Min}_k \lfloor z_k/n_k \rfloor$ represents the number of sets formed by a non-defective multi-type wafer. Due to the complexity of this problem, the author has developed a heuristic procedure for finding a near-optimal chip allocation.

(c) Custom Orders Consider that a custom order consists of d_k non-defective chips of type k, $k=1,...,K$. Then it is desirable to minimize the number of multi-type wafers needed to satisfy the custom order. This problem is related to minimizing the maximum of the "first passage times," which is known to be complex. For this reason, the author present two heuristics for determining near-optimal chip allocation. The first heuristic is based on chance-constrained programming while the second heuristic is based on Markov Chain approximation.

The models developed by Singh et al. (1990) and Tang (1990) assumed that the yield rate depends only on the chip type but not on the chip site. This assumption does not capture the "radial dependence" mentioned by Stapper (1982). This observation has motivated Avram and Wein (1990) to develop a different model for determining chip allocation when the yield rate depends on the chip type k and chip site s. We now describe their model briefly.

The fabrication facility is modeled as an open quasi-reversible queueing network that possesses a stationary distribution. It is assumed that the multi-type wafers arrive at the system according to a Poisson process with rate λ. In multi-type wafer manufacturing, let x_{sk} be the fraction of chip type k to be allocated to chip site s. Let Y_{sk} be a random variable that takes on the value 1 if a type k allocated to site s is non-defective, and takes on the value 0, otherwise. The random variables Y_{sk} are assumed to be mutually independent. The expectation and the variance of Y_{sk} are denoted by $E(Y_{sk}) = p_{sk}$ and $V(Y_{sk}) = v^2_{sk}$, respectively.

Avram and Wein have shown that the throughput rate (in terms of number of sets) increases as the value of the term $n_k^{-2} \Sigma_s x^2_{sk} v^2_{sk}$ decreases. This observation has led to the following surrogate objective functions for determining a good chip allocation: (a) Min $\Sigma_k \{ n_k^{-2} \Sigma_s x^2_{sk} v^2_{sk}\}$ and (b) Min $\text{Max}_k \{ n_k^{-2} \Sigma_s x^2_{sk} v^2_{sk}\}$. Notice that each feasible chip allocation satisfies the constraints (4.2) and (4.3). According to their simulation results, it seems that the throughput rate corresponds to the optimal chip allocation with respect to (a) is superior to that of (b).

4.3. Comments on models for wafer design

We now comment on the models presented in this section. In the models presented by Singh et al. (1990), Tang (1990), and Avram and Wein

(1990), there are three remaining issues to be considered. These issues are:

(a) Chip Size: The size of each type of chip has been considered to be homogeneous among chip types. In general, different types of chips may take up different amounts of space on a wafer. In that case, it is of interest to extend the models to the cases in which the chip size varies from type to type.

(b) Chip partitioning: All the models presented in this section assumed that the number of sites S is greater than the number of chip types K. In practice, it is quite possible to have a situation in which K > S. In that case, one needs to partition the chip types into groups, where each group of chip types form a multi-type wafer.

(c) Dynamic Design: The models presented in this section concentrate on static designs in which the composition of the chip types to be produced on the same wafer does not change over time. However, there are situations in which it is necessary or desirable to modify the wafer design. For instance, when some of the chip types become obsolete, then it is necessary to modify the wafer design. In addition, when the cost of producing mask sets becomes low or when electronic beam lithography becomes cost effective, then the cost of changing from processing a multi-type wafer to a different multi-type wafer becomes low. In that case, it is worthwhile to develop models for analyzing the tradeoffs between static design and dynamic design.

Avram and Wein (1990) present an interesting model for evaluating the benefit of multi-type wafer fabrication. However, there are some remaining issues that deserve special attention. In their model, the facility is assumed to be a quasi-reversible queueing network with wafers arriving at the facility according to a Poisson process. Thus, the departure process is also Poisson and is independent of the processing time of the system. This model is adequate for long-term planning, and the Poisson arrival is reasonable when the manufacturing lead time is long and there is no information about the yield until the final test. However, in a competitive market like integrated circuits, intermediate range planning is also critical. In that case, it is necessary to develop models that address the following issues:

(a) Processing time: When comparing the throughput rates for single-type wafer and the multi-type wafer, one should account for the discrepancies of the processing times. This is because the processing time tends to depend on the complexity level of the chip types to be produced on the same wafer (c.f., Connors (1991)).

(b) Production control: Most wafer fabrication facilities have developed new methods for early detection of defects on a wafer. (See Graf (1984), Healy (1981), and Marcus (1981) for detailed description of different inspection and testing techniques.) This development has helped management to establish inspection procedures for monitoring yield at earlier stages of the manufacturing system. In that case, the yield information should be used to develop mechanisms for determining the type and the quantity of the wafers to be released to the system.

5 Conclusion

In this paper we have review some mathematical models that focus on the issues of yield management and product design in the context of IC manufacturing. We have also commented on the existing models and indicated potential research areas in IC manufacturing. We hope that this review will generate some research interests in this area.

References

Albin, S.L. and D.J. Friedman, (1989), "The Impact of Clustered Defect Distributions in IC Fabrication," Management Science, 35, 9, 1066-1078.

Ansley, W.G., (1968), "Computation of integrated-circuit yields from the distribution of slice yields for the individual devices," IEEE Trans. Electron Devices, vol ED-15, pp 405-406, June.

Avram and Wein, (1990), "A Product Design Problem in Semiconductor Manufacturing," Working Paper, Sloan School of Management, MIT.

Balcer, Y., and S. Lippman, (1984), "Technological Expectations and Adoption of Improved Technology," Journal of Economic Theory, pp. 292-318.

Burman et al., (1986), "Performance Analysis Techniques for IC Manufacturing Lines," AT&T Technical Journal, pp. 46-57.

Chen et al., (1988), "Empirical Evaluation of a Queueing Network Model for Semiconductor Wafer Fabrication," Operations Research, pp. 202-215

Dennard, R.H., (1966), "Cost Study for Integrated Circuits with Many Logical Decisions per Chip," Research Report RC-1552, IBM Thomas J. Watson Research Center, Yorktown Heights, NY.

Glaser and Dobberpuhl, (1985), Design and Analysis of VLSI Circuits, Addison-Wesley.

Graf, (1984), "Testing - A major concern for VLSI," Solid State Technology, Jan, pp. 101-107.

Hamilton and Howard, (1975), Basic Integrated Circuit Engineering, McGraw-Hill Book Company.

Healy, (1981), Automated Testing and Evaluation of Digital Integrated Circuits, Reston Publishing.

Lawson, R.T., Jr., (1966), "A Prediction of th Photoresist Influence on Integrated Circuit Yield," Semicond. Prod. Solid State Technol. **9**, 22.

Marcus, (1981), "Diagnostic Techniques," in VLSI Technology, edited by Sze, McGraw Hill.

Meyer, F., and Pradham D., (1989), "Modelling Defect Spatial Distribution," IEEE Transactions on Computers, pp. 538-546.

Moore, G.E., (1970), Electronics, vol 43, p 126, Feb. 1970.

Murphy, B.T., (1964), "Cost Size Optima of Monolithic Integrated Circuits," Proc. IEEE, 52, pp. 1537-1545.

Paz, O., and T.R. Lawson, Jr., (1977), "Modeling Defects Induced by Diffusion," IEEE J. Solid-State Circuits, SC-12, 540-546.

Reinhard, (1987), Introduction to Integrated Circuit Engineering, Houghton Mifflin Company.

Seeds, R.B., (1967), "Yield, economic and logistic models for complex digital arrays," in 1967 IEEE Int. Conv. Rec., part 6, Oct. 1967, p. 60.

Shimkunas, A., (1984), "Advances in X-Ray Mask Technology," Solid State Technology, pp. 192-199.

Singh, M., C. Abraham, and R. Akella, (1990), "Planning for Production of a Set of Components when Yield is Random," IEEE Transactions on CHMT, March issue.

Stapper, C.H., P. Castricci, R. Maeder, W. Rowe, and R. Verhelst, (1982), "Evolution and Accomplishments of VLSI Yield Management at IBM," IBM J. Research & Development, Vol. 26, 5, pp. 532-545.

Stapper, (1985), "The Effects of Wafer to Wafer Defect Density Variations on Integrated Circuit Defect and Fault Distributions," IBM Journal of Research and Development, pp. 87-97.

Stapper, C.H., (1986), "On yield, fault distributions, and clustering of particles" IBM J. Res. & Development, Vol. 30, 3, pp. 326-338.

Stapper, (1987), "Correlation Analysis of Particle Clusters on Integrated Circuit Wafers," IBM Journal of Research and Development, pp. 641-650.

Tang, C., (1990), "Composing Batches with Yield Uncertainty," Working Paper, UCLA, also to appear in European Journal of Operational Research.

Tang, C., (1991), "Designing an Optimal Production System with Inspection," European Journal of Operational Research, pp. 45-54.

Warner, R.M., (1974), "Applying a Composite Model to the IC yield problem," IEEE J. Solid-State Circuits, SC-9, 86-95 (June).

Wilson, A., (1986), "X-Ray Lithography: can it be justified?" Solid State Technology, pp. 249-255.

Yanagawa, T., (1972) "Yield degradation of integrated circuits due to spot defects," IEEE Trans. Electron Devices, vol ED-19, pp 190-197, Feb.

A HIERARCHICAL APPROACH TO DESIGN, PLANNING, AND CONTROL PROBLEMS IN ELECTRONIC CIRCUIT CARD MANUFACTURING

Reza H. Ahmadi
John E. Anderson Graduate School of Management at UCLA
Los Angeles, California 90024-1481

I. Introduction:

The competitive nature of the electronic industry has greatly intensified in recent years. The capital investment in this segment of industry constitutes a large portion of current industrial investments. Printed Circuit Boards (PCBs) are an important module of electronic parts, and consequently, highly automated manufacturing systems have been designed to produce the required circuit cards at the desired volume. The design, planning, scheduling, and control problems which arise in this environment are challenging and multi-faceted.

Automated electronic assembly lines are typically designed for high volume production. Operationally, these lines have serial structures that are connected via a versatile material handling system. The operations in these systems consist of a number of assembly, inspection, and test operations. Machines employed are high precision pick-and-place robots, visual inspection systems and automatic test fixtures. An important characteristics of these production systems is the very large number of different operations, processes, and test stages required in assembling a board type which may require up to 400 components. The assembly process involves the placement of a large number of different components/devises (IC's, capacitors, resistors, etc.) on the circuit card. This activity, typically referred to as "populating" or "board stuffing", takes place in a number of stages. The devices installed at each stage are dictated by the bill of components for the products. The manufacturing line can be characterized as a coupled serial line, and resembles a continuous flow manufacturing system. Although line operations are discrete in nature, there is a strong requirement for continuity of the flow of the work units both from the technological restrictions and from the managerial desire to operate under a low work-in-process environment because of competitive consideration. Therefore, work initiated at the beginning of the line must proceed without any significant delay at or in between operations. The work-in-process is very sensitive to the work environments. A method of assembly, refereed to as Surface Mount Component (SMC) technology, is wildly used in large scale PCB environments. In this method, solder paste is applied to the surface of the board, at which components are to be mounted. Considerable effort is made to avoid contamination of the work unit before they are populated. The time to dry imposes strict limitation on the work unit's throughput time. In order to achieve the required high volume production capacity, a number of identical machines are used for the same operations as parallel processors. The electronic assembly lines differ from flexible manufacturing systems because of the high set up time of the machines employed in assembly operations. This lack of flexibility is associated with both: (1) the limited capacity of the machinery in handling the multitude of the components and the tools that are required in the assembly operations, and (2) the arrangement of the feeders on the machine. As an example, in populating PCB's the number of unique components that are placed on a board may exceed one hundred. The components that are loaded on the machine via individual feeders occupy limited available space on the machine. In most cases it is not possible to simultaneously load component requirements of several products, unless a high degree of commonality exists among board types. Products are produced on a regular periodic basis and are delivered to the customers in periodic shipments. Products mix changes are faster due to the short life cycle of the electronic parts.

Recently, there has been a growing interest in the modeling and analysis of the optimization problems that are crucial to the improvement of productivity in the PCB assembly operations. For examples of the research in this area the reader is referred to Ahmadi (1992), Ahmadi (1986), Ahmadi et al (1988), Ahmadi and Ali (1988), Ahamdi and Matsuo (1991a, b), Ahmadi and Kouvelis (1990a, b), Ahmadi et al (1990), Ball and Magazine (1988), Grotzinger (1988), Leipala and Nevalainen (1989), Lofgren and McGinnis (1987a, b), Lofgren et. al. (1991), Chen (1982), Chow (1985a, b), Chow (1986), Drezner and Nof (1984), Hu and Canter (1989), Hu and Kuh (1985), Gavish and Seidmann (1987), Soukup (1981), Magirou (1986), McKenna (1985), Pukanic (1985), Wittrock (1985, 1988), and Tang and Denardo (1988a, b).

In this paper we provide a hierarchical framework for addressing the design, planning, and control problems that arise in the assembly of the PCB's. We are interested in answering the following questions which encompass the set of problems encountered:

1. System design Problem: How many standard and special pick-and-place machines are required to meet the projected demand?

(1a). How many mini-lines should be formed?

(1b). What is the capacity of the system for a given product mix?

2. Process Configuration Problem: How many production stages are required for special and standard machines?

(2a). How should the feeder components and tools be partitioned among the production stages?

(2b). What are the process routings for the different PCB's?

3. Product Grouping Problem: How should we group the part types and form product families?

(3a). How should we assign the families to the mini-lines?

4. Mini-Line Production Problem: How many machine should be assigned to each mini-line at each production stage?

5. Scheduling and Control Problem: How should we schedule the production of the families in mini-lines given the technological constraints and managerial desire to operate under a low work-in-process environment?

6. Staging Problem: How should the feeders be positioned on the feeder carriers of each pick-and-place machine?

7. Placement Sequencing Problem: What should be the pick-and-place sequence of the components at each stage and for each board type?

The decisions in this environment are further complicated by the lack of complete knowledge of the technical data required for the decision making. The difficulty in determination of technical data is based on the performance sensitivities of the pick-and-place machines to their staging and arrangements. For example, the processing time to complete a PCB at any stage, which is required to make decisions regarding the planning and scheduling, is not known until the detailed questions about the machine staging and the sequence of the pick-and-place operations are addressed. Therefore, contrary to the classical models which assume that the parameters of the problems remains invariant, the hierarchical models developed have to produce the appropriate technical data for addressing the planning problems. Also, as the complexity and scale of the environment has grown, the interface and the multi-faceted nature of the problems have not been properly addressed. The interaction of the various problems and the subsystems requires an integrated view of the manufacturing environment.

Hierarchical Structure:

The objective of this paper is to devise a hierarchy of decision problems in which each level addresses a segment of the issues and constraints of the system and to provide the information to make decisions in the subsequent level. The problems in this environment are divided into three blocks. In the first block a mathematical model which deals with the process configuration is formalized. The input to this segment are the bill of components, tooling requirements, machine tooling capacities, and the demand rate of the components for each product type. The output of this segment will be the number of productions stages and the machine staging specifications. Given the solution of the first segment the process planners will be able to define the process routing of different product types in the system. The technical data required for the planning and scheduling purposes is produced in the second block. The input to the second block is the set of feeder components and the tools required to dispense them, i.e. the solution of the first module. Then for each product type at any production stage we need to find out the optimal positioning of the feeders on the feeder carrier and the sequence of the pick-and-place operations. The feeder reel positioning and pick-and-place sequence will enable us to generate the processing time of each product at any stage. Switching from one product type to another requires a set up change for the feeders' position on the pick-and-place machine. The set up change is necessitated due to different feeder reel positioning solutions for each PCB type. All the technical data as well as the solution of the first segment are used in the third block to generate the solution to the grouping and mini-line production problem. The three blocks together provide means of answering the questions posed above. The models in each segment provide the foundation for developing a hierarchical system to address the problems that arise in electronic card manufacturing. Figure 1 depicts an schematic diagram of the proposed hierarchical structure.

The contributions of this paper are the devising of the hierarchical system and the development of models which formalize the decision making in the PCB manufacturing. In the next section we present a mathematical program to address the Process Configuration Problem. Section 3 develops models to optimize the operational problems at the machine level and discusses the Staging Problem and Placement Sequencing Problem. In section 4 we present models to answer the system design, Product Grouping, and the Mini-line Production Problem. Some concluding remarks are discussed in section 5.

II. Process Configuration Problem:

The PCB manufacturing environment is constantly changing. Design

FIGURE 1

Hierarchical Structure

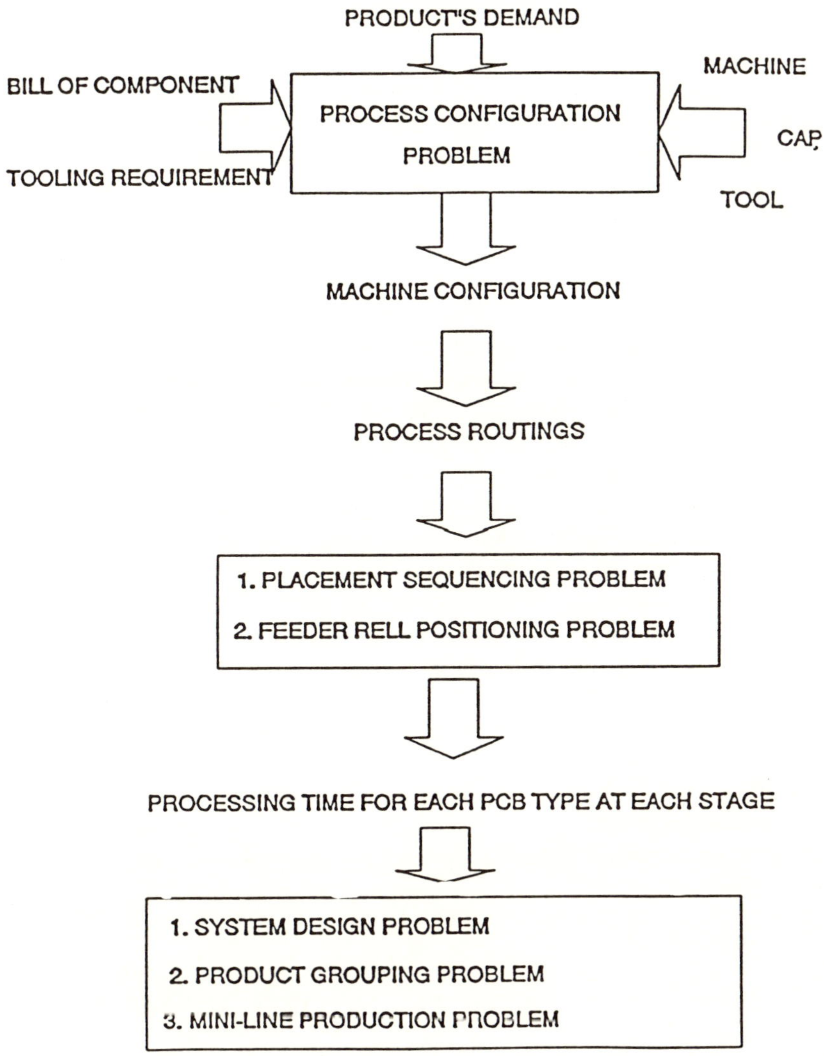

changes of the board are quite common and life cycles of the PCB's are relatively short. Demand for particular PCB types varies significantly. This dynamism in the industry indicates that the manufacturing system has to be redesigned each time that the product mix and their volume change considerably. However, we have observed that, for certain classes of electronic packaging systems, even though the product set and product volume manifest high volatility, the component used to produce the different PCB's show relatively more stable demand. This is conceivably due to the phenomena that technological changes and development in the component production (chips, capacitors, resistors) are moderately slower than the changes in the types of the PCB's and their variations. This observation suggest that the tooling requirement of the different product sets may not be substantially different from one product mix to another, and the system planners can simply reconfigure the system for a new product mix.

In this section, we formalize a mathematical program addressing the Process Configuration Problem. The model to be presented defines the machine staging problem (i.e. the components and tools assigned to each machine) for the manufacturing line. We are assuming that the demand rate for the product set and the tooling requirement for populating the circuit boards are known. The problem can be simply stated as follows: What is the best assignment of the tools and feeder reels to the pick-and-place machines at each stage such that the number of production stages is minimized and an apriori balance for each product set is achieved at each stage. We should mention that most of the previous research on this area has focused on the specific aspects of the placement or insertion machines in isolation. In fact it is not only the use of the machine but also the processing configuration of the system which considerably impacts the performance of the manufacturing line. The following notation is used to formulate the problem:

$p \in P$ = index for components used by all the products, referred to as bill of components.

$k \in K$ = index for production stages. $|K|$ is an apriori bound on the number of stages.

$j \in J$ = index for the tools used for dispensing the components on the boards. Sufficient number of any tool types are assumed to be available.

$i \in I$ = index for the product types or families.

W = capacity of the feeder carrier of pick-and-place machine.

L = capacity of the tool magazine of the pick-and-place machine.

σ = a bound on the total number of tools to be used.

$a_{p,j}$ = 1 if component p can be dispensed by tool j.

0 otherwise.

$c_{p,k}$ = cost of assigning component p to stage k. It exponentially increases as the number of stages goes beyond an apriori lower bound.

d_{ip} = demand for component type p induced by product i over the planning horizon.

$X_{p,k}$ = 1 if component p is assigned to stage k. 0 otherwise.

$N_{j,k}$ = 1 if tool j is assigned to stage k. 0 otherwise.

The Process Configuration Problem (PCP) can be formulated as follows:

Process Configuration Problem:

(PCP) $\quad v = \text{Min} \quad \Sigma_{p\in P}\Sigma_{k\in K}\, c_{pk}\, X_{pk}$

s.t.

$\Sigma_{k\in K}\, X_{pk} = 1$	$\forall p\in P$	(1)
$\Sigma_{p\in P}\, X_{pk} \leq W$	$\forall k\in K$	(2)
$X_{pk} \leq \Sigma_{j\in J}\, a_{pj}\, N_{jk}$	$\forall p\in P,\ \forall k\in K$	(3)
$\Sigma_{j\in J}\, N_{jk} \leq L$	$\forall k\in K$	(4)
$\Sigma_{j\in J}\Sigma_{k\in K}\, N_{jk} \leq \sigma$		(5)
$\Sigma_{p\in P}\, d_{ip}\, X_{pk} \leq \mathcal{L}_i$	$\forall k\in K,\ \forall i\in I$	(6)
$X_{pk} \in (0,1)$	$\forall p\in P,\ \forall k\in K$	(7)
$N_{jk} \in (0,1)$	$\forall j\in J,\ \forall k\in K$	(8)

where $\mathcal{L}_i$ is a managerial parameter which indicates the maximum load (i.e. the total number of components to be dispensed) at each stage for product type i. Constraint (1) ensures that each component type is assigned to a stage. Feeder carrier and tool magazine capacity are captured by constraints (2) and (4). Constraint (3) ensures the compatibility of the assignment of the feeders and the tools to the machine at each stage. Constraint (5) imposes a bound on the total number of tools used. Since tool changes are time consuming operations in pick-and-place machines, it is desirable to minimize the total number of tool changes. Tool changes can be reduced by limiting the total number of tools assigned to the different stages. Constraint (6) assumes that the load assigned to each machine for each product type i∈I is additive and can't exceed a desired level. This constraint serves as a balancing measure of the load assigned to each stage. Here we are implicitly assuming that the amount of time required to place the components of each product type is proportional to the number of components required by each product type. Balancing the processing time of a product over the production stages reduces the cycle time of the production and increases the efficiency of machine utilizations. The objective function tries to find the minimum number of stages required for the assignment of components and tools, which is indicated by the positive X_{pk}'s.

The rationale for the objective function of the (PCP) is twofold. By minimizing the number of production stages, one implicitly minimizes the number of machines needed to meet the system requirement. By minimizing the number of manufacturing stages, the capital investment in the system is also minimized. Also, since there is a limitation on throughput time of the PCB's in the system (the paste applied at the beginning of the line for mounting the component dries after certain number of hours), it is desirable to limit the number of stages. Minimizing the number of stages would reduce the potential waiting time between stages and thereby will tend to lower the throughput time of the work units in the system.

Problem (PCP) is a difficult large scale pure integer programming problem, which suggests that the problem is difficult to solve optimally. For a non-binding value of C, problem (PCP) reduces to a multiple choice multiple knapsack problem which is NP-hard (Garey and Johnson 1979). For the solution approaches to this problem the reader is referred to Ahmadi (1992).

The solution obtained in this segment, the assignment of feeder reels and tools at each stage, will provide the necessary information for defining the feeder reel positioning and placement sequencing of the components for each product family at each stage.

III. Feeder Reel Positioning and Placement Sequencing Problems

In this section we focus on the problems of optimizing the productivity in the PCB assembly operations. The output of this model is used as input data in the subsequent level of the hierarchical system. To proceed, we assume that each machine is staged for dispensing a subset of the bill of components such as chips, capacitors, and resistors, and that the tools required are allocated to each machine. The problem that we are interested in is: Given the feeder components, and the tools to dispense them, how should the machine be staged. I.e. what should be the position of the feeder reels on the carrier and what should be the sequence of the placement of the components?

The Essential Features of Pick-and-Place Machines:

Below, the essential features of the pick-and-place machines which are important from a modeling perspective are outlined. In presenting this overview, the objective is to develop a framework for analyzing this class of machines. In addition, the goal is to create an off-line method for deciding on the operational characteristics of the machine so that the concurrent operations are fully utilized.

Pick-and-place machines using Surface Mount Component technology is widely used in electronic packaging industry. These machines are commonly used to meet high volume production objectives. Dual delivery pick-and-place machines (DYNAPERT MPS500) are currently used in IBM's SMC line in Austin, Texas. Other types of placement machines used in assembly operations are discussed by Ball and Magazine (1987), and Hu and Canter (1989). These machines vary in their operating characteristics. As a result, different models are used to characterize the machine performance and productivity. In this section we describe the general characteristic and key operational parts which are common to most machines. The models constructed will consequently be applicable to most environments. The first step is the characterization of the machine's functional operations. Characterization should determine the crucial process parameters and their relationships.

Each pick-and-place machine includes a work board table, the placement arm, pick-and-place heads, component delivery (feeder reel) carrier, and tool magazines. The feeder carrier and the work board are capable of independent movement which is driven by a separate motor. In our generic machine, the work board moves in the x and y directions simultaneously and positions itself under the head to perform the placement operation. The carrier, which includes the feeder reels and the tool magazine, moves in the x direction only. The head is attached to a fixed length arm, which moves between two fixed positions, namely the pick and placement positions for the components. Tools must be changed if a different component type is to be placed on the board. Given that independent motors move the work board and the feeder carrier, it is clear that the efficient utilization of the machine is a function of the ideal exploitation of the concurrent movements involved in pick-and-place operations. The placement time of a component is given by the maximum of the work board movement, for positioning the board under the placement position of the head, and the sum of the feeder carrier movement and the head movement to placing position after the pick operation is performed. It should be noted that the head movement to pick position is performed concurrently with the feeder carrier movement. The greater speed of the motor used for the head movement is sufficient to complete this operation during the feeder rack movement. The crucial moves which are subject to optimization are the work board and feeder carrier movements. The head movements from pick position to place position and vice versa are constant and therefore not the subject of the optimization models to be presented. Also, as described earlier, a change of tools is usually required for the pick-and-place operations when the components belong to different tool groups. Such tool changes require a considerable amount of time which reduces the system productivity significantly.

Significant Factors for Modeling Purposes:

Of primary importance in the modeling of the Staging and Placement Sequencing Problem are:

(i) Tool Change Time: Tool changes represent unproductive time and degrade the throughput rate of the system considerably. During the tool changes all other functions of the machine come to a halt. This factor suggest that components belonging to a tool group should be inserted in sequence, avoiding unnecessary delays. (ii) Feeder Carrier Movement: The pick-and-place machines are high precision equipment and limiting their mechanical movement should improve their reliability. Accordingly, reducing the absolute movement of the feeder carrier is a desirable characteristic. Excess movements results in wear and tear that could lead to misalignment of the feeder locations on the carrier, and should therefore be avoided. This engineering consideration motivates the development of the specific solution approach for positioning of the feeder reels on the feeder carriers referred to as "oriented" positioning. An oriented positioning of the feeder reels is one in which all the required movements, for components in the same tool group, are accomplished by changing the direction of feeder carrier movement only once. For further detail the reader is referred to Ahmadi et al (1990).

(iii) Multiple Placement from a Feeder Reel: Quite often a PCB requires more than one placement from a given feeder reel, i.e. multiples of a component type must be placed on the board. To reduce the feeder carrier movements it is desirable to dispense these components in series, thereby eliminating the need for the extra feeder carrier movements. This in turn places a restriction on the placement sequences of the components on the board, which we will address later.

(iv) Feeder Carrier and Work Board Speed: Compared to the work board, movements of the feeder carrier are moderately slower. This is due to the much larger weight that is moved by the carrier, i.e. the movement of up to 60 feeders which may be positioned on the carrier. Also, the size of the boards which are populated in this process are much smaller than the length of the feeder carriers. Typical circuit board are 10"x14", whereas feeder carriers are more than 60" in length.

These observations motivate the development of the following models for (a) placement sequencing and (b) feeder reel positioning.

The Placement Sequencing Problem:

In this formulations we are interested in finding a placement sequence of the components with the following characteristics. For each product type $p \in P$, at each stage $k \in K$, and each tool type $j \in J$, we would like to find the shortest time cycle such that the components belonging to the same feeder reel are placed consecutively. Note that the cycles for each tool type can be obtained independently, since during the tool change operations all the other functions of the machine stop, and the board positions itself under the placement location. The sequence of the tool changes will be determined by the feeder positioning problem. The following notation is required to formulate the problem adequately.

$i,u \in I$ = index for all the components to be placed on the board using tool type $j \in J(k)$, where $J(k)$ is the set of tools assigned to stage k.

Γ_{iu} = time for the work board to move from position of component i to u, on the board.

δ = time for the head to move from pick to place position or vice versa.

$m,n \in M$ = index for the feeders assigned to a pick-and-place machine at a stage.

$I(m)$ = set of components dispensed from feeder m.

ζ = time required for the feeder carrier to move one unit.

ω_{iu} = $Max(\Gamma_{iu}, 2\delta)$, if i and u are dispensed from the same feeder
$Max(\Gamma_{iu}, \zeta+\delta)$, if i and u are dispensed from different feeders.

G_{iu} = 1 if component u is placed immediately after component i
0 otherwise.

The Placement Sequencing Problem (PSP) can expressed as:

(PSP) $\quad \text{Min } \Sigma_{i\in I}\Sigma_{u\in I}\, \omega_{iu}\, G_{iu}$

s.t.

$$\Sigma_{i\in I}\, G_{iu} = 1 \qquad \forall u\in I \quad (9)$$

$$\Sigma_{l\in I}\, G_{iu} = 1 \qquad \forall i\in I \quad (10)$$

$$\Sigma_{i\in I\backslash I(m)}\Sigma_{u\in I(m)}\, G_{iu} = 1 \qquad \forall m\in M \quad (11)$$

$$\Sigma i\in I(m)\Sigma_{u\in I\backslash I(m)}\, G_{iu} = 1 \forall m\in M \quad (12)$$

$$G_{iu} \in S \qquad \forall i\in I, \forall u\in I \quad (13)$$

$$G_{iu} \in (0,1) \qquad \forall i\in I, \forall u\in I \quad (14)$$

where S is the set of spanning trees. Constraint (9) and (10) ensure that each component position has to be visited once and only once. Whereas constraint (11) and (12) ensure that once a component of a given feeder reel is placed all other components using that feeder reel are also dispensed before any components from another feeder reel is placed. The cycle restriction of the tour is captured by the set of the spanning trees S, which guarantees that sub-tours are not formed. The objective function of the problem (PSP) seeks to find the minimum time cycle. The coefficients ω_{iu} indicate that if i and u are dispensed from the same feeder the time to place a component is the maximum, of the concurrent movements of the head and the work board. If i and u belong to the different feeders then ω_{iu} is the lower bound on the time to place the component u which is bounded by the maximum time of the work board move and the time that it takes for the head to go to the pick position and for the feeder rack to move to pick position.

In the context of Traveling Salesman problem, (PSP) may be described as State-TSP, where a salesman has to visit "cities" belonging to different "states" given that each state and each city should be visited once and only once. This problem is discussed in detail in Ahmadi and Mamer (1992) and Gavish and Seidmann (1987). Finally, note that the (PSP) has to be solved for each stage, each product type, and each tool group. Next we propose our model for the Feeder Reel Positioning Problem.

Feeder Reel Positioning Problem:

The second problem which needs to be addressed at this stage involves identifying the position of the feeders on the feeder carrier such that the total movement of the carrier is minimized. The positioning problem has to be solved for each product type and at each stage. We refer to this problem as the (FRPP). In the discussion that follows we assume that feeder reels have been re-indexed according to the order in which they are visited in the Placement Sequencing Problem. Now, recall that the carrier and work board move concurrently. Let G^*_{iu} denotes the solution of the Placement Sequencing Problem and let Γ^*_{iu} be the corresponding time that it takes for the work board to move from position of component i to u on the board. We are interested in movements among the "states", or moves which require a carrier movement. Therefore, let $\mu_n = \Gamma^*_{i'u'}$ be the time that it takes for the work board to move from the last "city" in "state" n-1 (i.e. component i') to the first "city" in "state" n (component u'). μ_1 is the move from the head place position to the first city in "state" 1, and $\mu_{|M|+1}$ is the move from the last city in "state" $|M|$ to the place position. Our objective is now to determine the feeder reel positions on the feeder carrier such that the total carrier movement in excess of $\Sigma_{n=1 \text{ to } |M|+1}\, \mu_n$ is minimized. We introduce the following notation to describe the problem (FRPP):

ξ_i = width of the feeder reel i. All feeder reels are assumed to have the same width.

R_n = Starting position of feeder reel n, reel n will occupy slot n. (The tool magazine is assumed to be in position 0 and $R_0 = R_{|M|+1} = 0$)

S_{mn} = an indicator variable that takes the value 1 if feeder reel m precedes n and 0 otherwise.

The component pickup by the head requires that the center of the reel should be aligned in the pick position therefore, the time for the carrier movement between feeder n and n+1 is given by the time for carrier to move the distance $|R_n+0.5\xi_i - R_{n+1}-0.5\xi_i|$. Thus the time for picking component n+1 is $\max\{\mu_{n+1}, |R_n+0.5\xi_i - R_{n+1}-0.5\xi_i|\}$. The problem (FRPP) for any tool group can be formulated as follows:

(FRPP) $$v = \text{Min } \Sigma_{n=1 \text{ to } |M|+1} \max\{\mu_n, |R_{n-1}+0.5\xi_i - R_n-0.5\xi_i|\}$$

s.t.

$$R_n + \xi_i \leq R_m + O.S_{mn} \qquad \forall m\in M, \forall n\in M, m\neq n$$

$$S_{mn} + S_{nm} = 1 \qquad \forall m\in M, \forall n\in M, m\neq n$$

$$R_0 = R_{|M|+1} = 0$$

$$\xi_0 = \xi_{|M|+1} = 0$$

$$S_{mn} \in (0,1) \qquad \forall m\in M, \forall n\in M, m\neq n$$

where O is a large number. Observe that the problem formulation (FRPP) can be extended easily to multiple tool group case, referred to as (MFRPP). This is accomplished by introducing, for each tool group, two dummy components with feeder reel width of zero that are positioned at location (R = 0). The two dummy reels correspond to the tool magazine and represent the first and the last movements in the placement sequence. This modification will assure that tool changes occur following the completion of the placement from the feeders in one tool group. Next we show that the oriented solutions constitute the class of dominant solutions.

The optimal solution to (FRPP) can be obtained as a shortest path in a layered network. For further details of the algorithm refer to Ahmadi et. al. (1990). To solve (MFRPP) we propose the following procedure.

Algorithm to Solve MFRPP:

Step 1. Solve (FRPP) for each tool group independently.

Step 2. Sort the tool groups according to non-decreasing order of the number of feeders in each group. The sequence of the feeder reels in each group and the ordering of the tool groups provides an initial "nested" solution to (MFRPP). In "nested" solutions, components within a tool group are placed consecutively on the carrier.

Step 3. Mesh the first two tool groups in the sequence given in step 2. The resulting solution is considered as one group and then it is meshed with the third tool group in the sequence. Repeat this step until meshing of the tool groups is considered. For details of the "meshing" algorithm the reader is referred to Ahmadi et al. (1990).

Given the sequence of the component placement and the feeder reels position on the pick-and-place machine, we can now completely generate the technical data (i.e. the processing time for each board type at any machine) required to address the System Design, Product Grouping, Mini-line Production, and Scheduling problems of the electronic manufacturing line. Before we discuss the remaining issues, we would like to draw some guide-lines for the design of the pick-and-place machines.

Clearly such systems can achieve their peak performances when there is no delay in the concurrent operations of the machine. Recall that the concurrent processes are: head movement, work board movement, and carrier movement. An interesting question which arises, is how should the manufacturing engineers design and set (reset) these systems for maximum performance? Since the concurrent operations are performed by independent motors, it might be possible to adjust and coordinate the speed of motors driving these operations such that the delay in the concurrent operations is minimized. This may be achieved even after the machine is staged and the placement sequences are identified. As a guide-line for the machine designers, observe that among concurrent operations, the head movement from pick to place position and vice versa is fixed. Therefore, considering the size of the circuit cards populated by these machines if the time required to travel the largest distance in the domain of the board is less than the head movement, we are assured that no delay is created because of these movements. Also if the maximum feeder carrier movement is less than the head movement, then all concurrent operations are done in a fixed time, which is defined by the head movement. Under this circumstance the time to place a component is determined by the head movement time and is fixed for all the component dispensement regardless of the feeder reel positioning and placement sequencing.

The output of this segment of the hierarchy generates the technical data, i.e. the processing time of the PCB's at each production stage, which are used as the parameters in the models of the next section.

IV. System Design, Product Grouping, and Mini-Line Production Problem:

In this section, we formalize an optimization problem to address the System Design, Product Grouping, and Mini-line Production Problem. The issues that we are interested in, cover a wide range, from system design to scheduling of the machines at the manufacturing floor. Among the decisions with which managers are faced are the following: • How many standard and special pick-and-place machines are required to meet the projected demand, given the capital constraint or the maximum number of machines that can be obtained? • How many mini-lines should be formed to manufacture different product groups? • How should we assign the PCB families to the mini-lines? • How many machine should be assigned to each mini-line at each stage? • How should we schedule the production of the families in mini-lines given the technological constraints and managerial desire to operate under a low work-in-process environment? All of these problems are highly interrelated and require an integrated approach in analyzing them. The analysis of these issues is further complicated due to the presence of set up time of the machinery employed in the pick and place operations. The lack of flexibility does not occur from operational limitations of the machinery, as far as populating circuit cards are concerned. The limitation as we have seen in section II and III is associated with limited capacity of the machinery in handling the multitude of sub-components that are required in the assembly operations, and the performance sensitivity of the pick-and-place machines to the arrangement of the feeders and the sequence of component placements on the board. Changing the product families on the lines also requires repositioning of the feeders on the machine. The magnitude of the set up changes for large assembly lines which we are envisioning could be up to one or two shifts. The production systems in the manufacturing line must also be tightly coupled. That is, only small buffer space is allowed at each machine. This limitation is due to the technological constraint discussed earlier. A circuit board is scrapped if the throughput time of the board exceeds a certain critical level. Below, we outline our approach in developing the decision models in this segment of the hierarchy.

Modeling Considerations:
(i). To achieve internal focus, reduce the set up changes, minimize work-in-process, and reduce throughput time of the work units, we organize the system as a collection of mini-lines or manufacturing cells (Ahmadi and Matsuo 1991). By forming several mini-lines and dedicating product families to them, we capture the benefits of group technology (Hyer and Wemmerlov 1984). Since set up times are high, batch production policy will be used for demand during the shipping cycles for families assigned to each mini-line. Furthermore, a pull production system which uses the kanban system for loading and controlling work units can be implemented to control the flow of material in each of the mini-lines. Kanban systems require concurrent dedication of the set of machines at each stage to the product family under

production, and have traditionally been used in environments where the production volume is high and stable. However, systems organized as a collection of mini-lines can also benefit from the advantages of kanban production systems as a mechanism for coupling the successive production stages.

The model to be presented used the following additional notations:

$i,h \in I$ = index for product families.
$g \in G$ = index for mini-lines.
$k \in K$ = index for production stages.
S_i = set up time for family i.
P_{ik} = total processing time for family i at stage k.
R = total number of machines available.
T_i = variable time assigned to the production of family i.
Z_{ig} = 1 if family i is assigned to mini-line g
0 otherwise.
Y_{gk} = number of machines allocated to mini-line g at stage k.
C = Total time to complete all jobs.

The Mini-Line Design Problem (MLDP) can be formulated as follows:

Mini-Line Design Problem:

(MLDP) $v = \text{Min } C$

s.t.

$$\Sigma_{g \in G} Z_{ig} = 1 \qquad \forall i \in I \qquad (15)$$
$$\Sigma_{i \in I} (T_i + S_i) Z_{ig} \leq C \qquad \forall g \in G \qquad (16)$$
$$\Sigma_{g \in G} \Sigma_{k \in K} Y_{gk} \leq R \qquad (17)$$
$$T_i Y_{gk} \geq P_{ik} Z_{ig} \quad \forall i \in I, \forall g \in G, \forall k \in K \qquad (18)$$
$$Z_{ig} \in (0,1) \qquad \forall i \in I, \forall g \in G \qquad (19)$$
$$Y_{gk} \geq 1 \text{ and integer} \qquad \forall g \in G, \forall k \in K \qquad (20)$$
$$T_i \geq 0 \qquad \forall i \in I \qquad (21)$$

The objective of the problem (MLDP) and constraint (16) minimize the makespan for the completion time of the PCB families. Constraint (15) ensures that each family is assigned to one mini-line. Constraint (17) is a surrogate for the capital constraint and provides a bound on the total number of machines allocated. Since the kanban implementation for a family requires a concurrent use of all machines in a mini-line, thus the variable time assigned to PCB type i is $T_i = \max_{k \in K} P_{ik}/Y_{gk}$. For a sufficiently large $|G|$, (MLDP) problem decides the optimal number of mini-lines to be formed, the grouping of the families, the configuration of each mini-line, and the timing for the production of each family.

Problem (MLDP) is a large difficult non-linear integer program, and we approach it hierarchically. We first solve the family grouping problem, that is the assignment of the families to mini-lines for a fixed $|G|$. Next we solve the mini-line production problem. The grouping problem should be such that it results in small makespan for the mini-line problem. Ahmadi and Matsuo (1990) presented a methodology for grouping families based on developing the similarity index between any pair of families, which we apply here to our problem. The proposed grouping problem uses the derived indexes to form the groups. We propose the following problem for generating the similarity indexes.

Similarity Index Problem:

$$(SIP_{ab}) \quad S_{ab} = \text{Min } (T_a + T_b)$$

$$\text{s.t.} \quad T_i Y_k \geq P_{ik} \qquad i=a,b; \quad \forall k \in K \qquad (22)$$

$$\Sigma_{k \in K} Y_k \leq R \qquad (23)$$

$$Y_k \geq 1 \text{ and integer} \qquad \forall k \in K \qquad (24)$$

$$T_i \geq 0 \qquad \forall i \in I \qquad (25)$$

The similarity index problem in a sense is an adaptation of the (MLDP) problem if families a and b are grouped together. It provides a measure for grouping of a and b if all the machines are allocated to these two families. (SIP_{ab}) can be solved optimally using the proposed greedy algorithm.

Algorithm for (SIP_{ab})

Step 1. Set $Y_k = 1$ for $\forall k \in K$ and $R = R - |K|$.

Step 2. For l=1 to R do

- identify the stage l such that additional machine will reduce the objective function the most. Break the ties arbitrarily.
- Set $Y_l = Y_l + 1$ and $R = R - 1$.

Continue.

The problem addressed here is a variation of the knapsack sharing problem. Federguen and Gronevelt (1986) have shown that the above greedy algorithm solves the similarity index problem optimally. Given S_{ab} for all $a \in I$ and all $b \in I$, we can define a regret measure for grouping a and b as follows:

$$R_{ab} = R.(S_{ab}) - \Sigma_{k \in K} P_{ak} - \Sigma_{k \in K} P_{bk}$$

The regret measure R_{ab} is an indicator for the amount of idle time which may be incurred if a and b are in the same group. The objective of the grouping problem is to form groups with minimum total regret. We also want to impose a restriction on the amount of total processing time assigned to a group. The grouping problem can now be formalized.

The Grouping Problem:

(GP) $V = \text{Min } \Sigma_{g\in G} \Sigma_{i\in I} \sum_{h=i+1 \text{ to } |I|} R_{ih} Z_{ig} Z_{hg}$

s.t. $\Sigma_{g\in G} Z_{ig} = 1 \quad \forall i\in I \quad (25)$

$\Sigma_{i\in I} P_i Z_{ig} \leq \Delta \Sigma_{i\in I} \Sigma_{k\in K} P_{ik} \quad \forall g\in G \quad (26)$

$Z_{ig} \in (0,1) \quad \forall i\in I, \forall g\in G \quad (27)$

where $\Delta \Sigma_{i\in I} \Sigma_{k\in K} P_{ik}$ is a managerial measure for the amount of processing time that should be allocated to a mini-line. It provides a measure for the load partitioning among the mini-lines.

Problem (GP) is a variation of the quadratic transportation problem and is NP-complete. Several heuristic solutions have been developed to solve variations of (GP) which can be used here. Other algorithms or search procedures such as simulated annealing procedure could be used to solve the grouping problem heuristically. For further details the reader is referred to Kusiak and Heragu (1987). Given the assignment of the families to the mini-lines, the scheduling and machine allocation problem can be expressed as follows:

Mini-Line Production Problem:

(MLPP) $v = \text{Min } C$

s.t. $\Sigma_{i\in I(g)} (T_{ig}+S_i) \leq C \quad \forall g\in G \quad (28)$

$\Sigma_{g\in G} \Sigma_{k\in K} Y_{gk} \leq R \quad (29)$

$T_{ig} Y_{gk} \geq P_{ik} \quad \forall g\in G, \forall i\in I(g), \forall k\in K \quad (30)$

$Y_{gk} \geq 1$ and integer $\quad \forall g\in G, \forall k\in K \quad (31)$

$T_{ig} \geq 0 \quad \forall i\in I, \forall g\in G \quad (32)$

Where I(g) is the set of PCB families assigned to group g. Mini-Line Production Problem can be solved in polynomial time as follows:

Mini-Line Production Algorithm:

Step 1. Let C_l and C_u be the lower and upper bound on C respectively.

Step 2. Set $C^*=(C_l+C_u)/2$.

Step 3. For $g\in G$ solve

$V_g = \text{Min } \Sigma_{k\in K} Y_{gk}$

s.t. $\Sigma_{i\in I(g)} (T_{ig}+S_i) \leq C^* \quad (33)$

$T_{ig} Y_{gk} \geq P_{ik} \quad \forall i\in I(g), \forall k\in K \quad (34)$

$Y_{gk} \geq 1$ and integer $\quad \forall k\in K \quad (35)$

$T_{ig} \geq 0 \quad \forall i\in I(g) \quad (36)$

using a greedy algorithm similar to the one for (SIP_{ab}).

Step 4. If $\Sigma_{g\in G} \Sigma_{k\in K} Y_{gk} < R$ ($\Sigma_{g\in G} \Sigma_{k\in K} Y_{gk} > R$)

• then set $C_l = C^*$ ($C_u = C^*$) and goto step 2
• else stop. The optimum solution has been found.

Mini-Line Production Algorithm for each value of C^* finds the minimum number of resources required to meet the target makespan. By changing the value of C^* using a binary search technique the algorithm finds the optimal solution. The solution of the Mini-Line Production Problem is the last step in our approach to the decision problems that arise in the electronic circuit card manufacturing.

V. Concluding Remarks:

In this paper we have provided a hierarchical decision model to address the problems arising in electronic card manufacturing. The hierarchical framework developed is self-contained in terms of its ability to provide the proper technical data as parameters of the models. The decision models developed assist the managers to understand their system and helps as a guide in making decisions regarding the design, production planning, and operational problems which they are faced. The models proposed attempt to maximize the throughput of the system and capture the benefits of a pull production control system. These models could also be implemented in other technological environments such as pin-and-hole through. The model for forming mini-lines is of high value to managers in large electronic card manufacturing systems in partitioning their facility to gain the advantages of focused systems.

Acknowledgments:
The author is thankful for the careful reading of this manuscript by Professor Gordon Shirley. This research has been supported by the grant # 95.

References:

•Ahmadi, R., 1992. "Hierarchical Approach to Design, Planning, and Control Problems in Electronic Circuit Card Manufacturing," Working paper, AGSM, UCLA.

•Ahmadi, J., S. Grotzinger and D. Johnson. 1988. "Component Allocation and Partitioning for a Dual Delivery Placement Machine," Opns. Res., 36, 176-191.

•Ahmadi, J., S. Grotzinger and D. Johnson. 1986. "Emulation of Concurrency in Circuit Card Assembly Machines," Re. Report RC-12161, IBM T.J. Watson Research Center.

•Ahmadi, J. 1986. "Optimization Problems of Flexible Assembly Systems in Large Scale Electronic Parts Manufacturing," unpublished Ph.D. dissertation, The University of Texas at Austin.

• Ahmadi, J. and I., Ali, 1988. "Integrated Production Scheduling in Large-Scale Flexible Electronic Parts Manufacturing," Annals of Opns. Res., 39, 1, 42-55.

• Ahmadi, R. and H., Matsuo, 1991a. " The Line Segmentation Problem," Opns. Res., 39, 42-55.

• Ahamdi, R. and H., Matsuo, 1991b. " Mini-Line Approach for Pull Production," working paper, AGSM, UCLA.

• Ahmadi, R. and P., Kouvelis, 1991a. " Staging Problem of a Dual Delivery Pick-and-Place Machine in Printed Circuit Card Assembly," working paper, AGSM, UCLA.

• Ahmadi, R. and P., Kouvelis, 1991b. "Design of a Testing Fixture for Printed Circuit Boards," working paper, AGSM, UCLA.

• Ahamdi, J., R., Ahmadi, H., Matsuo, and D., Tirupati, 1990. " Component Fixture Positioning for Printed Circuit Board Assembly with Concurrent Operations," working paper, AGSM, UCLA.

• Ahmadi, R. and J. Mamer, 1991. "Routing Heuristics for Pick and Place Machines," working paper, AGSM, UCLA.

• Ahmadi, R. and C. Tang, 1989. " An Operations Partitioning Problem for Automated Assembly System Design," Opns. Res. Vol. 39, No. 5, pp-824-835.

• Ball, M.O. and M.J. Magazine, 1988. " Sequence of Insertion in Printed Circuit Board Assembly," Opns. Res. 36, 192-201.

• Chen, P.H. and J. Talavage, 1982. "Production Decision Support System for Computerized Manufacturing Systems," Journal of Manufacturing Systems, 1, 2, 157-168.

• Chow, W. 1985. "Head and Disk Assembly Line Performance Characteristics, Part I: Analysis of Existing Lines," IBM Technical Report, TR 02.1256, IBM General Products Division, San Jose,
California.

• Chow, W. 1986. "Design for Line Flexibility," IEE Transaction, 18, 1, 95-103.

• Drezner, Z. and S.Y. Nof. 1984. "On Optimizing Bin Picking and Insertion Plans for Assembly Robots," IIE Transactions, 16, 3, 262-270.

• Federgruen, A., and H. Groenevelt, 1986. "The Greedy Procedure for Resource Allocation Problems: Necessary and Sufficient Conditions for Optimality," Opns. Res. 34, 909-919.

• Fisher, M.L., 1981 "The Lagrangian Relaxation Method for Solving Integer Programming Problems," Mgmt. Sci. 27, 1, 1-18.

• Garey, M.R., and D.S. Johnson, 1979. "Computers and Interactibility: A Guide to the Theory of NP-Completeness'" W.H. Freeman and Co., New York.

• Gavish, B. and A. Seidmann. 1987. "Optimizing the Retrieval Sequence for Component Insertion in Printed Circuit Boards," Working paper, Simon Graduate School of Business Administration, The University of Rochester.

• Grotzinger, S. 1988. "Positioning for a Dual Delivery Placement Machine," Research Report RC 13908, IBM Watson Research Center.

• Hyer, N.L., and U. Wemmerlov. 1984. "Group Technology and Productivity," Harvard Business Review, 140-149.

• Hu, X. and M.W. Canter. 1989. "Optimal Component Sequencing to Minimize Total Production Time," presented at CORS/TIMS/ORSA Meeting, Vancouver.

• Hu, T.C. and E.S. Kuh. 1985. "Theory and Concepts of Circuit Layout," in VLSI Circuit Layout: Theory and Design, T.C. Hu and E.S. Kuh (eds.) IEEE, New York.

• Liepala, T. and O. Nevalainen. 1989. "Optimization of the Movements of a Components Placement Machine," Eur. J. Open. Res., 38, 167-177.

• Lofgren, J.K. and L.F. McGinnis. 1986a, "Dynamic Scheduling for Flexible Printed Circuit Card Assembly," in Proceedings of the 1986 IEEE International Conference on Systems, Man and Cybernetics (October, Atlanta, Ga.

• Lofgren, J.K. and L.F. McGinnis. 1986b, "Soft Configuration in Automated Insertion," in Proceedings of the 1986 IEEE International Conference on Robotics and Automation (April), San Francisco.

• Lofgren, C.B., L.F. McGinnis, and C.A. Tovey. 1991, "Routing Printed Circuit Cards Through an Assembly Cell," Opns. Res., Vol. 39, No. 6, pp 992-1004.

• Soukup, J. 1981. "Circuit Layout," Proc. IEEE 69, 1281-1304. (Reprinted in VlSI Circuit Layout: Theory and Design, T.C. Hu and E.S. Kuh (eds.) IEEE, New York.)

• Magirou, V. 1986. "The Efficient Drilling of Printed Circuit Boards," Interfaces, 16, 4, July.

• McKenna, A. 1985. "Conceptual Model Aids in Development of Manufacturing System for PC Assembly Shop," I.E. April.

• Pukanic, R. 1985. "Flexibility and Integration are Key Material Handling Concepts in Electronic Assembly Environment," I.I.E., April.

• Tang, C. and E.V. Denardo, 1988a. "Models arising from a Flexible Manufacturing Machine, Part I: Minimization of the Number of Tool Switches," Opns. Res., 36, 767-777.

• Tang, C. and E.V. Denardo, 1988b. "Models arising from a Flexible Manufacturing Machine, Part II: Minimization of the Number of Switching Instants," Opns. Res., 36, 778-784.

• Wittrock, J. R. 1988. "An Adaptive Scheduling Algorithm for Flexible Flow Lines," Opns. Res., 36, 3, 445-453.

• Wittrock, J. R. 1985 "Scheduling Algorithm, 1985 " IBM Journal of Research and Development, 29, 401-412.

FUTURE DIRECTIONS

WHAT IS OPERATIONS MANAGEMENT?
A PERSPECTIVE FROM THE PAST
A PROJECTION FOR THE FUTURE

Jeffrey G. Miller, Boston University

In 1966, Elwood Buffa defined the boundaries of operations management in his book of selected papers, Readings in Production Operations Management. (John Wiley, 1966). He reached broadly into the the literature to develop this collection, which stimulated his peers to examine the field with a new perspective. In its subsequent use as a foundation text in operations management doctoral programs it influenced the thinking of many of today's leading scholars.

In 1991, Buffa again stimulated an examination of the field. The papers presented at the conference in honor of his career, and subsequently reproduced in this book, redefined the boundaries of operations management 25 years after the publication of Readings in Production Operations Management. A cursory examination of the two selections provides ample evidence that the answer to the question, "What is operations management?" has changed substantially.

The purpose of this chapter is to examine the changes in scope and definition of the operations management field over the last 25 years, and to project how the field might evolve in the 21st century. The social science methodology of content analysis is applied to the contents of Readings in Production Operations Management and to the contents of this book to reveal the underlying intellectual structure of the field over time. These structural elements are then used to develop definitions of operations management in 1966 and 1991. A comparison of the two definitions is used as the basis for the development of several scenarios for operations management in the year 2020.

Methodology

The specific methodology applied to analyse the contents of the 1966 and 1991 readings is known as the KJ© technique, after Jiro Kawakita, a noted anthropologist. The technique marries two separate techniques, affinity diagrams, and interrelationship digraphs,. These and related methods have been used extensively in theory development (see for example The Basics of Qualitative Research, Strauss & Corbin, Sage, 1990), in context analysis (Introduction to Qualitative Research Methods: The Search For Meanings, Taylor & Bogdan, Wiley Interscience, 1984), in problem solving for total quality management, (see The Memory Jogger Plus, GOAL/QPC, Methuen, MA, 1989. as well as to organize artifacts from anthropological studies.

To develop affinity diagrams, the titles of the papers in Readings in Operations Management and Perspectives on Operations were used as observations relevant to the content of each book. (See tables 1 and 2 for a list of titles.) An iterative process of grouping was then applied

Table 1
Observations from
Readings in Production Operations Management

1.) The Production Process: Operating Characteristics-Abruzzi
2.) Evolving Concepts in Production Management-Starr
3.) A Proposal for a General Model of a Production System-Alcalay & Buffa
4.) A General Model for Production & Operations Systems-Reisman & Buffa
5.) Linear Programming Approach to Production & Employment Scheduling-Hansmann & Hess
6.) Linear Programming in The Packing Industry-Green, Chatto, Hicks & Cox
7.) Study of a Textile Mill With the Aid of Linear Programming Eisenman, Williams, & Young
8.) An Application Of Linear Programming to Cupula Charging-Metzger, Schwarzbeck
9.) The Development of Factory Simulation Using Actual Operating Data, LeGrande
10.) Daily System for Sequencing Orders In a Large Scale Job Shop-Steinhoff
11.) Load Forecasting, Sequencing & Simulation- Bulkin & Steinhoff
12) Cost Models for the Application of Waiting Line Theory to Industrial Applications-Hillier
13.) Application of Queuing Theory For Determining Manpower Requirements For an In-Line Assembly Inspection Department-Sespeniak
14.) Waiting Line Models in Materials Handling-Fetter & Galiher
15.) A Job Vacancy Queue-Peterson

Observations from
Readings in Production Operations Management

16.) Transaction Oriented Information Handling Systems-Thompson
17.) Allocating Facilities With CRAFT-Buffa, Armour & Vollmann
18.) A Heuristic Method of Assembly Line Balancing-Kilbridge & Wester
19.) COMSOAL: A Computer Method of Sequencing Operations For Assembly Lines-Arcus
20.) Questions For Solving the Inventory Problem-Morgan
21.) Guides to Inventory Policy-Magee
22.) A Linear Decision Rule for Production & Employment Scheduling-Holt, Modigliani, Simon

Table 1 (Continued)

23.) Consistency & Optimality in Managerial Decision Making-Bowman

24.) Application of a Technique for Research & Development Program Evaluation-Malcolm, Roseboom, Clark & Fazar
25.) An Analytical Study of the PERT Assumptions-MacCrimmon & Ryavec
26.) Some Properties of Schedules for Large Projects With Limited Resources-Weist
27.) How Management Can Use the Improvement Phenomenon-Carlson
28.) Management in The 1980's-Leavitt & Whisler
29.) Management & The Black Box-Anshen

to each set of observations. The basis for subgroup formation was the conceptual affinity between a subset of observations that separates them from the other observations. After a subgroup was formed, a new name describing the collective content of the group was developed. Both the grouping and the group naming procedures are intuitive, and of course, researcher dependent. The reader can assess the face validity of the subgroups developed in this analysis by examining Tables 3 and 4. The process of grouping observations by the affinity of their contents can be compared to reverse outlining, where the subheads and major heads are deduced from the bottom up based on the data (observations), as opposed to inductively assigning data on the basis of an a priori set of organizing constructs. It should be noted that only preliminary titles of the papers in this book were available as data at the time this paper was written.

Table 2
Observations from Perspectives on Operations Management

1.) Product Standardization and the Strategic Management of Technology-Rosenbloom
2.) Manufacturing /Marketing Interfaces - Hausman
3.) Linking Technology and Business Strategy - Hax
4.) Creating Value Through Operations in the Global Enterprise - Kleindorfor and Cohen
5.) Models of Tightly Coupled Production Systems - Baker
6.) The Role of P/OM as Educator for Achieving World-Class Systems - Starr
7.) The Strategic Role of Information in Services - Fitzsimmons
8.) Management Issues in Flexible Manufacturing Systems - Seidman
9.) A Review of Variable Yield Problems - Bitran
10.) Design for Supply Chain Management: Methods and Examples - Lee
11.) Mathematical Models of Retailer Inventory Systems: A Review - Nahmias and Smith
12.) Stockless and Fast Production: Review and Research Agenda - Thomas

Table 2 (Continued)

13.) Service Quality Deployment: Quality Service by Design - Chase
14.) Operations Management: The Time for Meta Research is Now-Reisman
15.) The Information/Control/Buffer Portfolio-Schwarz
16.) Research In Manufacturing Strategy- A Cross Functional Perspective-
Karmarkar
17.) Precision of Aggregate Planning Models- Wagner
18) Modeling in Support of World-Class Operations - Silver
19.) TBA - Federgruen
20.) Some Thoughts on Inventory Modeling and Diagnostics - Graves
21.) Optimization Models for Electronic Card Manufacturing - Ahmadi
22.) Dynamic Control of Operating Systems - Dyer and Larsen
23.) TBA - Erlenkotter
24.) A Current Perspective on Layouts - Johnson
25.) Competing Through Quality: Perspectives and Research Directions -
Moskowitz
26.) A Decision Analysis Approach for Coordinating Design, Manufacturing,
and Marketing - Sarin
27.) Manufacturing Strategy - Skinner

Interrelationship digraphs represent the relationships between groups formed in the affinity grouping process. They are formed by drawing arrows to indicate hypothesized causal or other relationships between one or more groups. Only the names of the major groups are considered as data in developing the potential linkages at this point. The completed digraph thus represents all of the data that the theory builder has at his disposal at a high level of abstraction and generality. In this paper, the completed digraphs are also referred to as a "knowledge maps".

What is Operations Management 1966?

Table 3 shows the five major affinity groups formed around the observations in Readings in Production Operations Management, along with the names given to the subgroups and their parents The numbers under each group or subgroup name indicate the number of the paper from the book.(See Table 1.)

Table 3
Affinity Groups: Readings in Production Operations Management

A.) Production Systems Theory
+ Production Management (1,2)
+ Systems Analysis (3,4)

Table 3 (Continued)
B) Optimal Resource Allocation (5,7,8)
C) Production Process Design
+ Process (Systems) Analysis Tools (9,12,14,17)
+ Division of labor (18,19)
D) Factory Planning & Control
+ Shop Floor Scheduling (10,15)
+ Inventory Management (20,21)
+ Production Planning
+ Capacity Analysis & Forecasting (11,13)
+ Aggregate Planning (22,23)
E) Managing Change in Production Systems
+ Technological Change In Production (16,27,28,29)
+ Project Scheduling (24,25,26)
Numbers in () refer to observations in Table 1 that compose subgroup

Figure 1 is the interrelationship digraph (knowledge map) created to illustrate the relationships between the affinity groups defining the operations management construct in 1966.

Figure 1
Knowledge Map for
<u>Readings In Production Operations Management</u>

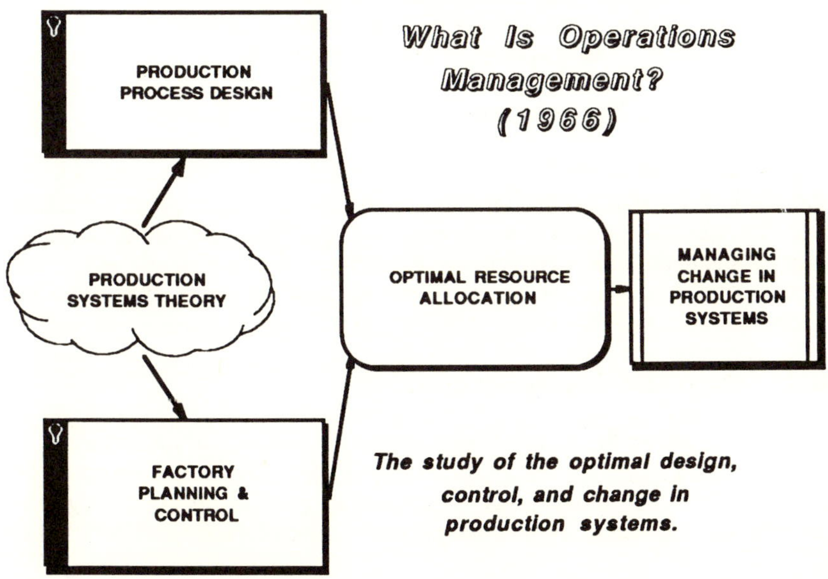

Production systems theory forms the core theoretical base for the field in the interpretation in Figure 1. The papers by Abruzzi, Starr, Alcalay and Buffa and Reismann and Buffa outline key elements of this theory as it existed in 1966. The arrows in the digraph represent the relationship between the theory and its application to problems in production process design and factory planning and control. The papers by Fetter & Galliher, Hillier, Buffa, Armour & Vollmann, Arcus, and Kilbridge and Wester are excellent examples of production process design applications. Steinhoff, Morgan, McGee, Holt, Modigliani, Muth & Simon, Bowman, and Hansmann & Hess, among others, illustrate applications to the problems of factory planning and control.

The titles by Eisenmann, Williams, and Young, Green, Chatto, Hicks, and Cox, and Metzger, and Schwarzbeck in Readings in Production Operations Management illustrate the application of linear programming to production design and control problems. However, the arrows linking this grouping to others, and its prominence in the center of the digraph, are indicative of a dominant feature of the Operations Management paradigm in 1966: a preoccupation with optimal resource allocation as an objective.

The last major element of the 1966 concept of Operations Management is represented by the group of papers in Readings in Production Operations Management that are concerned with the management of change in production systems. Thompson, Anshen and Leavitt and Whistler are concerned with the management of technological change, and in particular radical changes in information technology. Malcolm, et al, MacCrimmon & Ryavac, and Weist are also concerned with radical change through the management of large scale projects. Carlson, heralding what is to become an important theme in the 1990's, focuses on gradual, or incremental, changes in productivity.

What is Operations Management? The data from 1966 suggests that it was the study of the optimal design, control and change in production systems. Though the title of Buffa's book employs the then inclusive language "production and operations management" in recognition of the newly recognized importance of the service sector, there are no service sector articles in the collection.

What is Operations Management 1991?

Table 4 shows the major affinity groups formed around the observations (titles) in Perspectives On Operations Management. Five major groups were created from the 1991 observations, roughly corresponding to the five groups formed with the 1966 data. Few of the papers in the collection were focused on theory per se. But, the titles of the papers by Lee, Nahmias and Smith, and Reisman, among others, indicate that the theory being applied has a broader scope than production systems theory as it was represented in 1966. The term value chain theory is used to describe the larger set of multi stage system constructs being tapped by the researchers of 1991.

Table 4
Affinity Groups for Perspectives on Operations Management

A.) Operating Systems Design (5, 15, 21,24)
B.) Operating Systems Management (8, 12, 17, 20, 22)
C.) Value Chain Systems Theory (10, 11, 14)
D.) Winning in Global Competitiveness
- + Technology Strategy (1,3)
- + Operations Strategy (4,27)
- + World Class Operations (6,18)

E.) Organizational Process Integration
- + Information Systems Integration (7,8)
- + Cross Functional Processes (2,16)
- + Quality Management (9,13,25)

Numbers in () refer to observations in Table 1 that compose subgroup

The papers of Amahdi, Johnson, Baker, and Schwarz, clearly reflect the continuing application of theory to design problems. However, it can be argued that the titles of their works are more in keeping with the the 1966 orientation towards production design. One could draw a similar conclusion about the titles in the operating systems management subgroup. The titles employed by Siedman, Thomas, Wagner, Graves, and Dyer & Larsen, are all suggestive of factory applications in planning and control. However, the inclusion of other papers on the service sector in the collection, and the use of service sector examples in some of the titles in these subgroups, justifies the use of the broader term "operations." in these subgroup titles.

Three subgroups formed the basis for the major subgroup entitled "Winning in Global Competitiveness". The common denominator implicit in the names of each of the three subgroups was a focus on competitiveness in global markets. Rosenbloom and Hax deal with the management of technology in this context, Kleindorfor and Cohen, and Skinner deal with global operations strategy, and Starr and Silver's titles reflect attention to the emergence of world class standards. Competitiveness is a significantly more sophisticated and complex objective criterion to deal with than optimal resource allocation, the central focus of the 1966 collection. A global market view is significantly broader and more inclusive than a factory view. More than any other portion of the 1991 collection, the titles in this and the next major group reflect the greatest changes in the field since 1966.

The organizational process integration group also represents an expansion in the scope of the field from "change in production systems" in 1966. The names of the subgroups comprising this larger group suggest that the range of change issues in 1991 encompasses the development of integrated information systems (Fitzsimmons and Siedman), cross functional business processes (Hausman and Karmarkar), and quality management (Bitran, Chase, and Moskowitz).

Figure 2 shows the knowledge map for the 1991 collection.

Figure 2
Knowledge Map of Perspectives On Operations Management

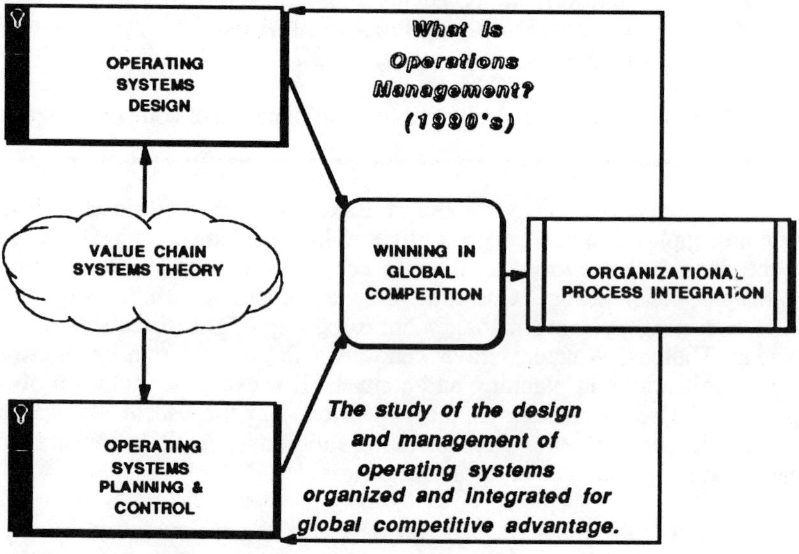

The similarity between Figure 1 and Figure 2 suggests that there was little change in the structure of the knowledge in the field, but a great change in its scope. Figure 2 does not simply replace Figure 1, it overlays it. Everything in the 1966 knowledge map (digraph) is subsumed in the 1991 map. Production systems theory is a special case of value chain systems theory. Optimal resource allocation is one factor to be considered in world class competition. Organizational process change represents a more general class of issues than production process change. We see a great deal of evidence in the 1991 titles that the 1966 paradigm lives on as a part of a larger definition of the field.

There is one major structural difference between the two knowledge maps, and that is the inclusion of the two feedback arrows emanating from the organizational process integration group. The intent of these arrows is to signify the implicit continuous improvement theme that ran through many of the papers in the 1991 collection, and that promises to accelerate the rate of change and improvement in the field in the future.

What is operations management in 1991? It is the study of the design and management of operating systems organized and integrated for global competitive advantage.

What is Operations Management 2020's?

Figures 1 and 2 represent two snapshots of the field taken twenty-five years apart. They are similar in kind to the time photos of the galaxy that led physicists to posit the "big bang" theory of the development of the universe. In both physics and operations, the scope of the observed object is expanding at a dizzying rate. But, since neither a universal theory of physics or operatons yet exists, our ability to explain what fueled development in either instance is poor. This does not mean that a thorough exploration of the causes of change in the operations management field does not deserve further analysis in other papers. In the balance of this paper, three future scenarios for the field will be explored, discontinuous change, collapse, and continued expansion.

Discontinuous Change - Hardest to predict, but the perhaps the most probable course in the long run is discontinuous change. We would not see a simple expansion or contraction of the basic knowledge maps of 1966 or 1991 here, but a dramatic alteration in their structure and content reminiscent of the basic paradigm shifts that Thomas Kuhn has observed in the world of science. (See his Structure of Scientific Revolutions). The easiest thing to predict about the field of operations management under this scenario is that it would no longer exist in any recognizable form, other than as a historical footnote in the development of some other new field. If this be the case, those involved in Operations Management in 1966 and 1991 should be able to look back from 2020 with pride in the contributions they made.

Collapse-The collapse scenario is somewhat more likely in the short term, at least for some schools of management. It presumes that the field has already exceeded its sustainable limits, and that its scope will begin to contract, perhaps to an enhanced version of Figure 1. Here, we might see operations departments and industrial or manufacturing engineering departments joining forces to develop deeper insights into production theory, and its applications to process design and operations control. The collapse scenario would represent a decision to invest in depth, rather than breadth.

Continued Expansion-The continued expansion scenario is based on the continued expansion of the universe of operations management in a manner similar to that which occurred between 1966 and 1991. Figure 3 might resemble a knowledge map for the future based on a simple projection of what has already occurred. From 1966 to 1991, the core subject matter of the field changed from production, to the more inclusive operations. In 2020, we might see the even more inclusive term "business processes" replace the term operations. Business process management would include the management of any purposive activity (including projects) in a manufacturing, service, or professional enterprise. Common operations case examples would include analyses of production lines and fast food, and other business processes like product development, order entry, or the development of a legal brief. Production systems theory and operating systems theory would then be considered a special case of business systems theory.

Figure 3
Interrelationship Digraph for 2020's?

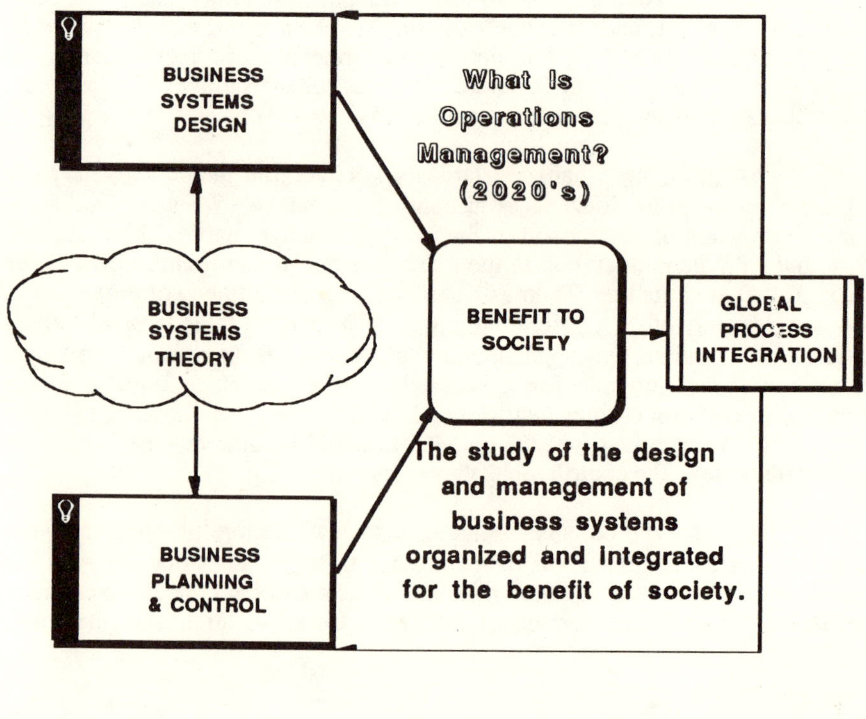

In a similar expansion of scope, competitiveness and optimal resource allocation as central objective criteria might be replaced by the more inclusive "benefit to society". We already see an increase in attention to societal issues, such as environmental impact and sustainable growth in the analysis of operating systems by major corporations. (See for example the annual reports of 3M and Dow Chemical). Organizational process integration and management, which concentrates on interfunctional activities within an enterprise, would be upgraded to interbusiness process integration and the management of alliances on a global level.

Though Figure 3 depicts but one possible scenario for expansion, it raises a critical question common to all: How much can the boundaries of the field expand before the field loses meaning and relevance? Will the scope of the field reach a point where the distance between the interests and scope of individual scholars is so great that meaningful discourse and interaction can no longer take place? Operations Management scholars are already challenged to communicate across a wider array of subjects, issues, and methodologies than they ever have been.

Other fields, such as economics, have been able to maintain their integrity despite their breadth of perspective. The devices they have used to maintain their common ground have included the development of a strong common theoretical base and language, and subdivision into related groups such as micro and macro economics. Assuming that further expansion of the boundaries of the field is the most likely near term scenario for Operations Management, the challenge for the community of scholars in it is to consider how we can use these or similar devices to maintain our own ability to contribute to the development of knowledge.

EDUCATING MANAGERS TO COMPETE: THE ROLE OF OPERATIONS MANAGEMENT

Martin K. Starr
Graduate School of Business
Columbia University
New York, New York

ABSTRACT

Conventional methods for managing firms - which worked for U.S. firms from the 1950s through the 1980s - are not so competitive in the final decade of the 20th Century. Traditional schools of business - which evolved from the 1950s through the 1980s - provide the MBAs that practice the not so competitive, conventional methods of managing.

Some widely accepted tenets of economics, finance, marketing and management science appear flawed for application in the 90s. Perhaps they were acceptable in the 1960s, but they are suspect now. The fact that they are still being applied without the validation of hypothesis testing is indicative of how serious a problem exists in our understanding of business as a system.

A driver of change has been the continuously improving computer facilities and their ability to provide a skein of interconnected information that reveals the structure of the operating systems. The systems point of view which this approach requires is neither well-understood nor widely-practiced. Under these circumstances, it is not surprising that management training is failing to address managerial quandaries.

To begin with, this paper identifies the competitiveness problems facing firms that are attempting to become global winners. Then, some of the causes of the problems are examined. Finally, educational-based solutions are suggested. The notion is developed that P/OM must begin to play a major role in problem-solving. P/OM-sight provides the systems point of view which helps business to achieve competitive world-class performance.

I. COMPETITIVENESS PROBLEMS

There have been two major interrelated problems which account for the competitive malaise of Europe and North America. First, is the deterioration of managerial competitiveness. This is particularly noticeable in the U.S.A. which after World War II was the unquestioned world leader in production, marketing and distribution. To show how much times change, while this article was being completed, the New York Times characterized the American President's trade mission to Japan as "disastrous" and the Japanese government was talking about various forms of aid to assist the ailing American economy.

Second, is the failure of the management education system to halt the deterioration of managerial competitiveness in the U.S. from the 1960s onward. The role of management education can be viewed in two parts. One was the 1940-1950 evolution of Industrial Engineering (I.E.) training which ultimately partnered up with Operations Research (O.R.) and Management Science (M.S.) in the 1950s. By the 1960s, there were functional field problems that transcended engineering considerations. A few academics[1] wrote basic texts that combined the materials in a new way which helped to solve the problems and launch the P/OM field.

Now, in the 1990s, to solve competitive problems requires transcending the traditional curriculum of business schools. This observation applies to both MBA and Executive Programs. It is not clear who will lead the way. To what extent can business schools change, and even if they do, to what degree can they lead, rather than follow the development of business practices.

As we proceed, there will be occasions to examine portions of the history of managerial developments with respect to both management itself and the system for educating managers that has evolved since the 1950s. We will be able to examine directions and stumbling blocks that slowed the development of competitive systems. We will also be able to speculate about the present competitive conundrums and what might be done about them.

A. DETERIORATION OF MANAGERIAL COMPETITIVENESS

Conventional methods for managing U.S. firms are not proving to be competitive in the final decade of the 20th Century. Many executives have been puzzled by the fact that the same methods were successful after World War II. At that time, companies made lots of money and returns on investment were exceedingly good. No one made a big deal about product quality and customers were happy to receive deliveries when they got them. By the mid-60s, signs of deterioration of managerial competitiveness were surfacing.

The effects of deterioration of competitiveness for U.S. firms became transparent (starting in the 1980s) in terms of seriously increasing trade deficits, budget deficits, and unemployment. The overall decrease of global market shares for many different U.S.-based industries was pronounced. Entire industries that produced radios and TVs literally disappeared. On an individual firm basis, return on assets, investments and equity declined. Productivity fell to levels that made little sense in view of major investments in computer and related technologies. Profitability gave way to incurring substantial losses as a result of tough new competitors that made use of advanced technology and methodology. These competitors could emerge anywhere in the world.

U.S. firms had significant sunk costs in old technology, as well as difficulty in changing from established conventional management methods. Manufacturers proceeded to rid themselves of outdated systems by engaging in numerous plant closings. Following extensive downsizing, firms like IBM, General Motors and Kodak cut workforces substantially and upgraded their technologies. These moves affected consumers' pocketbooks and investor attitudes. The negative interactions reinforced each other culminating in multiplicative deterioration. Bankruptcies of prestigious companies like Pan Am and Eastern Airlines, R.H. Macy and Federated Department Stores, Southland (owner of 7-Eleven stores), and Jim Walter (a leading home builder) became commonplace as economic turbulence prevailed. Government takeovers of foundering banks resulted from bad debt associated with real estate which became overstocked when prosperity (which drove the engine of expansion) disappeared.

American firms have been trying to turn this situation around. They have adopted new methods for increasing productivity, attaining high quality, reducing costs, improving service, etc. The efforts to bring about change have been effective, in some instances, almost heroic, but they have not been sufficient to outperform (or leapfrog) the competition. From game theory we know that zero-sum competitive events can have only one winner.

There is only one world champion team. It doesn't matter that it only won by one point. So it is not enough that companies are performing better on quality and costs. The winner leads the field as the best there is available. If a firm is getting better at a faster rate than any of its competitors it has a chance to surpass all contenders. The rate of improvement is consequential. When improvement is based on imitation of the leaders rather than innovating to surpass them, then the game is lost. In definitive terms, research is not required to prove the point that winning is based on surpassing capabilities.

B. FAILURE OF THE MANAGEMENT EDUCATION SYSTEM

Traditional methods of education fail to develop student awareness of many factors which count in gaining competitiveness in the present-day world. Among the more important factors we want to list are:

1. coordinating interfunctional systems dependencies,
2. recognizing technological development cycles,
3. knowing the criteria for technological investments,
4. using time-based management methods,
5. achieving innovative changes with quality improvements,
6. understanding consumer dynamics,
7. getting P/OM to play facilitator in achieving 1-6.

The traditional methods of education are associated with the POSDiC framework which is planning, organizing, staffing, directing, and controlling. These management of organization terms which are the backbone of a business education have been designed to support the conventional methods of managing. As a consequence, management teaching cannot help but fail to solve the present problems of competitive malaise. The functional isolation that permeates what is taught and the models and methods used for teaching are responsible. As an example, consider the propensity of typical company bureaucracies to protect themselves by segmenting business functions.

It is not a surprise that the typical curriculum of business schools mimics the practice of functional separatism. As long as business schools implicitly endorse current business practice through what they teach and how they teach it, the schools cannot remedy the competitive shortfalls that business is experiencing. We would like to encourage teachers to begin to lead and not follow business practice.

II. CAUSES OF PROBLEMS

The industrial world is undergoing enormous change which is characterized by the wearing out and replacement of old technologies that cease to be competitive. They no longer provide attractive returns on investments. Replacement technologies are often connected to new computer technology and information systems. An immediate effect of this change is that the technology of manufacturing, and doing work, in general, has been irreversibly altered. Understanding what is happening provides distinct competitive leverage. Not comprehending causes and impact puts one at a distinct competitive disadvantage. What should be taught in business schools must be congruent with the new technologies - and their applications - to be valid and applicable. The cooperative relationship of business and engineering is no longer a luxury. It is a necessity.

A. ACADEMIC AND INDUSTRY GOALS DO NOT ALIGN

Some say business schools are responsible for the problems of American industries. This seems a wasteful argument. What counts is that business schools do not appear to accept the responsibility for helping to solve the problems of industrial competitiveness. Industry and academia are, at best, casually bonded in a partnership of shared responsibilities, benefits and penalties. Some realignment is necessary whereby academia and industry respect and support each others contributions to theory and practice, respectively.

While engaged in a critical evaluation of the role of the schools, it should be noted that the failure of business and engineering school faculties to find the means for fundamental cooperation has been remarkably disappointing to both academics and managers in industry[2]. The failure underscores the inability of academic institutions to repair their own bureaucratic hangups. It is evident that criticism of industrial bureaucracies can also be applied to the organizational structure of universities. A unified technical-management educational focus might have a powerful influence on industry to repair the same kind of gap which divides management into financial as opposed to technical functions.

For the most part, the research efforts of university faculty members are not directed toward the near-term improvement of corporate performance. This is because academic requirements for tenure are best satisfied by contributing to the theoretical understanding of phenomena that are area-specific, quantifiable, and tightly focused. Fuzzy issues that reflect uncertainties associated with human behavior are not likely to lead to acceptable dissertations or articles that can be published in the leading technical journals.

Business people are seldom aware that the goals of academicians do not align with those of industry. Business schools are not powerless to change this situation. Faculties could champion an alliance of business schools dedicated to installing new curricula and rewarding applications-oriented research. The existing accrediting organization for business schools has been acting for years as though everything is just slightly less than fine. It is not alone. Business journalists do not perceive that problems exist. The entire educational system acts as if it does not have a clue about how to bridge the gulf between academic programs and industrial needs. The time has come; the opportunity is almost gone for bringing about those changes that would benefit industrial infrastructure. The hitch is that business schools must change their goals to align them with existing industrial goals.

B. THE 1950s P/OM-IDENTITY PROBLEM RESOLVED

P/OM was an invention whose mother was necessity. The field was created to blend together the mature I.E.methods, the upstart O.R./M.S. decision models, and the management heuristics and decision-modes based on instincts and experiences. The awkward name derived from the fact that production was identified with manufacturing and the plant floor whereas operations were identified with logistics (such as airline routing) and services (such as the backroom of banking). Operations were viewed as amenable to I.E. and O.R./M.S. improvements, but having a fuzzier structure and closer contact with the customer than manufacturing, often made it appear more like an art than a science[3].

Historical insight might help us here. The O.R./M.S. methods were so powerful that they created what could be called the 1950s P/OM-identity problem. Before the mid-1950s, production was taught and practiced as an outgrowth of I.E. with powerful roots going back to Frederick W. Taylor and the concepts of efficiency specialists. It is a mistake to believe that I.E. had no humanistic missions. For its time and place in history, early I.E. had many advocates of humane behaviors to develop motivational drives[4].

By present standards, the relationships were limited and stilted.

The problems of this era were related to the difficulty of bringing together disparate points of view and fitting them together in a single systematic framework. We can characterize these different points of view as follows:

1. the human realities of production,
2. the technological potentials and constraints of equipment and materials, and
3. the mathematical regularities of production logistics.

To solve the problem calls for rationalization of what C.P. Snow called the two cultures (viz., humanism versus science). Put another way, we are talking about blending the art and the science of management. In terms that are comfortable to both industry and academia, the gap was between theory and practice.

To bridge that gap it was necessary to find a way to combine the analytics of I.E. and O.R./M.S. with the knowledge base of Industrial Psychology and the pragmatics of labor-intensive production departments. For the new field of P/OM, what percentage of effort should be devoted to the qualitative body of knowledge as compared to the quantitative? There is no way in which a percent breakdown can be given. Rather, the approach of leading production thinkers like Elwood Buffa was to create a text describing all of the pieces and parts of the production and operation mix which needed to be considered for successful design, implementation and operation of the production system. As possible, employee relations issues and quantitative decision models were connected to the fabric of running a good P/OM department rather than to each other. It was a good solution which helped establish P/OM as a legitimate functional field of business.

C. OPTIMAL STRATEGIES CHANGE: 1990s CHALLENGE

In the 1950s and 60s, one of the main objectives of U.S. industries was to produce high volumes of goods which could satisfy the increasing demand levels. Producer resources were scarce and it was common to describe the situation as a seller's market. The scarce resources were materials, skilled and unskilled labor, plant capacity and management time.

Under such circumstances, the optimal allocation of scarce resources was a critical goal.

Implementation of O.R./M.S. models held the key to scarce resource allocation. O.R./M.S. groups, known by various names, were established in tens of thousands of U.S. firms. This was an extremely productive period for the causal-type, phenomenalogical modeling that distinguished O.R./M.S. from strictly statistical analysis. The new techniques were directed to scarce resource allocation which epitomized an era when there was not enough of everything to supply all demands.

The emerging O.R./M.S. field was ideally suited to balance resource constraints, minimum production requirements and the profitability or costs of the products that could be created by blending and mixing different amounts of the scarce resources. Manufacturing technology was mature and there was little choice concerning the best technological strategies. The fact that with proper technological design it was possible to reduce costs by taking advantage of economies of scale lead to further efforts to aggregate volume.

Those companies able to increase output without incurring additional costs developed a competitive advantage. This led them to take further steps to do more of the same. The big firms got bigger as they aggregated volume. Marketing departments grew in importance as it became clear that selling volume was the key to success. The first objective for production was to turn out the product in sufficient volume to match what was sold. Quality, which never had a front seat in American industry, took a seat even further back. There was little incentive to innovate processes or develop new products.

By the 1980s, technological changes in manufacturing capabilities put in question the basic assumptions concerning economies of scale which had been developed in the 1950s. A new type of competitor began to enter the global marketplace. These new entrants made the older companies (which had succeeded through volume output) extremely vulnerable. The new companies produced a more reliable and generally higher quality product.

Resource allocation models were not irrelevant but they provided secondary considerations. Quantitative models, acceptable to management, would help to improve quality, allow faster product development, increase the rate of innovation, and provide greater motivation for employee participation. The 1990s mission statement for P/OM would have to take into account the fact that major changes in business objectives had occurred.

III. EDUCATIONAL SOLUTIONS FOR THE 1990s

To what sources should managers who want to know how to become winners turn? They are so busy running the present systems and putting out fires that they have little time to invent their futures. There may be some very good answers available from various sources, but it seems clear that in the 1990s, industry is not turning to the business schools of America for advice on what to do and how to do it. There are prior precedents for industry calling on academia to provide appropriate educational assistance.

Because of our academic orientation, teaching and research experiences, and the resources which are available to us, we are prone to ask why are we not leading in research and teaching that can provide the kind of guidance that industry would support. If asked, what would we suggest for an educational solution to the competitiveness problem? The advice which follows is couched in terms that are applicable to the university as a whole, but the realistic hope is that business schools will be the change-agents.

The first step would be to get rid of the art vs science conflict which plagues university campuses in the U.S.A. What is needed is the respect of each culture for what can be learned from the other. The scientific culture must come to accept applications (of theory) as being relevant for true scholarship. Accordingly, academic tenure should be grantable for applications of theory. Conversely, university budgets could reward cooperation between the humanities and science instead of pitting them against each other.

The second recommendation is even more controversial than the first. The university system should reward individuals for working together in teams, rather than for competing with each other for grades[5]. Eventually, grade schools might emulate the academic procedures of colleges and universities. The ultimate objective would be to have firms recognize the advantages of team cooperation over adversarial behavior for competitive advantage.

The third recommendation is to increase rewards for intelligent risk-taking. This means that errors should not be penalized when they are logically incurred in pursuit of learning. Information systems must be developed for enhancing the memory of what works and what does not. We wish to avoid the repetition of errors, but we need intelligent experimentation to learn. Learning systems are the only way to adapt to global competition.

The fourth recommendation is to reward hands-on knowhow of technological systems. It is unfortunate that the financial management of systems encourages confidence in abstract parameters that have nothing to do with how to bake bread, make chairs, or mine coal. This encouragement is misleading both in the short-run and in the long-run. The managers of successful global companies know the nuts and bolts of their businesses. Those in the baking business know the specific parameters of baking bread. Baking bread really differs from making autos or mining coal. When MBAs are not aware of the physical process and its technological parameters, they are likely to make incorrect decisions.

The fifth recommendation is to design, install, and motivate the use of the systems point of view. To function properly the systems perspective must be developed through communication interchange capabilities that allow the systems participants to see and act on the whole picture. Schools in the U.S.A. are among the least likely organizations to understand, let alone instruct anyone in, the use of cooperative teamwork that epitomizes the systems points of view. Competitive grading and stress on each individual's achievements overrides the potential benefits to be gained from teamwork. In the business schools, competition between students abounds, which MBAs take as a sure sign of what they will be up against after they graduate and start working for a large organization that rewards its stars and passes over the team player types. Competition also exists between members of the faculty and their respective departments which are constantly engaged in a struggle to gain or at least maintain their share of the teaching and research budget. Removing competitive barriers and installing an interfunctional curriculum is of the order of an Herculean task.

IV. 1990s ROLE OF P/OM AS AGENT PROVOCATEUR

Things are not looking as good as they use to look in MBA-ville. Economic turmoil has cast out prosperity and replaced it with hard times. Companies are hiring fewer business school graduates. MBAs, in particular, are being scrutinized in terms of what they can contribute to a company's competitive capabilities. Students are questioning whether business schools can get them jobs. The costs of tuition, books and lodging are exceedingly high, reflecting galloping inflation of educational costs in an otherwise non-inflationary economy. The forecasts for future applications to MBA programs are not rosy. Schools are reexamining their business plans.

The reasons that a broad spectrum of companies have ceased to hire MBAs is threefold. First, the hire-on costs of MBAs seem out of line with the costs of hiring managers from other sources, including undergraduate schools. Second, companies require fewer middle managers since their operations utilize fewer workers. Flatter organizations are performing better than those with many-level hierarchical structures. Third, gnawing doubt is giving way to certainty that American business schools do not know how to train managers who can improve competitive performance. If these beliefs persist, there will be a substantial contraction in the number of business schools that will be around at the turn of the millennium.

The scholar's mission, at this time, is to point out that if business schools engage in serious and substantive change, then the millennium scenario is not likely to happen. The disassociation of practicing management from academic training in management does not take into account the future of business schools which have revised their agendas. What is the nature of the revised agendas? How can business school graduates be trained to help their firms become industry leaders? No one supposes that by altering the teaching and research agendas of business schools we can totally resolve the competitiveness problems of industry. At the same time, there is agreement that a new business curriculum which trains managers in globally competitive practices will be welcomed in the marketplace.

The products of schools are students (and research that converts as soon as possible to improve what students are taught). Students are analogous to the products and services produced by companies. Superior products are received in the same way as the output of any company producing a superior product. Curriculum to a business school is a close equivalent to product or service design. Teaching success with that curriculum is analogous to process effectiveness for the production of a product. But what should such a new curriculum look like? It is doubtful that many currently proposed themes for revising business curricula (e.g., ethics and internationalism) are likely to alter industrial competitiveness. Those proposed themes set conditions that are quite likely necessary, but not sufficient.

We have listed five general educational initiatives in Section III designed to revitalize the MBA degree. To some extent, each and every one of them is dependent upon elevating the P/OM function to have equal planning and decision-making status in the corporation. For business schools the challenge is to break down the functional compartmentalization of the faculty, the curriculum, the budget, and the state of mind that relegate P/OM considerations to second rate status. For successful global competitors, finance and marketing are now equal partners with production (technology and the control of operations). Organizational and educational systems and structures have to catch up with this realization.

It is particularly worrisome that from the educational perspective, a large number of traditional concepts that are regularly taught in business schools all over the worlds are incorrect and misleading. The new curriculum should root out these erroneous concepts that are part of the foundations of business teaching in the West. Many of them have occurred because P/OM representation has been missing from the board of decision-makers in the corporation and the business school. P/OM allows engineering and technology back in the boardroom. The erroneous concepts briefly discussed below engender non-competitive performance.

1. Let us correct the oversimplified price - demand elasticity model to include the P/OM parameters of quality and time. It is essential to teach that a relationship exists in which demand will increase as quality improves because that product becomes less substitutable. Time - quality - price elasticity refers to both the rate of innovations for new features and delivery due dates.

2. Let us correct the criteria for determining when to invest in new technology. Discounted cash flow (DCF) is insensitive to when the cash is on hand and unable to capture the value of technological knowhow. Cash flow control is a time-based management issue with which P/OM is deeply involved. Inventory and shop floor planning as well as model changeovers and conversion to altered processes involve exposure to cash outages. Cash drains from bottlenecks and

delays in value-adding are P/OM matters. DCF must be modified with P/OM inputs to reflect the value of technological learning.

3. Let us connect Design for Manufacturability with Design for Marketability. P/OM inputs about how to make the product so that it conforms to the highest standards of product design with near-zero defectives are essential. Materials, technological and training issues abound. Changes in product and process, including engineering design changes for incremental improvements should be reviewed by P/OM with respect to the effect on meeting standards and percent defectives.

4. Let business schools teach how production and marketing interact. The appropriate management model captures how P/OM considerations affect the dynamics of the market place. How do product qualities, such as reliability, affect customer loyalty? Is it worth spending more to improve reliability and thereby loyalty? What is it worth to improve service speed with respect to customer defections (switching to a competitor's brand) resulting from product failure? Can larger investments in the technology of the production process, or training of employees, reduce the switching rate to competitors' products?

5. Let business schools teach the new methods of project management which include concurrent engineering[6] and on-going or continuous project management. R&D and marketing are crucial compatriots with P/OM in these endeavors to achieve high quality innovative changes on schedule with niche marketing requirements and competitive offerings. P/OM is the logical candidate to play the facilitator in these efforts which involve the judicious use of incremental changes as well as occasional heroic breakthroughs.

We could go on at length with further examples of P/OM acting in its technology and training capacity to operate as a full fledged partner in the organization. P/OM is the best place to start when moving from independent functional compartments to the integrated non-compartmentalized systems perspective. It is the responsibility of the business schools to create the environment for this to take place. It is the opportunity for business schools to renew their importance to the university, industry and society.

Endnotes

1. Among these academics was Professor Elwood Buffa whose early P/OM books helped interrelate technology and training with qualitative and quantitative models.

2. A major initiative which was undertaken by a governmental science academy to promote cooperation between business and engineering schools was spectacularly unsuccessful.

3. A top-level bank manager who frequently sent his backroom personnel to Columbia's Executive Program in Production commented that bankers could learn more from talking to manufacturers than from other bankers.

4. For example, Lillian Gilbreth was a renowned industrial psychologist of that era.

5. This notion is inspired by the grading philosophy of Dr. W. Edwards Deming, as well as by the experiences of firms that reward teamwork. Dean's List should be reserved for group efforts.

6. Concurrent engineering has also been called simultaneous engineering. However, as the term is presently being used to describe interfunctional teamwork with the systems perspective, we should rename it concurrent management of projects.

THE INFORMATION/CONTROL/BUFFER (I/C/B) PORTFOLIO: AN OPERATIONS MANAGEMENT PARADIGM

Leroy B. Schwarz
Krannert Graduate School of Management
Purdue University

Many different paradigms are used in teaching Operations Management. Among these are the Industrial Engineering, Operations Research, and Product Life-Cycle paradigms. Each of these paradigms provides a different, but useful way of introducing students to what Operations Management is, and to what managing operations means.

The "Information/Control/Buffer (I/C/B) Portfolio"paradigm, described below, has proven itself very effective in introducing students to what Operations Management is, and to what managing operations means. And, in addition, it provides a useful framework for *comparing* different types of management systems (e.g., MRP versus JIT). Even more important, it is effective in motivating students to think creatively about designing and managing *new* management systems. It also provides a link between Operations Management, Technology Management, and Management Information Systems.

And, finally, although the I/C/B Portfolio was designed as a *teaching* paradigm, we have found it to be just as valuable in providing a perspective on current research in Operations Management, and a provocative research agenda.

THE INFORMATION/CONTROL/BUFFER PORTFOLIO

All operations management systems do basically the same thing: they all "plan" and "control" production. "Planning" means making decisions about what to do. "Controlling" means making these decisions happen.

From an organizational viewpoint, management systems include all of the people involved in planning or controlling production: all of the office personnel, plus anyone involved in supervising or monitoring the actual production process. From an equipment or facilities viewpoint, management systems include all of the computer software and hardware that's involved in planning and controlling production, plus all of the offices, office equipment, etc. that contain or support this activity.

From a functional viewpoint, management systems can be divided into three separate but related systems: (1) the **"information-system"**, which provides information for decision-making; (2) the **"control-system"**, which plans and controls production; and (3) the **"buffer-system"**, which compensates for less-than perfect information- or control-systems. We call the combination of an information-system, a control-system and a buffer-system an "Information/Control/Buffer Portfolio", or "I/C/B Portfolio".

THE CONTROL-SYSTEM

The "control-system", short for planning and control system, is the core, of the management system. The role of the "control-system" is to make and implement decisions. The control-system involves all of the equipment, facilities, people, hardware, software, and know-how involved in making decisions and then implementing them.

Control-System Decision-Rules. The control-system's decision-making role is played by what we call the "control-system decision-rules". Decision–rules might be informal, or intuitive, or they might be an expert system. They might be simple or mathematically sophisticated. The decision-rules might be formal and automated; or, they might be totally within the head of some individual or group of individuals. Two important characteristics about decision–rules is the "optimality" of the rules and the "decision-rule leadtime". "Optimality" means how good the decision–rules are at picking the correct decision *given* the quality of information provided by the information-system. The "decision-rule leadtime" is the amount of time required by the decision-rules to reach a decision; that is, how long it takes to make the decision once the appropriate information has been provided. This leadtime might be short or long; it might be a fixed amount of time, or variable; known or unknown. If leadtime is unknown, it may have a known or unknown probability distribution whose parameters are either known or unknown.

Implementing Decisions. Once a decision is made, it must be implemented. Implementation usually involves some "paperwork" to authorize or initiate activity. And often, some preliminary actions must be taken, like ordering materials or preparing equipment. Finally, production itself takes place.

Two important characteristics about implementation are the implementation leadtime and the accuracy-of-implementation. The "implementation leadtime" is the amount of time required to put a decision into action; in other words, the time between making the decision and having it implemented. Like decision-rule leadtimes, implementation leadtimes might be short or long, fixed or variable, known or unknown. And, if unknown, implementation leadtimes may have a known or unknown probability distribution (whose parameters are either known or unknown). The "accuracy–of–implementation" measures how closely the implementation of the control–system's decision matches the decision. Perfect accuracy-of-implementation means no difference. In practice, differences may be small, as in a tightly controlled JIT system, or large, as in silicon-chip fabrication, where yield losses are unpredictable and difficult to control.

Often trade–offs occur between the implementation leadtime and the accuracy–of–implementation. The phrase "quick–and–dirty" means short decision-rule and implementation leadtimes but non-optimal, or only locally-optimal decision with less than perfect accuracy–of–implementation. American managers working in Japan report that typically, American managers make decisions more quickly than Japanese managers do — that is, American decision-rule leadtimes are shorter. On the other hand, the Japanese usually get their decisions implemented more quickly and more accurately— that is, the Japanese implementation leadtimes are shorter and their accuracy of implementation is better.

The overall quality of a control–system is measured as some combination of the four characteristics we have identified: (1) the quality of the decision–rules, (2) the shortness of the decision-rule leadtimes, (3) the shortness of the implementation leadtimes, and (4) the accuracy–of–implementation. Overall quality can range from "crude" (a control–system whose rules make poor decisions, with long and unpredictable leadtimes, yielding goods or services of unpredictable quantity and quality) to "perfect". A perfect control-system would make perfect decisions (given the quality of the information provided to it) instantaneously, and then implement them instantaneously, and with perfect accuracy.

Another important characteristic of a firm's control-system is its cost; that is, the cost of the people, equipment, facilities, hardware, and software that make up the information-system. Generally speaking, the cost of an information-system will be an increasing, and marginally-increasing function of its overall quality.

THE INFORMATION-SYSTEM

The role of the "information-system" is to provide historical, status, and future-oriented information for decision making. This information might be about demands, about costs, about materials, capacities, etc..

The overall quality of an information-system is measured as some combination of: (1) it's accuracy, or the correspondence between past, present, and future *reality* and what the information-system says about it; (2) the horizon, or how far into the future (or past) the information-system looks; (3) aggregation level; and (4) the information-system's leadtime. The information-system's leadtime is the time between an event and when the information-system is capable of reporting (or otherwise taking advantage of knowing about) it.

At their worst, information–systems provide grossly inaccurate information, or even irrelevant information. "Better" information systems typically provide some historical information and, perhaps, some status information. Still better information-systems provide future-oriented information, such as demand forecasts and cost projections. Near-perfect information–systems provide a virtually unlimited amount of accurate information about the past, present, and future. A "perfect" information-system is the proverbial crystal ball.

Another important characteristic of a firm's information-system is its cost; that is, the cost of the people, equipment, facilities, hardware, and software that makes up the information-system. Generally speaking, the cost of an information-system, like the cost of a control-system will be an increasing, and marginally-increasing function of its overall quality.

THE BUFFER-SYSTEM

In a perfect world, information-systems would provide perfect information about all aspects of the past, present, and future. And, in this perfect world, the control-system decision-rules would always make the perfect decision, and the control-system would implement these decisions perfectly, too.

But what about the real world? What about Murphy's Law? Murphy's Law says: *whatever can go wrong, does go wrong.* Murphy's Law says that there's no such thing as a perfect information-system or a perfect control-system.

Management copes with imperfect information–systems and control–

systems by using "buffer–systems". Planned inventories or extra capacity are popularly-used buffers. Planned inventories, for example, can be used to satisfy demand when it is larger than it was forecast to be. Inventories also compensate for actual production that's less than planned. Extra capacity and over-stated leadtimes are also buffers.

And, of course, Murphy's Law applies to buffer-systems, too. In other words, no buffer-system provides perfect buffering — at least not every time. This leads to lost sales, backordering, expediting, rework, and a host of other consequences.

The most important characteristics of a buffer-system are: (1) the form of buffering involved (e.g., inventory, capacity, time); (2) its effectiveness (i.e., the scale of imperfections it can buffer against); and (3) its robustness (i.e., the variety of imperfections it can buffer against). And, of course, the cost of the buffer-system is important as well. The cost of the buffer-system is the cost of all the buffer capacity, buffer inventory, and buffer time, plus all the costs associated with imperfect buffering (e.g., lost-sale cost, expediting cost, etc.)

AN EXAMPLE: THE NEWSBOY INVENTORY MODEL

Although the newsboy inventory model is a model of a very simple management system, it provides a simple example of an I/C/B Portfolio.

The Information-System. In the newsboy management system the information-system provides the probability distribution of future demand, estimates of the inventory-associated (i.e., backordering and inventory-holding) costs, and the current status (i.e., on-hand inventory).

The Control-System. In the newsboy management system the control-system selects the optimal order-quantity then implements this decision by ordering that quantity. Usually, the control-system's "decision-rule" is chosen to minimize expected one-period costs. The implementation part of the newsboy model's control-system is seldom, if ever, even speculated about.

The Buffer-System. The buffer-system in the newsboy management system is, of course, the planned safety stock; that is, the difference between the optimal base-stock and expected one-period demand. The cost of the buffer-system is the total expected shortage and leftover cost

THE I/C/B PORTFOLIO

A management system is an Information/Control/Buffer *portfolio*, because, like stocks in a financial portfolio, it costs money to acquire an information-system, a control-system, and a buffer-system. And, dollars spent on one component (say, the information–system) use up dollars that might be spent on another component (say, the control-system). Further, like a financial portfolio, imperfections in one component — say, a high–risk stock or a low–quality information–system — can be compensated for by another component — such as low–risk government bonds or a high–quality buffer–system. Finally, like a financial portfolio, a firm's I/C/B Portfolio should be changed over time as circumstances or the environment change.

In selecting the "best" I/C/B Portfolio, the object is not to select either the best possible information-system, control-system, or buffer system, but to select the *portfolio* with the minimum possible total cost, where:

Total I/C/B Cost = Total Information-System Cost

\+

Total Control-System Cost

\+

Total Buffer-System Cost

For example, a near–perfect combined information- and control–system may have low total buffer-system cost but very high information- and control-system costs. On the other hand, a crude information- and control–system may be low-cost, but be so imperfect that an enormously expensive buffer system is required to compensate for it.

INTRODUCING STUDENTS TO "MANAGING OPERATIONS"

Given the I/C/B Portfolio paradigm "managing operations means two closely-related activities: (1) using the *existing* I/C/B Portfolio to make and implement decisions; and (2) developing *new* I/C/B Portfolios.

USING THE EXISTING I/C/B PORTFOLIO

Using the *existing* I/C/B Portfolio to make and implement decisions — to plan and control operations — means using information provided by the *existing* information-system, making decisions the *existing* control-system decision-rules, implementing decisions using the *existing* mechanisms for implementation, and buffering against imperfections in all of this with the *existing* buffer-system.

In one sense, there is nothing new here except a lot of new, initially confusing, nomenclature (e.g., control-system decision-rules, etc.) On the other hand, our experience with other paradigms and many generations of students, indicates that graduates are often confused and disappointed when they find that the prescriptive decision-rules, as taught in many good operations management courses, simply "don't fit" their company, and they are confused about why. Often, these graduates mistakenly discard or discount everything we taught them because of the mismatch, and blame their former professors for misleading them.

The I/C/B paradigm helps them to see what fits and what does not fit; and, to work as best they can within their firm's existing I/C/B Portfolio.

DEVELOPING NEW I/C/B PORTFOLIOS

The I/C/B Portfolio paradigm encourages students to think about managing operations in a dynamic way: by developing *new* I/C/B Portfolios. In particular, the focus shifts away from learning "if you find a management system which operates under the following assumptions.....use the following decision-rule" to encouraging students to think about developing new information-systems or new control-systems, or to buffer in different ways.

For example, in the newsboy management system, students are encouraged: (1) to consider information-systems which reduce demand uncertainty or eliminate it (2) to consider how newsboy decisions are implemented, and how implementation might be changed in order to reduce total I/C/B costs; and (3) to consider different buffering mechanisms that safety-stock inventory.

Students are also encouraged to think about tradeoffs in constructing new I/C/B Portfolios. For example, in any given management situation uncertainty might be managed by improving the accuracy of forecasting and reducing buffers, or by spending less money on forecasting and spending the savings on better buffers.

A FRAMEWORK FOR COMPARING DIFFERENT TYPES OF MANAGEMENT SYSTEMS

The I/C/B Portfolio paradigm provides a unified framework for comparing different types of management systems. In particular, different systems can be examined on the basis of how their information-systems differ, how their control-system decision-rules and/or implementation systems differ, and how their buffer-systems differ.

As an illustration, Table 1 compares the three most popularly implemented manufacturing inventory management systems: (Q,ROP), MRP, and JIT, on the basis of their information-, control-, and buffer-systems.

THE I/C/B PORTFOLIO AND OPERATIONS MANAGEMENT RESEARCH

Although the I/C/B Portfolio was designed as a *teaching* paradigm, we have found it to be just as valuable in providing a perspective on current research in Operations Management, and a provocative research agenda.

A PERSPECTIVE ON CURRENT RESEARCH

If past and present Operations Management research is viewed from the perspective of the I/C/B Portfolio paradigm, then the following general observations emerge:

A Focus on Optimizing the Decision-Rules to Minimize Buffering Costs. Generally speaking, the overwhelming majority of research published in Operations Management is focused on optimizing decision-rules in order to minimize buffer-system costs. An excellent example is, again, provided by the newsboy model. In the newsboy model the characteristics of the information-system are fixed. Specifically, perfect information about marginal costs is provided and a known probability-distribution provides the demand information. Implementation considerations are left out.

The goal is to derive a decision-rule for ordering that will minimize expected buffer-system (expected inventory-holding and shortage) costs.

Assumptions About Perfect Information-Systems. Much of current Operations Management research assumes a perfect, or nearly-perfect information-system. In particular, costs are usually assumed to be known,

Table 1: Comparing (Q,ROP), MRP, and JIT Manufacturing Systems

	(Q,ROP)	MRP	JIT
Information-System:	• Forecasted Usage • Inventory Status • Standard Leadtime	• Master Prod Schedule • BOM • Inventory Status • Standard Leadtime • Shop Status	[•Fixed Future Usage] • Inventory Status
Control-System:	• Order-Release Decisions • Pull • Continuous-Review • Centralized	• Order-Release Decisions • Push • Periodic-Review • Centralized	• Order-Release& Sequencing Decisions • Pull • Continuous-Review • Decentralized
	• Shortage List • Sequencing Decentralized	• Load Report with Due Dates • Sequencing Decentralized	• Kanbans [• High Bandwith]
Buffer-System:	• FG Inventory • WIP Inventory	• Overstated Leadtimes • WIP Inventory	• Capacity

although in practice these costs are usually quite difficult to estimate. Further, demand for the product or service being modelled is usually assumed to be known or accurately described by probability distribution whose parameters are known.

Assumptions About Perfect Control-System Implementation. Finally, virtually all current research in Operations Management assumes that implementation is perfect. In other words, that whatever decision values (e.g., amounts ordered) the decision-rules specify get implemented as specified. In addition, most current research assumes that decision-rule and implementation leadtimes are either zero or some other constant.

A PROVOCATIVE RESEARCH AGENDA

The I/C/B Portfolio paradigm suggests the following research agenda:

Expand the Boundaries of Modelling. Operations management researchers should expand the boundaries of the models they optimize to include the entire I/C/B Portfolio. For example, model boundaries should be extended to include as decision-variables that involve the accuracy and validity of the information-provided by the information-system, and the accuracy and leadtimes of the implementation process. These are variables that practicing Operations Managers deal with every day, with very little guidance from existing research.

Develop Models to Evaluate I/C/B Portfolios. It is also very desirable to have models for choosing between different I/C/B Portfolios; i.e., for choosing between different management systems. Most of the literature that compares JIT systems with MRP systems, for example, involves testimonials and war-stories. The literature which boasts the marvels of TQM, cycle-time reduction, and cross-functional integration is similarly based.

The I/C/B Portfolio paradigm provides a framework for evaluating different management systems, although optimization under this framework would be very difficult.

Develop Models to Evaluate Leverage Points in Current I/C/B Portfolios. Finally, it would be of enormous benefit to practitioners if models were developed which could guide their decisions about "leverage" points in their existing I/C/B Portfolios. In other words, most operations managers know that their existing I/C/B Portfolios are far from being optimal. What they don't know is: where to make the next improvement — where to spend the next dollar — to get the maximum benefit. For example, is the best course of action for the manager of an MRP system to improve forecasting accuracy to improve bill-of-material accuracy, to adopt better master-scheduling decision-rules, to shorten workcenter standard leadtimes, or to invest in more finished-goods inventory.

The I/C/B Portfolio paradigm again provides a framework for identifying these leverage points.

OM: TIME FOR META RESEARCH[1]

Arnold Reisman
Case Western Reserve University
Department of Operations Research
Cleveland, Ohio 44106

HISTORICAL AND PHILOSOPHICAL BACKGROUND

The debate which raged during the 60's, regarding differences between Operations Research and Systems Engineering, finally petered out to a standoff. However, the 80's have recorded a much more profound debate - a great deal of questioning as to what has become of OR and as to where it is heading [1, 3, 4, 9, 10, 11, 13, 14, 22]. The differences between OM and OR have certainly been recognized by some although they were not the subject of any vigorous debates. Yet, the differences were sufficient to spawn several new professional societies and a number of new journals. The subject matter of a majority of articles published under the rubric of OM are not that different from the majority of those published in the OR/MS literature during the sixties and, to a lesser extent, the seventies.

More specifically, the current OR/MS literature is predominantly by, and for, what Russ Ackoff [1], calls "input-oriented professionals, namely, those "defined by the physical and intellectual instruments (for example concepts, theories, tools, techniques and methods) that their members employ to solve problems". Among the examples he cites for input-oriented professionals are statisticians, computer programmers, accountants, and now operations researchers.

To a lesser degree the current OR/MS literature reflects the work of Ackoff's "output-oriented professionals": those who are "defined by their products, by the kind of problems they solve"; those who "generally employ a wider variety of instruments than do input-oriented professionals but contribute less to the development of such instruments". As examples of

---

1 This paper is based on the forthcoming book, MANAGEMENT SCIENCE KNOWLEDGE: ITS CREATION, GENERALIZATION AND CONSOLIDATION, Quorom Books Publishing Company, New York, 1992.

output-oriented professionals he suggests designers of incentive systems, production and inventory control systems, management information systems, etc.

On the other hand, the "new born" OM literature was ostensibly directed at, and for, Ackoff's "market-oriented professionals". According to Ackoff, these are "defined by the class of users they address. They attempt to deal with as many as possible of the problems that their users have. Such professionals use a wider range of instruments and consult with a wider range of input-oriented technicians and output-oriented specialists than professionals of the other two types. They are generalists in the same way that general practitioners of medicine treat anyone who is ill, whatever the illness. Medical GP's frequently use specialists. For example, "they use input-oriented technicians such as anesthesiologists".

In "OR, A Postmortem", Russ Ackoff [1], traces "the devolution of Operations Research from its original state as a market-oriented profession through the stage of output-orientation to its current state of input-orientation".

Editors of OM journals, organizers of OM Conferences, authors of OM texts and professors of OM subjects must stay very aware and always be on the alert against the creeping "devolution" - a natural tendency to move in the direction of the *output* or worse yet the *input* directed professionals.

Yes, it is a natural tendency for a field of knowledge to follow a learning curve [18], to progress analogously to Rostow's [20] stages of economic development. OM subdisciplines can be classified based on their current stage of development using the learning curve. This is shown in Figure 1. Alternatively, one can take a longitudinal, retrospective look at the progress of a single discipline as shown in Figure 2, for Search Theory. What is inevitable, however, is that disciplines create literature at an exponential rate. This is shown for Location Theory using actual data [5] in Figure 3.

Explicitly or implicitly OM still claims to be directed at the Ackoff's "market-oriented professionals" - at those concerned with solving real-world problems. It therefore behooves the OM profession to provide the problem solver all the descriptive tools (models, heuristics, algorithms) as well as the case histories in a manner that is both efficient and effective. Needless to say, this objective is not incompatible with the needs of the OM student nor that of his or her professor.

In 1964, as a "post doc" at the Western Management Science Institute (WMSI) at U.C.L.A., I assembled some thoughts on the need for unifying and consolidating the knowledge base of OR/MS so as to make it easier to teach, to learn, to comprehend and to use. The draft which lay dormant for some time [13] began with:

> Students of management beware, or you shall be buried in models! The time has come for all those interested in management science to reflect. We must take stock of where we have been, where we are, and where we will be if we keep going at the current rate and in the current manner.

It went on to say that:

> The development of specialized models in management science, industrial engineering, operations research, and econometrics, began out of necessity, is degenerating into a habit. The literature is replete with models that fit a specific type problem like a glove but are useless for problems that differ in type. All these works need synthesis. Model builders should generalize upon previous works to integrate, systematize or categorize the specialized models within their fields. The science of management needs the kind of integrative thought given the science of chemistry by Mendeleyev.

Since the 60's much work has been done along those lines. A long list of references to articles attempting to integrate the states of knowledge of OR/MS subdisciplines can be found in Reisman [13 and 18]. However, for every published article attempting to unify or consolidate knowledge of OR/MS and now OM, there must be literally hundreds of articles which report on some specific model, algorithm, solution technique, etc.

Starting with the takeoff or fast growth stage of Figures 1 and 2 there is a need in every OM subdiscipline to encourage what is known as *meta research*. The following two sections will summarize both the objectives and the means for Meta Research in OM.

THE OBJECTIVES FOR META RESEARCH IN OM

1) *Consolidation.* Making the knowledge base of OM more unified within each subdiscipline as well as across the many subdisciplines;
2) *User-friendliness.* Making OM knowledge more amenable for use by the researcher, the practitioner, the educator, and the student;
3) *Expansion.* Providing a vehicle for major strides forward in expanding the knowledge base both in theory and in practice.

THE MEANS FOR META RESEARCH IN OM

1) *Meta research*, or research on research as discussed in [6, 7]. This is a more systematic, less anecdotal, more rigorous version of the traditional literature review.
2) *Taxonomies*. This is both a means to several ends and an end in itself.

Taxonomies create a structure and/or a vehicle for doing meta research. They provide a vehicle for classifying the existing literature in a way that is efficient yet robust. Also, as shown in references [16, 17, 18], taxonomies can serve as a means for identifying literature voids.

3) *Generalization*. As shown in [2, and 19] many subdisciplines of Operations Management can be unified while, at the same time, expanded through work which generalizes upon existing models and thereby subsumes each and all of them into its structure as special cases.

4) *Expert Systems*. As shown in [21], with the advent of this technology it is possible, based on the characteristics specified by the inputer, to have a computer select and suggest the most appropriate model, algorithm, and/or heuristic from the entire universe of potential approaches.

The above objectives to be served by meta research in OM and the stated means for achieving them are used to create a two dimensional incidence matrix. The result, Figure 4, shows, at best, but one empty cell. Alternately stated, each of the means identified can serve toward fulfilling each of the objectives. Moreover, the cells of Figure 4 are not mutually exclusive. They are in many cases supportive of each other. For example, many an expert system may well benefit from having a taxonomy of the field. This is especially so if generalization of the field is also available. Research on Research, especially if directed at identifying voids, is obviously aided and abetted by the systematization possible via a good taxonomy. The availability of a generalization clearly does not hurt in such cases as well.

Using a sample of references Figure 5 validates the message of Figure 4. Clearly, this is not an exhaustive set of examples. It may be noted however, that with one exception, the examples were drawn from this author's own publications - a sign of much laziness and of some vanity.

Figure 6 - A Taxonomy of Taxonomies [12] summarizes the fact that taxonomies themselves can be classified in a number of ways. Some are purely descriptive; they provide a means to classify what exists in a given, well defined and bounded field. On the other hand some also serve prescriptive ends by pointing out, in efficient and in operationally meaningful terms, the voids that exist in the literature of that field.

Along another dimension Figure 6 differentiates a taxonomy as discussed above from one which includes and/or is part of a general model for the field of knowledge. The last dimension simply categorizes the different formats of taxonomies. These are amply discussed in Reisman [12 and 18].

Taxonomies can and have been used to classify mature as well as emerging OM subdisciplines and to pinpoint knowledge voids in same. In OM it is useful that knowledge and therefore voids therein be described within a three dimensional continuum, e.g. *real data*, *methodology* and the *problems addressed* [17,18].

SUMMARY

- Knowledge creation in any field (including all subdisciplines of OM) is a learning system.
- It follows a learning curve.
- Learning curves tend to saturate
    - Contributions are at best incremental and small
    - A "normal science" according to Herbert Kuhn
- Literature, however, expands exponentially as a field of knowledge passes through the structuring stage and goes into takeoff and beyond.
- There is a need for knowledge consolidation
    - systematization
    - generalization
- This need is felt by the
    - Researcher
    - Practitioner
    - Student
    - Teacher
- There are ways to satisfy such needs. These are
    - Research on Research
    - Taxonomies
    - Generalization
- Such approaches in addition to making the literature more "user-friendly" can be used to identify voids in the literature. Such voids may exist in the
    - Data Availability
    - Appropriateness of Available Methodology
    - Problems in need of addressing

CONCLUDING REMARKS

- Note the *Expert Systems* column cells of Figure 4 are not empty except for the one addressing *Expansion of Theory*.
- Yet, a draft of Figure 5 with many more references than this chapter allows had very few entries indicated in this column.
- This may be due to the fact that this author has not published in this area and has not even done the homework necessary to review the relevant

literature.

- It may also be due to the relative newness of *Expert Systems* and a time lag in bridging that technology with the various subdisciplines of OM.
- If the latter case is true then this delineates a great potential for productive research.
- Such research, in addition to opening up new vistas for publishing articles and creating new journals, may even do some good in making the existing models, algorithms, heuristics, etc. more *user friendly* for the researcher and more importantly for the *practitioner*, the *student*, and the *teacher*.
- If it is successful in this regard it may even contribute to the *Expansion of Practice* by improving the quality of solutions to problems currently within the purview of OM practitioners and by allowing "missionary work" [4] to proceed with greater confidence into economic sectors currently untouched by the OM professional.

REFERENCES

1. Ackoff, R., (1987), *OR, A Postmortem*, Operations Research, Vol. 35, No. 3, May-June, pp. 471-474.

2. Balinsky, W. and A. Reisman, (1971), *Some Manpower Planning Models Based on Levels of Educational Attainment*, Detroit, September; Management Science, Application Series, Vol. 18, No. 12, August 1972, pp. 691-705.

3. Barnett, A., (1982), *Misapplications Review: An Introduction*, Interfaces, Vol. 12, No. 5, pp. 47-49.

4. Blumstein, A., (1987), *The Current Missionary Role of OR/MS*, Operations Research, Vol. 35, No. 6, November-December, pp. 926-929.

5. Brandeau, M.L. and Chiu, S.S., (1989), *An Overview of Representative Problems in Location Research*, Management Science, Vol. 35, No. 6, pp. 645-674.

6. Cooper, C.M., (1988), *Organizing Knowledge Synthesis: A Taxonomy of Literature Reviews*, Knowledge in Society, Vol. 1, pp. 104-126.

7. Cooper, C.M, (1989), *Meta-Analysis and the Integrative Research Review*, a paper presented at the meeting of the Operations Research Society

of America/The Institute of Management Sciences, New York, October.

8. Escueta, E., K. Fiedler, A. Reisman, (1986), *A New Hospital Foodservice Classification System*, Journal of Foodservice Systems, Vol. 4, No. 2, pp. 107-116.

9. Halse, R. and Lilien, G., (1986), *The Ombudsman: Academic Research in MS/OR: Science or Trivial Pursuit?*, Interfaces, Vol. 16, No. 3, May-June, pp. 41-48.

10. Lilien, G., 1985), *OR/MS on Thin Ice*, Interfaces, Vol. 15, No. 4, July-August, pp. 12-13.

11. Magnanti, T.L., (1986), *Editorial*, Operations Research, Vol. 34, No. 1, January-February, pp. 2-3.

12. Reisman, A., (1987a), *Expansion of Knowledge Via Consolidation of Knowledge*, presented at the Second International Symposium on Methodologies for Intelligent Systems, Charlotte, N.C., October 14-18. ISMIS-87 PROCEEDINGS, pp. 159-172, ONRL-6417, Oak Ridge National Laboratory.

13. Reisman, A., (1987b), *Some Thoughts for Model Builders in the Management and Social Sciences*, Interfaces, Vol. 17, No. 5, October, pp. 114-120.

14. Reisman, A., 1987c, *Where Have We Lost Our Way*, OR/MS Today, October.

15. Reisman, A., (1988a), *On Alternative Strategies for Doing Research in* *ht* *Management and Social Sciences*, IEEE Transactions in Engineering Management, Vol. 35, No. 4, November, pp. 215-221.

16. Reisman, A., (1988b), *Finding Researchable Topics Via a Taxonomy of a Field of Knowledge*, Operations Research Letters, Vol. 7, No. 6, December, pp. 295-301.

17. Reisman, A., (1988-89), *A Systems Approach to Identifying Knowledge Voids in Problem Solving Disciplines and Professions: A Focus on the Management Sciences*, Knowledge in Society: An International

Journal of Knowledge Transfer, Vol. 1, No. 4, Winter, pp. 67-86.

18. Reisman, A., (1992), Management Science Knowledge: Its Creation, Generalization and Consolidation, Quorum Books Publishing Company, New York, 495 pages.

19. Reisman, A. and E. S. Buffa, (1962), *A General Model for Investment Policy*, Management Science, Vol. 8, pp. 304-310, April. Also, MODELS FOR PRODUCTION AND OPERATIONS MANAGEMENT, E. S. Buffa, Wiley, 1963, pp. 402-416; and MODELS AND ANALYSIS FOR PRODUCTION MANAGEMENT, M. P. Hottenstein, International Textbook Company, Scranton, PA., pp. 53-60, 1968.

20. Rostow, W.W., (1971), The Stages of Economic Growth: A Non-Communist Manifesto, University Press, Cambridge, MA.

21. Shah, V.C., G.R. Madey and A. Mehrez, (1991), *Scheduling Assistant: A Methodology for a Knowledge Based Model Management System for Job Shop Scheduling Decision Support*, forthcoming in Omega.

22. White, J.A., (1991), *An Existence Theorem for OR/MS*, Operations Research, Vol. 39, No. 2, March-April, pp. 183-193.

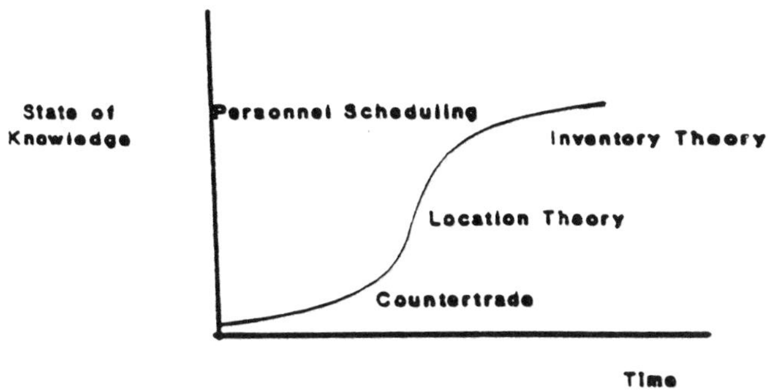

FIGURE 1

STATE OF DEVELOPMENT OF VARIOUS OR/OM SUBDISCIPLINES

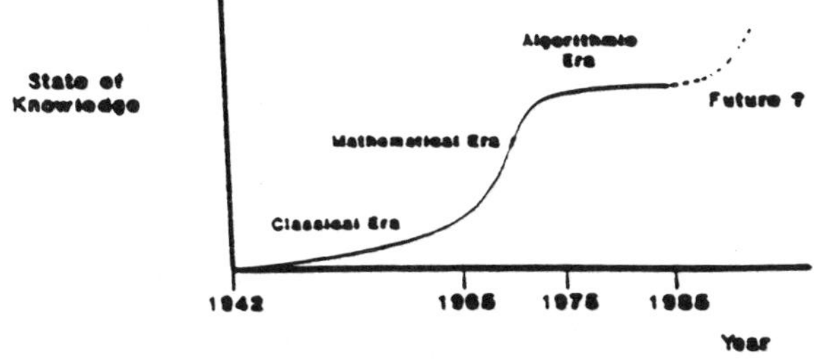

FIGURE 2

SEARCH THEORY:

STAGES OF DEVELOPMENT

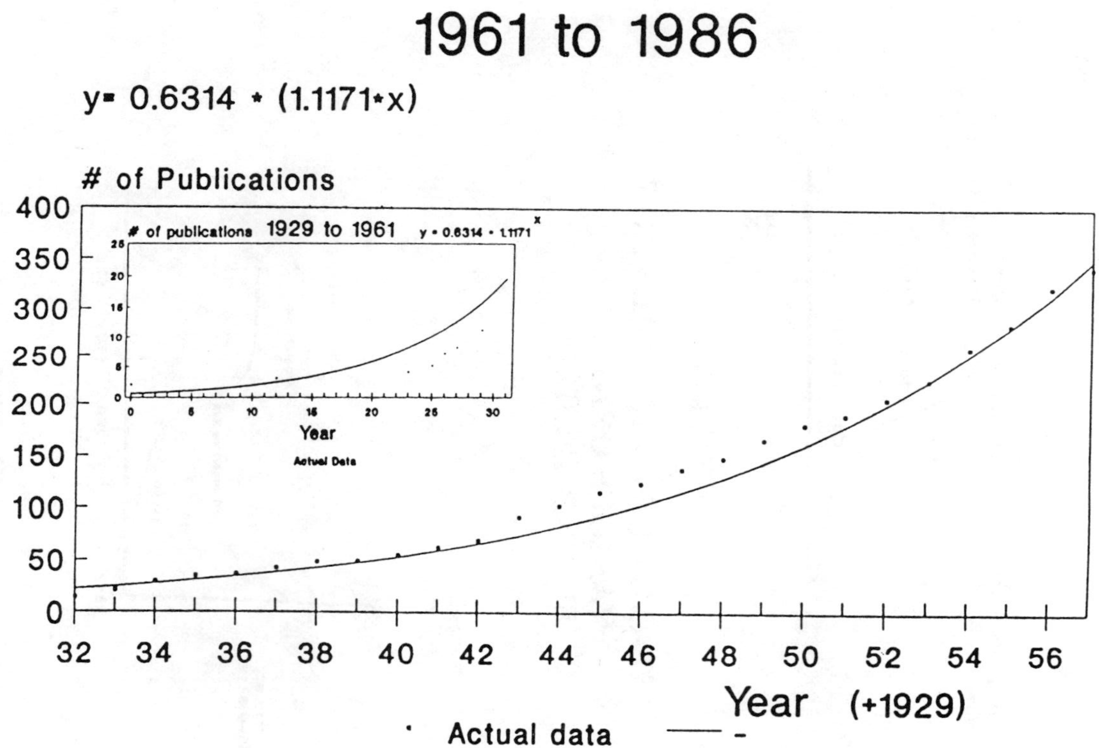

FIGURE 3

NUMBER OF PUBLICATIONS IN THE LITERATURE OF LOCATION THEORY

THE OBJECTIVES		THE MEANS			
		Research on Research	Taxonomies	Generalization	Expert Systems
	Consolidation	X	X	X	X
	User Friendliness				
	Researcher	X	X	X	X
	Practitioner		X	X	X
	Student/Teacher	X	X	X	
	Expansion of Theory	X	X	X	
	Expansion of Practice	X	X	X	X

FIGURE 4

THE OBJECTIVES AND THE MEANS FOR
META RESEARCH IN OM

THE OBJECTIVES		THE MEANS			
		Research on Research	Taxonomies	Generalization	Expert Systems
	Consolidation	**15,16,17,***	**8,16,19**	**2,19**	**21**
	User Friendliness Researcher	8,15,16,17,19	**8,16,19**	**2,19**	**21**
	Practitioner	8	**8,19**	**2,19**	**21**
	Student/Teacher	8,15,16,17	**8,16,19**	**2,19**	**21**
	Expansion of Theory	8,15,16,17,19	8,16,19	**2,19**	
	Expansion of Practice	2,8	**8,19**	**2,19**	**21**

FIGURE 5

SOME REFERENCES TO META RESEARCH IN OM:
THE OBJECTIVES SOUGHT AND THE MEANS USED

* **Numbers in this matrix correspond to references in the back of the paper.**

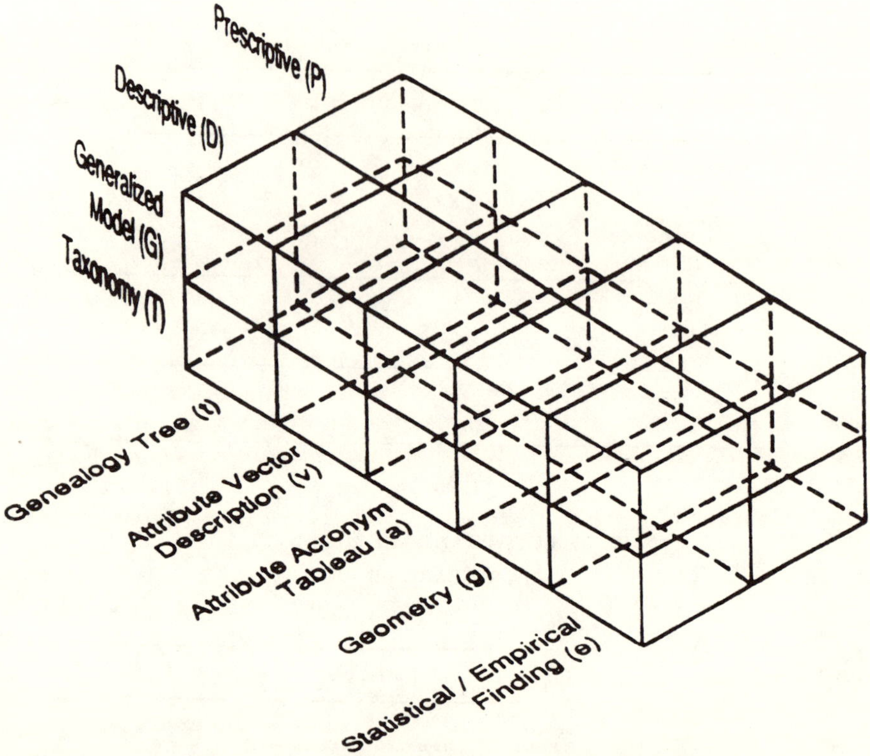

FIGURE 6
A TAXONOMY OF TAXONOMIES

REFINING OPERATIONS MANAGEMENT AROUND TOTAL QUALITY MANAGEMENT

Roger V. Johnson*
School of Business Administration
University of Michigan
Ann Arbor, Michigan 49103-1234

Failure to embrace total quality is a lot like failure to pay income taxes. Both are possible to postpone. But sooner or later it is clear that early attention is preferable. Of the two, total quality offers the better return. We gain the power to enhance operations and to provide new opportunities in the classroom and in research. But to obtain the full benefits, it is necessary to reexamine and correct some of the fundamental assumptions upon which operations management is built. What are some of those assumptions?

Consider **Corning's Policy**, typical of companies that practice total quality, by which all managers and employees are expected to know and live by:

> "It is the policy of Corning to achieve Total Quality performance in meeting the requirements of external and internal customers. Total Quality performance means understanding who the customers are, what the requirements are, and meeting those requirements, without error, on time, every time."

On time, every time is easily underestimated. It means that all orders should be on time, not just those that Corning feels they can conveniently fit into their schedules.

In contrast to Corning's policy, consider the myriad of models in the operations management literature that calculate the optimal level of not satisfying customers: optimal inventory safety stocks that determine how many customers will not be served; the project "least-costing" model that determines how late the project should be completed; backlogging in aggregate planning models to determine how many customers should be served late; balancing lateness costs with other costs in multi-criteria machine sequencing models. *All*

---

* A doctoral student of Elwood Buffa from 1968 until 1973, and a visiting faculty colleague of Elwood from 1982 until 1986.

imply that failing to satisfy some customers is normal and optimal. To be repetitious, Corning people strive for "meeting customer requirements, without error, on time, every time".

Models that implicitly aim to not always satisfy the customer have no relevance in the new world of total quality. And the growing chorus of industry spokespersons, each telling his company's story of total quality, compels us to incorporate total quality into our curriculum and to do it now.

Moving into total quality is still difficult for academics, since total quality research and teaching success models are rare, although many academics instinctively practice some aspects of total quality. In the absence of a school-wide total quality program that penetrates the teaching and research functions, each operations management academic pursuing total quality might move through six intellectual stages before achieving major benefits from a fully implemented total quality mode of operation. During these stages, changes in personal behavior, management of the class, course content, and perhaps research agenda, will occur. These six stages are:

1. Teach quality control and quality assurance.
2. Elevate quality to have a vitally important status.
3. Embrace quality over cost minimization.
4. Redesign courses around total quality.
5. Practice total quality in the classroom.
6. Identify new research opportunities.

A discussion of the six stages follows.

Stage 1. Teach quality control and quality assurance.

Until total quality had an impact, most courses and most textbooks devoted at most two class sessions or chapters to quality, usually the mathematics of statistical process control and perhaps of quality acceptance levels. These topics were usually positioned towards the end of the semester or text, implying quality is a useful but not an integral part of operations management. This reflected company practice.

Stage 2. Elevate quality to have a vitally important status.

At some point the operations management academic realizes that quality is vitally important. This realization might be prompted by visits to companies that practice total quality, listening to industry speakers at conferences, attendance at a Xerox or Procter and Gamble Quality Forum, or other means. The academic also realizes that the agendas of practicing operations managers are very different today from what we have traditionally taught in operations management. Usually managers place quality at the top.

To reflect the vital importance of quality, the one or two sessions on quality are typically expanded to three or four positioned at the beginning of the course. The new topics might include: quality function deployment, total quality, or the Baldrige criteria. Quality management is virtually ignored for the rest of the semester or text, implying that quality is not important enough to impact traditional approaches to operations management.

Stage 3. Embrace quality over cost minimization.

A troubling issue in teaching total quality is the role of costs. This occurs because operations management has implicitly been a cost minimizing discipline. In contrast, a basic tenet of total quality is to lower costs by increasing quality or reducing lead times, moving away from directed cost minimization. By practicing total quality, cost savings occur, for example, when internal or external suppliers and customers identify, agree upon, and satisfy mutual needs, which largely eliminate misunderstandings, thereby reducing delays, rework, and waste. But more important is the faith -- as it initially must be -- that when suppliers do the things that matter to their customers, market share will build, and so, therefore, will profits[1].

The better companies had realized by the early 1980s that total quality must permeate everything they do. In order to incorporate this new view, total quality must also permeate every topic in operations management courses. This is easily stated, but the impact on preparation of classroom material is easily underestimated. The entire traditional operations management curriculum

1. Former Chairman of Xerox David Kearns and Procter and Gamble President John Pepper espoused this chain at Xerox's Quality Forum II in Leesburg, Virginia, in August 1990.

needs to be reassessed and modified to ensure consistency with total quality objectives. The same applies to the interface between Operations Management and other functional areas, particularly marketing.

Many traditional operations management topics are not in line with total quality in their usual form. Some topics, such as inventory models, can be modified and expanded to incorporate total quality. Others, such as management of programs that focus on cost-cutting and productivity improvement, are obsolete. We should replace these with ideas that companies are finding effective, such as: total quality programs, lead time reduction programs, concurrent engineering, and design for manufacturability, serviceability, and logistics.

The question is, how do we alter this traditional theory to incorporate total quality? The following two sections outline examples of this alteration as it pertains to inventory management and project "least costing".

Inventory management

Much inventory management theory is based upon cost minimizing models such as the fixed order-interval model or the fixed order-quantity model. Some cost models are easily adapted to focus on customer satisfaction. A service level of 99.999%, for example, might be established to ensure a reliable supply. This replaces the cost minimizing "optimal safety stock" formula. An important subtlety here is that the 99.999% is selected simply to ensure reliable service, not to account for the high implied cost of possible future lost sales. This model improvement is a simplification.

Models can also be enhanced by focusing on degrees of freedom rather than specifying a predetermined objective function. The fixed reorder interval model, for example, has three linked variables for an anticipated demand distribution: the order interval, the buffer stock, and the order quantity. Fixing any two determines the third variable's value, implying two degrees of freedom. The first degree is normally used to set the buffer stock to attain the desired service level. The second degree is used to satisfy other needs of the internal or external customer. For example, if the item is delivered along with other items, the order interval is fixed to conform, thereby implying what the order quantity should be. If instead the item is delivered alone, economies of delivery might be the objective, which might be obtained by selecting a mean order quantity to match the capacity of the delivery vehicle, thereby implying what the order interval should be. Specification of which variable drives what is not defined in the model: this is driven by customer requirements.

The SOLA/Corning Electronic Link Team[2]

Total quality demands a broader view of inventory management than current inventory models offer. Part of the total quality approach is for supplier and customer to identify needs, agree on what will be done, and agree on a performance measurement. The work of the SOLA/Corning Electronic Link Team is cited as a successful inventory management example. Here are pertinent details.

Corning sells lenses to SOLA Optical (nee Coburn Optical Industries). Prior to the formation of the SOLA/Corning Electronic Link Team, orders and releases were transmitted verbally by phone between Coburn and Corning every two weeks. This took two to three hours.

In response to Corning's total quality initiatives in April of 1986, Coburn outlined several improvements they would like Corning to implement. Coburn wanted shorter lead times, reduced minimum order quantities, improved on-time delivery, and improved order accuracy. Corning agreed to address these issues, and from 1987 through 1989 they automated relevant internal processes with impressive results.

In 1988 Coburn was purchased by SOLA Optical, a major manufacturer of plastic lenses. Because these lenses competed with Corning's glass lenses, some Corning personnel thought they would lose this customer. However, Corning by this time had established itself as a reliable and cooperative supplier that added value to Coburn's operation, and the relationship survived the ownership change. It appeared to these personnel that the results of the total quality program saved this very important customer.

Nevertheless, SOLA had different priorities than Coburn, and Corning's focus was changed to match SOLA's focus on simplifying the process of doing business. In 1990, it was proposed that an electronic link be provided. Since November 1990, a Corning-based computer has allowed SOLA direct access to significant amounts of information, some previously

---

2 Information was kindly provided by Corning.

unaccessible, including SOLA's average weekly usage, Corning's inventory on hand, order status, and a product dictionary. SOLA people can request a delivery date, and Corning responds with a promised date. This link saves SOLA six to eight hours of calculations each week, and can now perform on-line computations in planning, forecasting, and scheduling processes that were previously unavailable. The system also consolidates shipping, saving SOLA about $1000 per shipment. These improvements allow SOLA to keep lower inventories and serve their customers better. The Corning system also provides electronic messaging, thereby eliminating telephone tag and resulting delays.

The project has virtually eliminated non-value added steps for both SOLA and Corning in producing and communicating an order. The time it takes Corning to confirm an order has been reduced from three to four days to a few minutes. It saves Corning about six hours per week.

By expanding the scope of inventory management to include relations with customers, its power is increased from a cost minimizing tool to a means of capturing more business.

Project Management Least-Costing

A popular enhancement of the critical path method is "least-costing" or "crashing". This is often taught as a technique to select the duration of the project which minimizes total costs. The implication is that if accelerating the project by a week costs more than the resulting savings, the project should not be accelerated. But from a total quality perspective, the major issue is "meeting the customers requirements on-time, every time". This is in conflict with the cost focus of the least-costing model. The model is not discarded, however. By requiring the project to be completed on-time, the model can be used to identify the lowest cost way of meeting the customer's due date.

Stage 4. Redesign courses around total quality.

In pursuing further benefits of quality, it is realized that the most fruitful sources of quality stem from the design of products, services and processes. In developing a teaching framework for quality materials, there is a realization of the logical consequences of this truth. It means that if we are to successfully teach the management of operations, it is necessary to include the management of design of products, services and processes as an integral part.

In order to place the management of design into the natural flow of operations management, it might be helpful to adopt another framework to organize materials around. W. Edwards Deming provides the basis of a useful model in Out of the Crisis. Deming views production as a system. Improvement envelops the entire system, from incoming materials and equipment, through receipt and testing of these, followed by production, assembly and inspection, after which comes distribution to the consumers. From consumers is obtained consumer research, which is fed back to the design and redesign of products. Other actions on the system include tests of processes, machines methods and costs. In this view, there is no start point to the improvements of the ongoing system.

Deming's diagram is not linear in time since it emphasizes repeated movement around the various stages of the system. For course organization, it is necessary to place the topics in linear sequence, perhaps from design through ongoing operations to suppliers to distribution. The rationale used for the core MBA O.M. course offered at the University of Michigan in the Winter 1992 semester is as follows, in which sections II, III and IV of the course match closely with Deming's framework.

Course Organization

The course is organized into six sections. The purpose of each is provided below.

I. Course introduction (4 sessions)

We probe the way that most successful companies are today organizing their operations: around total quality programs. Their realization is that to please the customer and to increase profitability, it is necessary to pay attention to quality: not just of the output, but of how things within the organization are performed. In the introduction we also recognize the operations portion of the value chain: inputs are converted by processes into outputs which marketing people sell. Understanding the flows through the processes and their capacities is fundamental to designing and understanding operations.

II. Design of products, services and processes (4 sessions)

Before the value chain can operate it is necessary to design the products and services offered to customers -- and design the processes that provide them. We do not delve into the engineering and styling issues, but we do examine management processes that are most likely to lead to superior products and services. These include quality function deployment, design for manufacturability, and concurrent engineering teams. All are used within the context of a total quality program.

III. Ongoing operations (10 sessions)

After the processes that provide the products and services are in place, it is necessary to manage the ongoing operations. In pursuing quality we spend more time on the improvement of the processes that provide the products and services than on the products and services directly. And we examine planning and control systems that ensure the services and products are provided in appropriate quantities and times. These include inventory control, aggregate operations planning, master production scheduling, just-in-time, and materials requirements planning.

IV. Distribution and suppliers (2 sessions)

Suppliers and customers share an organization's interests in providing the customer with value. A good organization therefore sees its suppliers and customers as partners rather than adversaries. We delve into developing beneficial relationships. We also look at distribution systems.

V. Strategic operations initiatives (4 sessions)

Total quality management demands continual improvement in everything everyone does. Strategic operations initiatives are put in place by top managers to galvanize improvements around a common theme. This might be to achieve a ten-fold reduction of rejects within eighteen months, or to reduce product development time by thirty percent within three years. These initiatives are simultaneously focusing and implementing operations strategy.

VI. Global citizenship (3 sessions)

> The course ends where the academic year began: with global citizenship, the focus of the orientation to our MBA program. We focus on these themes: strategic initiatives that reduce toxic emissions and thereby reduce costs, new product development that integrates environmental impact, strategies to reduce environmental impact, global warming, and projects done during your first academic year in the MBA program.

Once the framework is agreed upon, a renewed look at topics is required. Teaching the management of design calls for material such as concurrent or simultaneous engineering, quality function deployment, and other "design for" approaches: design for manufacturability (DFM), for assembly (DFA), for serviceability (DFS), and for logistics (DFL). And as the potential of good design is recognized to cover more and more areas, "DFX", to use an abbreviation devised in General Motors, where the "X" stands for anything relevant.

The course should address other burning issues of today's operations managers, which might be seen as mechanisms to ensure continual improvement, focus on improving quality, better service as defined by the customer, lead time reduction, sourcing, supplier relations, international operations, management of creativity and innovation, ethics, labor relations, minority representation, and, for many manufacturers, environmental concerns. Topics such as MRP systems, which MBAs will use much less than they will total quality programs, are moved into elective courses.

On realizing that focus on the process of achieving objectives is more helpful than focus on the outcome, it becomes obvious that it is important to teach the means to improve the process. Process mapping and analysis, and problem solving techniques provide this. The latter usually includes a standard set of problem-solving steps that might be either the Shewhart (or Deming) Plan-Do-Check-Act cycle, or a set that starts with forming a team, identifying the customer, identifying the root cause, and only then searching for a solution, and implementing it. There are a number of problem solving tools, including the "seven tools" and the seven "new tools". Students need to be taught these in or prior to their initial operations management course. Many industry people say 90% of their process improvements come from problem solving with basic tools.

It is also important that we academics do not fall into the trap that many companies have: some have marshalled all their creative resources to implement the total quality program, but failed to make necessary product and process innovations. In our case, we should not be diverted from making desirable innovations in other areas by focusing only on total quality.

Stage 5. Practice total quality in the classroom.

The total quality process can be used to facilitate classroom learning. Discussions that start with "what do you want?" take place, and students make requests. At first the emphasis is the teacher satisfying the students. After a time the classroom attitude evolves, or should be steered, to a supplier/customer relationship in which both parties contribute to their own satisfaction by helping the other party. This might include the teacher making more, not less, demands on the students.

The continuing dialog with students should include a formal component if the effort is not going to die. A student team might meet with the instructor each week to discuss what they want from each other, and how this might be achieved. Part of the formal dialog should be feedback as to how each party is succeeding in finding and implementing solutions to problems. The team should include someone with and someone without work experience, and someone with and someone without a numerate background. Students not in the team should also have an opportunity to meet with the instructor in order that both sides can gain a better understanding of the other's positions.

We now know that most cost minimizing and productivity programs have not worked, yet many quality programs do; and they can excite people. The same can happen in the classroom. Teaching how to achieve quality and increase market share is more interesting to students than how to shave costs. But as with quality programs, a number of things must be in place in the classroom before it works. Credibility is necessary, and industry can help here: speakers, perhaps taped, quotes and articles from industry leaders, and lots of success stories.

Stage 6. Identify new research opportunities.

Research opportunities abound in a changing field. Two groups are identified. First, all operations management material needs to be reassessed and in many cases modified or extended, with particular attention paid to criteria which are consistent with total quality. Some were mentioned earlier.

A second area extends from identified best practices. Several good companies now focus on processes. Xerox, for example, has identified and pays attention to over sixty processes that affect their customers. These range from the way Xerox responds to a service call to the way their warehouse picking operations are performed. Process benchmarking is used as a means of identifying, measuring, and analyzing a best practice of a particular process, wherever it occurs. Then it is improved and installed. This opens research opportunities that are particularly valuable to businesses. Once a best practice is identified, the researcher knows it is a process that is important to companies, and of which the important managerial variables are readily identified. The academic improves the best practice. To use warehouse picking operations as an example, part of a Xerox best practice is to cluster and re-position items that the computer identifies are to be picked in the next several hours in order to create shorter picker routes. This saves considerable picking time so orders are shipped to customers sooner. Building from this, academics can provide refinements, extensions, or algorithms that businesses want.

In conclusion, total quality provides lots of opportunities for us to improve and work on. Those that fail to take advantage of total quality will be penalized just as surely as those who do not understand that our taxation requirements should be met without error, on time, every time. However it would be overly optimistic to expect that the adoption of quality principles will result in substantial immediate improvements. It takes time and effort over and above normal teaching preparation.

[redacted]

[redacted]

Perspectives in operations management.

21019920

6-24-93

[redacted]

[redacted]